EIGHTH EDITION

CONTEMPORARY

ECONOMIC

PROBLEMS

AND

ISSUES

THOMAS J. HAILSTONES

Distinguished Professor Emeritus of Economics

Xavier University, Cincinnati

FRANK V. MASTRIANNA

Dean of the College of Information Sciences

and Business Administration

Slippery Rock University of Pennsylvania

Published by

H53 **SOUTH-WESTERN PUBLISHING CO.**

CINCINNATI   WEST CHICAGO, IL   CARROLLTON, TX   LIVERMORE, CA

PREFACE

Students successfully completing a good introductory course in economics develop an understanding of the workings of the American economic system and the use of basic tools and techniques in economic thinking. But because of the comprehensive scope of a principles text, only a relatively small segment of the text can be devoted to socioeconomic problems and issues. In many texts, treatment of such issues is confined to mini-readings or concluding comments within each chapter. To the students, readings and abbreviated applications within the text often seem superficial and lacking in depth. When one considers that most students' formal education in economics begins and ends with the principles course, all too often the unfortunate result is a lack of insight into the major problems of the day.

The basic goal of this text is to fill this need by presenting fifteen major problems and issues in a manner that the student finds both interesting and enlightening. Our approach is to present the chronological development or evolution of each problem along with the current facts necessary for proper understanding of the problem. The text contains an economic analysis of the alternate solutions that have been suggested. In most cases, however, the decision as to what should be done is left to the reader. It is hoped that after the reader has marshaled the facts, analyzed the alternatives, and weighed the merits of proposed solutions, decisions can be rendered and supported on what should be done about the various problems and issues.

For those familiar with the previous edition, it should be noted that two new chapters are presented with this eighth edition. Chapter 4 covers the beginning, continuing, and present day problems in agriculture. It deals not only with the current farm crisis but also with the widespread problems among financial institutions serving farmers.

Chapter 5 highlights and analyzes the current upheaval in the banking industry. It traces the impact of recent legislation to improve competition among financial institutions. Moreover, it explains the rash of mergers and bank failures taking place in the industry.

The other thirteen chapters have been extensively revised and updated to give the reader the most current information available as it pertains to both general and specific problems and issues. Chapter 2, for example, now treats problems caused by disinflation and deflation as well as inflation. In addition, Chapter 14 gives some examples of hyper-inflation and currency changes that have recently taken place in Third World countries.

In each chapter, the student should recognize the application of important economics concepts such as supply and demand, opportunity cost, and cost-benefit analysis. To avoid unnecessary repetition, graphic models are used selectively with the belief that students can readily apply these tools in chapters where such concepts are described. The main thrust of each chapter, however, is to emphasize the trade-offs individuals and society as a whole must make to achieve a desired end.

In evaluating economic policy, students should keep in mind the political realities of a democratic society such as ours. Frequently it may appear that solutions proposed by economists are both efficient and equitable and thus worthy of implementation. But public policy is largely determined by elected representatives who, in addition to seeking efficiency and equity in our economic system, are also very much concerned with the basic task of being reelected. Keeping this in mind, students can better understand why a significant number of public programs may provide substantial benefits in the present but even greater costs in the future.

Feedback from professors using previous editions indicates a variety of ways in which this book can be effectively utilized. In many cases the book is used as a supplementary text for the full-year principles course, while in others it is used for the one-term problems and issues course. In still other cases, the text serves as the basic learning tool for upper-division courses in current economic problems and issues, with greater emphasis on student research and subsequent written or verbal presentations on particular aspects of a given problem.

The authors are again indebted to the many individuals who aided in the development of the eighth edition of this book. Three members of the Economics Department of Xavier University, however, deserve special mention. We are particularly grateful to Dr. David Weinberg for accepting the responsibility for researching and writing the chapter on natural resources, to Dr. Harold Bryant for his work in providing extensive research and writing assistance for the chapter on Third World debt, and to Professor John Rothwell for writing the chapter on the banking industry. An expression of thanks is also due to Mrs. Marjorie Schmidt for typing various parts of the manuscript.

Thomas J. Hailstones
Frank V. Mastrianna

ABOUT THE AUTHORS

Thomas J. Hailstones is Distinguished Professor Emeritus of Economics and former Dean of the College of Business Administration at Xavier University in Cincinnati, Ohio. Before teaching at Xavier, he taught at the University of Detroit, Notre Dame, and for the U.S. Air Force. He received the Ph.D. in Economics from St. Louis University.

Dr. Hailstones has lectured throughout the United States and abroad. He is a member of the Board of Directors of the Clopay Corporation; Gradison Cash Reserves, Inc.; The Ohio National Fund; and the Student Loan Funding Corp. He has been an economic consultant to many companies, both large and small. In addition to having written numerous articles for academic journals and trade periodicals, Dr. Hailstones has written or co-authored a number of college textbooks. His current texts include ones on introductory, managerial, and supply-side economics.

Frank V. Mastrianna is Professor of Economics and Dean of the College of Information Science and Business Administration at Slippery Rock University in Pennsylvania. Prior to this present position, he served in a similar capacity at Xavier University. He received the Ph.D. in Economics from the University of Cincinnati.

Dr. Mastrianna has been a visiting professor at a number of universities. In addition, he has conducted workshops and been a featured speaker for many government, business, professional, and civic groups. He currently serves on the Board of Directors of the Union Central Life Insurance Company as well as numerous other boards and councils. As founder of his own consulting firm, he has engaged in extensive work involving economic evaluations of businesses and human capital. Dr. Mastrianna has co-authored textbooks covering a wide range of economic topics. He has also served as consulting editor and reviewer of several others.

CONTENTS

1

UNEMPLOYMENT
WHY AND
FOR HOW LONG?

Domestic goals have been cited for our economy by Presidential commissions, Congressional committees, business organizations, private economic groups, and labor unions. The three goals that are most prominent are full employment, economic growth, and stable prices. Serious unemployment must be avoided because of its economic consequence for the total economy and the hardship it brings to individuals and families.

The double-digit levels of unemployment associated with the recession of 1982 and carrying over into mid-1983 were the worst since the Great Depression when unemployment reached 25 percent. These high levels of unemployment exacerbated the strong feeling about the problem of unemployment.

Since the end of World War II the American economy has experienced intermittent periods of full employment. But even during periods of prosperity, unemployment has stayed at undesirable levels. After a 3½-year period of full employment in the late 1960s, nagging unemployment recurred in 1970 and averaged between 5 and 6 percent in 1971 and 1972. It eased below 5 percent in 1973, but was aggravated by the recession of 1974–1975 and reached 9.2 percent in the spring of 1975. During the economic recovery, unemployment dropped to 7 percent by 1977, 6 percent in 1978, and averaged 5.8 percent in 1979. With the recession of 1980, however, unemployment rose to over 7.5 percent by mid-1980 and over 8 percent toward the end of 1981. It then peaked at 10.7 percent in December, 1982, and remained at double-digit levels until June, 1983, when it dropped to 9.8 percent. By early 1987 it was still 6.6 percent.

The problem, then, is that in spite of highly prosperous conditions and substantial growth in our economy, we have had hard-core unemployment in many years. Naturally, questions arise. Why have we not had full employment? What do we mean by full employment? Who are the unemployed? How serious is the unemployment? Who are the hard-core unemployed? How do we compare with other nations in this respect? Is anything being done about it? What is the outlook for the future? Before analyzing the problem of unemployment, let us look at the definition, size, and composition of the labor force.

THE LABOR FORCE

In January, 1987, the total population in the United States was 242.4 million. Of this total, 183.6 million were in the category known as the *noninstitutional population;* that is, all persons 16 years of age or older, including members of the resident armed services, but excluding persons in institutions and the armed forces overseas. In January, 1987, of the noninstitutional population, 120.8 million were in the total labor force. The *total labor force* includes those in the noninstitutional population who are working or looking for work. Thus, it includes the unemployed as well as the employed. Furthermore, it includes proprietors, the self-employed, and members of the armed forces. The labor force, however, excludes all persons engaged exclusively in housework in their homes or attending school. Students, for example, are not members of the labor force unless they are working in addition to attending school. If they work or look for work during the summer vacation period, however, they become members of the labor force. Likewise, when they graduate, they generally become members of the labor force.

The Civilian Labor Force

If we subtract the number of persons in the resident armed forces from the total labor force, the remainder is known as the civilian labor force. By definition, the *civilian labor force* consists of all persons in the total labor force except members of the resident armed services. Since there were

1.8 million persons in the resident armed services as of January, 1987, the civilian labor force amounted to 119.0 million. Of the total labor force, 8.0 million, or 6.6 percent, were unemployed. The *unemployed labor force* includes all persons in the labor force seeking work.

Employment and Unemployment

The *employed labor force* is the difference between the total labor force and the unemployed. It is composed of all employed workers—including persons who did not work at all during the census week because of illness, bad weather, vacation, or labor disputes, but who had a job or business. It includes part-time as well as full-time employment. In January, 1987, the number employed was 112.8 million. Of this total, 3.1 million were in agricultural work, while 107.9 million were in nonagricultural employment.

There were 62.8 million persons in the noninstitutional population who were not in the total labor force. Homemakers comprised 29.3 million of this group. Another 9.2 million of those not in the labor force were in school. There were 1.5 million who stated that they were unable to work. The remainder, 21.8 million, was composed of those who had retired, individuals who did not want to work, those who did not have to work, and those having other reasons for not working. A breakdown of the population and the labor force is shown in Table 1.1.

Table 1-1

Population and Labor Force, January 1987

Category	Millions
Total population	242.4
Noninstitutional population	183.6
Total labor force	120.8
Armed forces	1.8
Total civilian labor force	119.0
Unemployed labor force	8.0
Employed labor force	111.0
Agricultural employment	3.1
Nonagricultural employment	107.9
Persons not in the labor force	62.8
Keeping house	29.2
In school	9.1
Unable to work	2.7
Other reasons	21.8

SOURCE: *Employment and Earnings* (February 1987).

Unemployment—Why and for How Long?   **3**

According to payrolls the bulk of the labor force is engaged in nonagricultural employment. The largest number, 24.2 million, is engaged in wholesale and retail trade. The second largest category, services, has 23.7 million workers. There are 19.2 million engaged in manufacturing, and government employment is fourth with 16.9 million workers, of whom 14.0 million work for state and local government. The sources of nonagricultural employment are shown in Table 1.2. From another point of view, 75.4 percent of today's workers are employed in service-producing industries, compared to only 24.6 percent in goods-producing industries.

Labor Force Participation Rate

The labor force has been growing in size because both the population and the labor force participation rate have been increasing over the past few decades. The *labor force participation rate* is the ratio of the labor force to the population. It can be calculated as a percentage of either the total population or the noninstitutional population. Since 1950, the percentage of the noninstitutional population in the labor force has been between 59 and 64 percent. The labor force participation rate compared to the total

Table 1-2

Employment in Nonagricultural Payrolls
by Industry Division, January 1987

Industry	Millions	Percentage
Goods-producing industries	25.0	24.6
Manufacturing	19.2	18.9
Mining	.7	.7
Construction	5.1	5.0
Service-producing industries	76.7	75.4
Transportation and public utilities	5.4	5.3
Trade (wholesale and retail)	24.2	23.8
Finance, insurance, and real estate	6.5	6.4
Services	23.7	23.3
Government (federal, state, and local)	16.9	16.6
Total nonagricultural employment[a]	101.7	100.0

[a]Total derived in Table 1-2 is not comparable with total nonagricultural labor force reported in Table 1-1. Table 1-1 includes proprietors, self-employed persons, domestic servants, and unpaid family workers; counts persons as employed when they are not at work because of industrial disputes, bad weather, etc.; and is based on a sample of the working-age population. Table 1-2 is based on reports from employing establishments.

SOURCE: *Employment and Earnings* (February 1987).

population has risen from about 40 to 48 percent, as shown in Table 1.3. The fact that the total labor force participation rate does change over the years indicates that the labor force has some degree of elasticity. The participation rate jumped noticeably in the 1970s due to the entry of the large number of women into the labor force.

The size and participation rate of the labor force also fluctuate annually. They have seasonal peaks and troughs. Consequently, actual labor force figures are usually adjusted for seasonal changes for purpose of analysis. It is a well-established pattern, for example, that rises in the size and participation rate of the labor force can be an aftermath of the pickup in the November–December Christmas rush. It must be remembered that these changes are not always evident in total labor force figures because the cyclical movements are superimposed on seasonal changes in the labor force. Seasonally adjusted data are necessary, therefore, for a proper interpretation of what is happening to the size and participation rate of the labor force as well as to the level of employment and unemployment. Seasonally adjusted figures show what is happening to the labor force exclusive of the seasonal changes that are taking place. Table 1.4 presents seasonally adjusted labor force and employment figures.

Age and Sex of Labor Force

Before we leave the structure of the labor force, the age composition of the labor force should also be noted. Table 1.5 shows that 7.1 million people

Table 1-3

Labor Force as a Percentage of Total Population

Year	Total Population (Millions)	Total Labor Force (Millions)	Total Labor Force Participation Rate
1950	152	64	42.1%
1955	166	68	41.0
1960	181	72	39.9
1965	195	77	39.6
1970	205	85	42.0
1975	214	95	45.3
1980	227	109	47.8
1985	239	117	49.0
1986	242	120	49.6

SOURCE: *Employment and Earnings* (February 1987).

Table 1-4

Status of the Labor Force, 1980–1987

(Thousands of persons 16 years of age and older, except where noted)

	Noninstitutional Population[a]	Total Labor Force[a]	Total Employment[a]	Civilian Employment			Total Unemployment	Unemployment Rate
				Total	Agricultural	Nonagricultural		
1980	169,349	108,544	100,907	99,303	3,364	95,938	7,637	7.0%
1981	171,775	110,315	102,042	100,397	3,368	97,030	8,273	7.5
1982	173,939	111,872	101,194	99,526	3,401	96,125	10,678	9.5
1983	175,891	113,226	102,510	100,834	3,383	97,450	10,717	9.5
1984	178,080	115,241	106,702	105,005	3,321	101,685	8,539	7.4
1985	179,912	117,167	108,856	107,150	3,179	103,971	8,312	7.1
1986	182,293	119,540	111,303	109,597	3,163	106,434	8,237	6.9
1987 (Jan.)	183,575	120,782	112,759	111,011	3,145	107,866	8,023	6.6

[a]Including resident armed forces

SOURCE: *Economic Indicators*, (February 1987).

in the civilian labor force are in the 16–19 age bracket, while at the other extreme 2.9 million are in the over-65 age bracket. The largest number, 80.8 million, are found in the 25–54 age group.

Although not shown in the table, it might be noted for future reference that of the total 52.4 million women in the civilian labor force approximately 60 percent are married. Of the married women in the labor force, 88 percent of them have husbands who are working.

THE MEANING OF FULL EMPLOYMENT

It is obvious that when we talk about full employment for our economy, we are not referring to a situation in which there is a job for everyone. Many people are too young to work, and others are too old. Some do not want to work, and others are physically or mentally incapable of work. At the beginning of 1987 there were about 29 million persons too busy with housekeeping chores to enter the labor force, and a large number of young people are still in school. Thus, full employment is not a condition in which the entire population is employed. In fact, in our dynamic economy with its mobile labor force, it cannot be expected that everyone in the labor force will be working. There will always be some workers quitting, others being discharged, and some moving to other positions. Furthermore, many persons upon completing vocational or skilled training are unable to find a job immediately at the particular occupation they desire and will refrain from accepting another position until they can find the type of work they want. In addition, we have a number of persons who want to work but have difficulty obtaining or holding a job because of physical or mental incapacities. Consequently, we can always expect some unemployment in our labor force.

Table 1-5

Civilian Labor Force by Age and Sex, January, 1987 (In millions)

Age	Total	Male	Female
16–19 years	7.1	3.7	3.4
20–24 years	15.0	7.8	7.2
25–54 years	80.8	45.0	35.7
55–64 years	11.9	7.0	4.9
65 and over	2.9	1.8	1.2
Total	117.7	65.3	52.4

SOURCE: *Employment and Earnings* (February 1987).

For nearly two decades it was generally accepted by reliable authorities, such as the President's Council of Economic Advisers, that full employment existed whenever 96 percent of the civilian labor force was employed. This allowed for 4 percent frictional unemployment, which was held to be consistent with full employment.

Today there are in the labor force a larger number and percentage of young people, women, and minority workers than existed in the mid-1950s when we came to accept the 4 percent unemployment figure as being consistent with full employment. These groups historically have had higher unemployment rates than the labor force as a whole. Consequently, when more statistical weight is given to these categories in establishing a full-employment unemployment rate today, it will yield a figure in excess of 4 percent, perhaps something in the range of 4.5–5.5 percent. The 1987 *Economic Report of the President,* for example, projected unemployment to remain between 6.7 percent and 5.5 percent until 1992 in spite of a predicted good economic recovery.

It has been suggested by others that full employment should be near the point where the number of job vacancies is about the same as the number of unemployed. Unfortunately, up to the present time this is a difficult measurement, since the Department of Labor's estimate of unfilled job orders is less than precise.

Although there are some sources that think the full-employment rate should be a lower or higher number, it is reasonable to define full employment as a condition in which 5 percent or less of the U.S. labor force is unemployed.

FULL EMPLOYMENT AND BALANCED GROWTH ACT OF 1978 (HUMPHREY-HAWKINS ACT)

After nearly a year of debate, revision, and compromise, Congress enacted the controversial Humphrey-Hawkins Bill under the official title of "Full Employment and Balanced Growth Act of 1978." The Act amended and embellished the Employment Act of 1946. In particular it included specific numerical targets and timetables regarding full employment and inflation rates. The Humphrey-Hawkins Act also required the President to spell out in the *Economic Report* measures designed to accomplish the objectives of the Act. This must be done by January 20 of each year.

The Act set a goal of 3 percent or less unemployment among adult workers (20 years of age or older) and 4 percent unemployment for the total labor force by the end of 1983. It required, too, that the rate of inflation be reduced to at least 3 percent by that time. In achieving these goals, the Act did give some preference to full employment insofar as it stated "the policies and programs designed to reduce the rate of inflation shall not impede the achievement of the goals for the reduction of unemployment." The long-term goal for reasonable price stability, however, was to reduce the rate of inflation to zero by 1988.

According to the Act, a balanced budget and balanced economic growth are to be sought after the goals concerning unemployment are reached. Moreover, the Act called for a narrowing of the differences in unemployment rates between various categories of the unemployed. Very little progress has been made toward this goal, however.

Two major items were included in the original Humphrey-Hawkins Bill which did not survive until the final version of the Full Employment and Balanced Growth Act of 1978. First was the concept of using the federal government as an "employer of last resort" for the unemployed if unemployment were not reduced to target levels within the five-year period. This item would have required the federal government to provide or find jobs for those who could not find employment. The second item was the establishment of a broad and specific planning system for the U.S. economy.

RECENT PROBLEMS OF UNEMPLOYMENT

Unfortunately, the endeavors of the Nixon Administration to effectively combat inflation in 1969 and 1970, along with other factors, led to slow-down in the economy. Unemployment, which had averaged 3.5 percent of the civilian labor force in 1969, rose to 6.2 by the end of 1970. Thus, in 1970 we experienced the anomaly of having both inflation and recession simultaneously.

In spite of the several measures tried by the Nixon Administration to reduce unemployment in 1971 and 1972, unemployment for 1971 averaged 5.9 percent and was 5.6 percent in 1972. The unemployment rate did drop to 4.9 percent in 1973. As a result of the 1974–1975 recession, it rose again to 5.6 percent in 1974 and, after peaking at 9.2 percent in May, 1975, averaged 8.5 percent for that year. With economic recovery, unemployment fell to 6.1 percent in 1978 and 5.8 percent in 1979.

Recession occurred again in 1980, and unemployment rose to 7.8 percent by July of that year. It remained at or near 8 percent until late 1981, when it rose above 9 percent. It continued to rise in the recession of 1982, finally peaking at 10.7 in December of that year. By the end of 1983, a year of economic recovery, unemployment was down to 8.2 percent and had fallen to 6.9 percent by December 1985. But by early 1987 it was at 6.5 percent.

We might reflect upon the current problems of unemployment in many ways: Why has unemployment stayed high among certain groups, such as teenagers and nonwhites, even in a full employment period? How can hard-core unemployment be eliminated or alleviated? What is causing the present state of unemployment? How is it possible to have unemployment and inflation simultaneously? On the other hand, we might ask, just how serious is unemployment?

WHO ARE THE UNEMPLOYED?

The question has often been asked in recent years: Who are the unemployed? Are they members of a particular group? Are they older or younger people? Are they skilled or unskilled? Are the same people continuously unemployed, or is there a turnover in the ranks of the unemployed? Let us try to find a few answers.

Types of Unemployment

Frictional unemployment arises from the normal operation of the labor market. It will occur even in periods of full employment. Workers are constantly being hired, fired, quitting, withdrawing from the labor force for special training, taking their time on entering or reentering the labor force to search for the right job, and relocating, often interstate, from one job to another. This unemployment is usually short term in nature. Much of the full-employment unemployment can be attributed to frictional unemployment.

Structural unemployment is caused by an imbalance between the skills possessed by workers and those demanded in the labor markets. Structural unemployment tends to be long term in nature. It is argued that a substantial portion of our present unemployment is structural in nature insofar as we have imperfect adaptation of workers to jobs. In short, unemployed workers are often not qualified to fill available job openings. For example, workers are rapidly displaced because of technological development and automation. New skills arise and old skills are no longer as important (or the demand for them no longer as great) as they once were. Consequently, displaced workers, usually 2 million or more annually, have difficulty finding new jobs. Some workers are unemployed because they lack proper training, others because they lack the geographic mobility to take advantage of job opportunities. As a result, we always have unfilled jobs on the one hand and jobless workers on the other.

Current data from the Department of Labor indicate that the number of vacancies is substantial. Moreover, it has been suggested by some groups that perhaps in our growing, dynamic economy we ought to expect a higher rate of structural unemployment than we have been accustomed to in the recent past.

Technological unemployment is a term describing a particular form of structural unemployment. It occurs when workers are displaced because of the introduction of modern labor-saving machinery and equipment.

Cyclical unemployment is the result of a less than full use of productive capacity due to a recession or depression. It is due to insufficient aggregate demand in the economy. If the aggregate demand can be strengthened by an increase in consumption, investment, or government spending, the level of business activity can be increased and the cyclical unemployment perhaps reduced. Much of the high rate of unemployment associated with the recessions of 1974–1975, 1980, and 1982 was cyclical in nature. Cyclical unemployment may be short term or long term, depending on the length of the recession.

Induced unemployment is a consequence of subsidies provided by public socioeconomic programs. One source of induced unemployment is the unemployment compensation system in the United States. Although it serves as an automatic stabilizer and has many other benefits, it tends to increase unemployment. The tax levied on the employer, for example, is not in direct proportion to the employer's layoffs. Therefore, it may be easier for a firm to lay off workers than it would be if it had to pay the full cost of the unemployment benefits that result. Secondly, in many cases the minimal difference between unemployed persons' unemployment benefits and what they may be able to earn on other jobs may deter their search for a new job. This is especially true where laid-off workers, such as those in the auto industry, receive supplementary unemployment benefits (SUB) from their employers. Unemployment benefits, plus SUB, provide the workers with an income equal to 90–95 percent of their regular take-home pay. Other programs such as welfare payments, food stamps, and rent subsidies can cause a similar upward bias on the rate or duration of employment.

Duration of Unemployment

The average length of unemployment for an idled worker generally varies with economic conditions. The period of unemployment is longer in a time of economic sluggishness or recession than in a period of prosperity. In the recession of 1982, it reached 20 weeks. In 1985, it averaged 15.6 weeks. Long-term unemployment, persons seeking work for 15 weeks or more, generally tends to change correspondingly.

Long-term unemployment is usually more prevalent among certain groups in the labor force, such as older workers, nonwhites, and workers laid off in industries manufacturing durable goods. Although stable employment seems to be more characteristic of white-collar workers, unemployment is found among professional and technical workers, craftsmen, clerks, and salesworkers, as well as skilled and unskilled workers. This was especially true in the recession of 1982 and into 1983.

Characteristics of the Unemployed

An analysis of unemployment data points out some of the characteristics of the unemployed which will add to our understanding of the problem.

Age and Race. The incidence of unemployment is usually high among teenage workers who are likely to change jobs frequently. The rate of unemployment among nonwhite male workers has been twice the rate of white male workers. This, of course, is due in large part to the fact that the nonwhite workers are concentrated rather heavily in the unskilled and semi-skilled occupations where unemployment rates are generally high. The number of unemployed by age and race can be seen in Table 1.6. Note that 35.7 percent of our total unemployed was in the 24 years of age or under category, and that

Table 1-6

Unemployment by Race and Age, January, 1987

	Millions	Percentage
Total Unemployed	8.6	100.0%
Race		
White	6.6	76.7
Black	1.8	20.9
Other	.2	2.4
Age		
16 to 19 years	1.3	15.4
20 to 24 years	1.7	20.3
25 to 54 years	5.0	58.1
55 to 64 years	0.5	5.2
65 years and over	0.1	1.0

SOURCE: *Employment and Earnings* (February 1987).

15.4 percent was in the 16–19 age bracket. In the 25–54 age group it was 58.1 percent.

At the other end of the spectrum, nearly 3 million elderly persons on pensions or Social Security worked and were counted as employed. On the other hand, about 750,000 people in this age group, were seeking work and listed among the unemployed.

Table 1-7 shows the unemployment rate for various categories of people in the civilian labor force. The lowest unemployment rate, 4.5 percent, was among married men. Teenage unemployment was nearly 18.0 percent. The unemployment rate for black teenagers was over 39 percent.

Sex and Marital Status. Of those unemployed, 3.5 million, or 41 percent, are female. The unemployment rate among women in January, 1987, was 5.9 percent, which was below the national average. Other unemployment rates are given in Table 1-7.

Since nearly 60 percent of the women in the labor force are married, it can reasonably be assumed that married women will comprise a substantial percentage of our unemployment. Data for January, 1987 show this figure to be 19 percent. In fact, 77 percent of the 1,338,000 unemployed wives

Table 1-7

Unemployment Rates within Various Categories
in the Civilian Labor Force, January, 1987

All workers			6.7%
White		6.5%	
Male	7.0%		
Female	5.9		
Black		13.9	
Male	14.1		
Female	13.8		
Hispanic origin		10.6	
Married men			4.5
Experienced workers			6.3
Teenagers (16–19 years)			17.7
White	16.2		
Black	39.1		
Labor force time lost[a]			7.6

NOTE: Not seasonally adjusted.
[a]Total hours lost by the unemployed and persons on part-time employment for economic reasons as a percent of potentially available labor force hours.

SOURCE: *Employment and Earnings* (February 1987).

in the civilian labor force had husbands who were currently employed, and 54 percent of the unemployed husbands had wives who were working. In only a small number of the households, 194,000, were both husband and wife unemployed.

The Hard-Core Unemployed. Within the hard core are illiterates, the chronically ill, those physically and mentally incapacitated, and the like. The hard core generally is so demoralized that individual rather than mass treatment by job placement services is needed. Many existing government and other retraining programs are far beyond the capabilities of the hard core. These people form a highly singular class of unemployed and unemployables, socially and economically isolated.

It is suggested that even when the full employment goal of 5 percent unemployment is realized, there still is at the bottom 1 percent unemployable. Studies indicate that the hard core has developed into a class of social outcasts, characterized by very low incomes, by residence in blighted areas, by living in isolation from the mainstream of life, and by their "attitude." The attitude found includes: (1) no feeling of obligation to the family; (2) deep dejection at inability to find work; (3) general loss of self-respect; and (4) mental imbalance. It was further revealed that many of these people do not know how to look for a job, fill out an application form, or market any skills and experience they may have.

MEASUREMENT OF UNEMPLOYMENT

There are many ways in which the employed and the unemployed can be measured. Usually employed figures are more reliable than those for unemployment, and much controversy can arise regarding the method used to count the unemployed.

The Survey Method

Information on the labor force is obtained from the Bureau of Labor Statistics (BLS) Current Population Survey and is reported in the Department of Labor publications, such as the *Monthly Labor Review* and *Employment and Earnings.* Figures are gathered through a survey taken the week ending nearest the twelfth of the month. The sample is made up of 59,500 households in 629 areas throughout the nation, a sampling ratio of 1 in every 1,200 households in the United States. According to the survey, persons are classified into three basic groups: employed, unemployed, and not in the labor force. The survey is made by well-trained interviewers. Questions are asked carefully and skillfully in order not to influence the response of the interviewee. Any responsible person, usually the homemaker, may answer the questions concerning the working status of other members of the household who are not at home during the time of the interview. The BLS also conducts a monthly survey of 200,000 business establishments that employ about 35 million people.

Employed persons are (1) those who did any work at all as paid employees during the survey week or who worked in their own businesses, professions, or farms or who worked 15 hours or more as unpaid workers in family businesses; (2) all those who were not working but who had jobs or businesses from which they were temporarily absent because of illness, bad weather, vacation, labor disputes, or personal reasons; and (3) members of the armed forces stationed in the United States.

Unemployed persons are all persons who did not work during the survey week, who made specific efforts to find a job within the prior four weeks, and who were available for work during the survey week (except for temporary illness). Also included as unemployed are those not looking for work, but (1) waiting to be called back to a job from which they had been laid off or (2) waiting to report to a new job within 30 days.

Not in labor force includes all civilians 16 years and over who are not classified as employed or unemployed.

One group classified as not in the labor force under present definitions consists of *discouraged workers*. These are persons who believe they cannot get a job because no jobs are available or because some personal factor, such as age, lack of skill or training, or some sort of discrimination would prevent them from finding work. A special survey in 1981 indicated that two-thirds of the 1.1 million discouraged workers that year cited job-market factors as the reason they could not find work. A substantial controversy exists among labor leaders, business leaders, and government officials as to whether discouraged workers ought to be counted as unemployed rather than as not in the labor force. The number of discouraged workers reached 1.2 million during the 1982 recession.

The household survey is the major source of information about the total labor force as well as about the employed and unemployed. Although the survey method has some weaknesses, the Department of Labor prefers it to the count of state unemployment claimants because a large number of persons in the labor force are not eligible for unemployment benefits. In addition, claimants may exhaust their benefits, cease to be claimants, and therefore no longer be listed as unemployed in the state employment office count. According to the Department of Labor, the survey method is 95 percent certain of being correct. Nevertheless, the state BES (Bureaus of Employment Services) data have certain advantages because the states prepare them every week rather than monthly, as the BLS survey method does. Furthermore, the data can yield specific information about local labor market areas.

Recent experiences indicate that total unemployment can be estimated by dividing the number of unemployment claimants by a corrective factor of 0.44. Thus, average weekly unemployed claimants of 3.4 million in January, 1986, would be adjusted to 7.7 million unemployed in the total civilian labor force. This is close to the actual BLS survey count of about 7.8 million. But such a method of projecting total unemployment is not as accurate as the survey method, especially since the corrective factor used varies from state to state and changes over time.

Underemployment

For decades the BLS survey measured the employed and unemployed, but made no attempts to measure the *underemployed.* For this reason, some critics maintained that the survey did not measure the true level of unemployment since it neglected unemployment due to the short work week. So great was this concern that recording of underemployment was started in the mid-1960s.

Labor force time lost is a measure of worker-hours lost to the economy through unemployment and involuntary part-time employment and is expressed as a percentage of potentially available worker-hours. It is computed by assuming: (1) that unemployed persons looking for full-time work lost an average of 37.5 hours, (2) that those looking for part-time work lost the average number of hours actually worked by voluntary part-time workers during the survey week, and (3) that persons on part-time work for economic reasons lost the difference between 37.5 hours and the actual number of hours they worked. In January, 1987, labor force time lost was 7.6 percent compared to the actual measure of 6.6 percent unemployment. The difference between these two figures is a measure of underemployment.

This procedure has some weaknesses, however. First, it is questionable whether 37.5 hours or 40 hours should be used as the norm. Second, if we incorporate underemployment in the rate of unemployment, should we adjust the unemployment rate downward when we have overtime employment, such as 40.5 average hours of work? Even using a 37.5-hour norm, that would mean hyperemployment of 8 percent ($3.0 \div 37.5 = 8$ percent) from overtime. Overtime then more than offsets the labor force time lost.

Multiple Job Holders

Some critics feel that the practice of *moonlighting,* the holding down of two or more jobs, should be curtailed in the interest of reducing unemployment.

Statistics indicate that in 1986, for example, 5.5 million persons, or 4.9 percent of the employed labor force, held two or more jobs. Most of these multiple job holders, 4.6 million, were in nonagricultural industries. Eliminating moonlighting would not increase employment proportional to the decrease in multiple job holders, of course, because most of the secondary jobs are part-time. Some of them amount to only a few hours a week, and frequently the rate of pay is less than that for the primary job of the worker. In some cases the secondary job is that of an unpaid family worker. In the case of two-thirds of a million persons whose primary jobs were in nonagricultural industries, the secondary jobs were in agriculture. Many of these were farmers working at city jobs.

Job Vacancies

Sometimes it is suggested that jobs are available if the unemployed would just get out and look for them. In many cases, however, the unemployed may not have the skill, aptitude, or mobility to take advantage of existing job vacancies. In other cases the job may be insignificant, vocationally and economically, compared to the usual line of work of the unemployed person.

Job vacancies, as measured by the BLS, are the stock of unfilled job openings for all kinds of positions, both full-time and part-time, permanent and temporary, for which employers are actively seeking workers. The job vacancy rate is computed by dividing the number of current job vacancies by the sum of employment plus vacancies and multiplying by 100.

In 1984 there were an average of 572,000 job vacancies per month reported by state employment services in the United States. About half of these were considered as long-term vacancies, those that remained unfilled for 30 days or more. At that time there were 8.4 million persons jobless. The unemployment rate was 7.4 percent, and the job vacancy rate was 0.6 percent. If the unemployed had filled the job vacancies at that time the rate of unemployment would have dropped less than one percentage point.

Criticism of the Survey Method

Although there is general agreement that the BLS survey count of employment is very good, numerous criticisms have been leveled at its method of counting unemployment. Some critics feel that we should not count secondary wage earners among the unemployed, others suggest that students should not be included in the ranks of the unemployed, and many critics object to the listing of retirees as unemployed. Others go so far as to say that we should not count anyone as unemployed who does not need a job. This does, however, set up a normative qualification which could be very difficult to measure. Furthermore, if people not working who do not need a job are not counted as unemployed, should the people who are working but do not need a job be excluded from the count of the employed and perhaps removed from the count of the labor force? In still other cases, it has been suggested that the degree of unemployment may be influenced by the enthusiasm of the census takers; the harder they look to find unemployment, the more they will find. On the other hand, some critics suggest that the unemployment figure is too low for a number of reasons. Among these they cite the fact that many job seekers may become discouraged and withdraw from the labor market. Then there are those who feel that we put too much economic, social, and political emphasis on the measure of unemployment. They contend that the important measure is that of total employment rather than that of unemployment.

Over the years the Department of Labor has admitted that many complexities exist in the measurement of the unemployed. It has staunchly defended the BLS survey method, however, explaining the purpose and reasoning behind many of the statistical calculations. The Department has stated many times that what is needed is not a rejection of the statistics, but provision for more detail and more meaningful breakdowns so that the data would be more useful for public policy decisions.

Evaluation of Data

The United States generally receives high praise from statisticians throughout the world for the methods, techniques, frequency, thoroughness, and integrity of its statistical data on such matters as unemployment, production, and prices. Nevertheless, in recognition of the many questions about the collection and measurement of employment and unemployment data, the President and Congress in 1976 established a National Commission on Employment and Unemployment Statistics to evaluate our present system of collecting, calculating, and disseminating employment and unemployment statistics and to make recommendations on methods of improvement.

After months of hearings, study, and deliberation, the Commission issued its final report, *Counting the Labor Force,* on Labor Day, 1979. The Commission stated that it was "reasonably satisfied that available data is used in appraising current labor market trends." But the Commission found that the "richness" of the data that describe an individual's current labor force status is not matched by information on how that status came about or under what conditions it would change.

The Commission proposed no major changes, but offered 88 recommendations to help the statistical system reflect changing economic conditions and policy needs, including a plan for expanding the size of the statistical sample and several minor changes in labor force definitions. No recommendation was made, however, for changing the definition of unemployment by including "discouraged workers" or for changing to 18 the age for inclusion in the labor force. Nor did the Commission recommend a much discussed "hardship" index linking income to unemployment data. It did, nevertheless, recommend some type of annual report on economic hardship.

U.S. UNEMPLOYMENT COMPARED TO THAT OF OTHER NATIONS

Critics of our economy have called attention at times to the fact that unemployment in some countries, such as Japan and West Germany, is lower than it is in the United States. After being adjusted for comparison purposes, unemployment rates for the United States and other major nations are shown in Table 1.8. The U.S. ranking improved after the recession of 1982. Moreover, the U.S. economy created new jobs at a faster rate than most other nations during the 1983–1987 recovery period.

Table 1-8

National Unemployment Rates
(Adjusted to the United States definition)

	1982	1984	1986
United States	9.7%	7.5%	7.1%
Australia	7.1	9.0	7.9
Canada	11.0	11.3	9.5
France	8.5	10.0	10.3
West Germany	5.9	7.8	7.5
Great Britain	12.3	13.0	13.1
Italy	4.8	5.6	6.3
Japan	2.4	2.8	2.8
Sweden	3.1	3.1	2.6

[a]Second quarter.

SOURCE: *Statistical Abstract of the United States: 1986 and Monthly Labor Review* (January 1987).

WHAT IS BEING DONE ABOUT UNEMPLOYMENT?

Under the Employment Act of 1946, the Administration in office has an obligation toward preventing, reducing, or eliminating unemployment. Section Two of the Act declares:

> The Congress hereby declares that it is the continuing policy and responsibility of the Federal Government to use all practicable means consistent with its needs and obligations and other essential considerations of national policy with the assistance and cooperation of industry, agriculture, labor, and state and local governments, to coordinate and utilize all its plans, functions, and resources for the purpose of creating and maintaining in a manner calculated to foster and promote free competitive enterprise and the general welfare, conditions under which there will be afforded useful employment opportunities, including self-employment, for those able, willing, and seeking work, and to promote maximum employment, production and purchasing power.

This Act, furthermore, requires the President to make an *Economic Report* in January of each year. The report, delivered to a Joint Congressional Committee, reviews economic conditions of the previous year, gives a preview of the current year, and makes recommendations for bringing about maximum or full employment. The Act likewise provides for a President's Council of Economic Advisers to aid and assist him in making the report and in making recommendations for implementing the Act.

The Humphrey-Hawkins Act of 1978, as mentioned earlier, expanded and strengthened the mandate of the Employment Act of 1946. It specified a procedure for the setting of numerical goals to be reached by 1983. The goals specified in the law are an adult unemployment rate of 3 percent, an overall unemployment rate of 4 percent, and a rate of inflation of 3 percent. The President, however, may, and has, recommended modification of the timetable for achieving these goals.

In addition to continuation of the annual Presidential *Economic Report,* the Humphrey-Hawkins Act requires the Federal Reserve Board to submit to Congress twice annually its objectives and plans for monetary policy and to discuss the relationship between those plans and objectives and the President's short-term goals contained in the *Economic Report.*

The Act also emphasizes the importance of reducing the differences in unemployment rates of specific groups within the labor force, such as black and Hispanic Americans, in the attempt to achieve full employment.

Measures to Alleviate Cyclical Unemployment

Many attempts have been made to lessen unemployment. The various measures attempted or suggested fit into two broad categories: those that try to reduce cyclical unemployment, and those that seek to correct structural unemployment. In the first category are the several Congressional acts and executive actions that aim at raising the total demand in the economy by encouraging higher consumption, greater business investment, or higher government spending. Easier monetary policies by the Treasury and the Federal Reserve have worked in this direction. Government deficits in excess of $100 billion in the past several years have been used with the notion that they might help increase employment or prevent unemployment.

During the period 1960–1986 many measures and programs were implemented to alleviate cyclical unemployment. In the period of nagging unemployment of the early 1960s, for example, personal and corporate taxes were reduced, an emergency public works program was implemented, a program was started to reduce poverty, tax credits were used to stimulate investment, unemployment compensation was extended for an additional 13 weeks, and excise taxes were reduced.

During and following the recession of 1974–1975 taxes were again reduced, unemployment compensation extended, large federal deficits tolerated, the money supply increased, interest rates lowered, tax credits increased, public employment programs expanded, and other measures taken to bolster the level of economic activity and to reduce unemployment.

In the early 1980s various supply-side measures were implemented to encourage savings, stimulate investment, motivate worker effort, and deregulate the economy. Many of these measures were contained in the Economic Recovery Tax Act of 1981. With the severity of the recession of 1982, unemployment benefits were extended from their normal 26 weeks to 52 weeks. Moreover, many workers who were laid-off or who lost jobs as a result of increased imports became eligible for additional unemployment programs under the Trade Adjustment Assistance Program that was established in 1974.

Measures to Alleviate Structural and Technological Unemployment

Measures mentioned in the previous section are attempts to increase over-all effective demand through increases in investment, consumption, and government spending. Some attempts have also been made to eliminate or alleviate structural unemployment. Structural and technological unemployment are conditions in which unemployed workers' skills do not fit available job opportunities.

Area Redevelopment Act (1961). This Act tried to bring industry to depressed areas and jobs to displaced workers. The main features of this Act were the financial aids provided for distressed areas or areas with labor surpluses. These aids took the form of loans and grants for the construction of community projects and loans for private industrial undertakings of various types that would help to lessen unemployment in the area. Included in the program was training to prepare workers for jobs in new and expanded local industries. During the life of the Act, over 1,000 projects involving 65,000 trainees in 250 development areas were approved.

Manpower Development and Training Act (1962). The primary purpose of this Act was to provide training for the unemployed and underemployed to qualify them for reemployment or full employment. The MDTA allocated funds among states on the basis of each state's proportion of the total labor force, its total unemployment, and its average weekly unemployment payment.

The Act established training courses in those skills or occupations where there was a demand for workers, and the trainees had a reasonable chance of securing employment upon completion of the training program. Such programs were set up through the local state employment service utilizing state and local vocational education institutions, although private schools and other training institutions could be used. During the lifetime of the Act, over 2 million enrollees received training under MDTA programs.

Economic Opportunity Act (1964). This Act was passed for the purpose of establishing several programs in an effort to eliminate poverty. Included in the war on poverty were numerous programs sponsored under the cooperation of several federal, state, and local agencies. Among other things, the Act provided for the establishment of youth conservation camps, work-training programs for unemployed youths, and work-study programs for high school and college students of low-income families. Included also in the poverty package were provisions for special programs to combat poverty in rural areas, loans to business to increase investment and raise employment, urban job centers for youth, literacy programs for adults, and a VISTA Corps (Volunteers in Service to America).

The Office of Economic Opportunity (OEO) was established for the implementation of the Act. Most programs were coordinated at the local level through Community Action Commissions set up in metropolitan centers.

Appalachian Regional Development Act (1965). As a result of the findings of President Kennedy's Appalachian Commission, Congress voted to enact the Appalachia Bill, which provided for various types of aid for a 13-state area extending along the Appalachian Mountains from New York State to eastern Mississippi. The program for this depressed area was aimed at developing an economic base to encourage subsequent private investment as a means of improving its economic level.

In the early stages major emphasis was placed on road construction, health facilities, land improvement and erosion control, timber development, mining restoration, and water resource surveys. The Act provided nearly $1 billion to improve the economic condition of the area in the hope of raising production, employment, and income of its inhabitants.

The JOBS Program (1968). Continued emphasis has been placed upon worker development and training. Programs involving more creative collaboration between private industry and the federal government have been attacking the problem of hard-core unemployment and poverty. President Johnson's Manpower message of 1968, for example, called for the establishment of Job Opportunities in the Business Sector (JOBS).

The program was built on a commitment by groups of business people in metropolitan areas to hire thousands of disadvantaged people and give them on-the-job training, counseling, health care, and other supportive services needed to make these individuals productive workers. The program was built on the premise that immediate placement on a job at regular wages, followed by training and supportive services, rather than training first in an effort to qualify for the job, would provide superior motivation for these disadvantaged workers. By 1978, several hundred thousand disadvantaged workers had been given jobs by individual company efforts and through Department of Labor contracts. Six of every eight workers hired on federally financed programs were blacks and one in eight was Hispanic American. About half of those hired were under 22 years old.

Public Service Employment. The Emergency Employment Act of 1971 authorized the establishment of a Public Employment Program (PEP). For this purpose Congress appropriated $2.25 billion for a two-year period to finance transitional public service jobs at state and local governments levels. Funds were allocated when national unemployment equaled or exceeded 4.5 percent for three consecutive months.

Although the Emergency Employment Act expired, the concept of public service employment was continued and incorporated into the Comprehensive Employment and Training Program. In 1980, over one million workers were engaged in public service employment.

Comprehensive Employment and Training Act (1973). A major step toward the decentralization and decategorization of employment programs was taken when President Nixon signed into law, on December 28, 1973, the Comprehensive Employment and Training Act (CETA). The goal of the Act was to transfer the responsibility and resources for many manpower programs from the federal government to states and localities. In doing so, an effort was made to reduce the fragmented efforts of nearly 10,000 programs throughout the nation. The Act made governors and elected officials of major cities and counties responsible for the planning and operation of comprehensive employment programs. In addition to regular funds for such things as public service employment programs, the Act provided special funds for areas where unemployment was 6.5 percent or higher for three consecutive months. Eighty percent of the available funds were directed to state and local governments. The other 20 percent were used by the Secretary of Labor to administer certain national programs such as the Job Corps. Many existing programs, such as JOBS and MDTA, were placed under the jurisdiction of CETA. Since the Act was passed, several states have implemented comprehensive manpower programs.

CETA was subsequently amended by the Emergency Jobs and Unemployment Assistance Act of 1974 and by the Emergency Jobs Program Extension Act of 1976. CETA was reauthorized by Congress in 1978. The purpose of CETA was to provide training, employment and other services leading to unsubsidized employment for economically disadvantaged, unemployed, and underemployed persons. The several titles of CETA authorized a variety of activities.

Title I, for example, established a nationwide program of comprehensive employment and training services administered by prime sponsors which, for the most part, were states and local government units. Titles II and VI provided programs of temporary public service employment during periods of high unemployment. Title III provided funds for supervised training and job placement programs for special groups such as young workers, ex-offenders, older workers, persons of limited English-speaking ability, Indians, and migrant and seasonal workers. Title IV authorized a program of intensive education and training, known as the Job Corps, for disadvantaged youths. Title VIII established a young adult conservation corps. In 1981, Congress appropriated $7.7 billion for various CETA programs, and 2.9 million persons participated in CETA programs.

CETA had approximately 60–70 percent "positive terminations" of participants in its programs. Positive terminations are defined as individuals who were placed either directly or indirectly in unsubsidized employment, found jobs through their own efforts, enlisted in the armed forces, or engaged in other activities that increased their employability.

In fiscal 1981, for example, 68 percent of the CETA participants received positive terminations. Thirty-seven percent were placed on jobs either directly or indirectly or through self-effort. Sixteen percent were considered positive terminations since they left CETA programs to enroll full time in school, to enroll in a program not funded by CETA, to enter the armed forces, or to engage in any other activity that increased their employability. Thirty-two percent of the participants were nonpositive terminations who refused to continue their participation in CETA programs or left for reasons unrelated to jobs or activities that increased their employability.

In addition to several special programs for American Indians, migrant workers, ex-offenders, and older workers, CETA provided funds for a summer employment program for economically disadvantaged youth. In 1981, $769 million was distributed to help provide short-term jobs for 774,000 youths aged 16–21. Participants worked in such places as schools, libraries, community service organizations, hospitals, and private nonprofit agencies. Typical positions included nurse's aide, typist, school maintenance aide, cashier, library aide, clerk, and nutrition and day-care aide.

The Jobs Corps, established by the Economic Opportunity Act of 1964, continued under Title IV of CETA. It was designed to assist disadvantaged youths aged 16–21 to become more responsible, employable, and productive. All participants are out of work, out of school, and in need of additional education, vocational training, and counseling. A total of 114,000 enrollees were in the Job Corps in 1981. Most of these were from low-income families. Fifty-five percent were black youths, and eleven percent were of Hispanic origin. Eighty-seven percent had less than a high school education. In fiscal 1981 the Job Corps had an overall placement rate of 86 percent. Of 41,500 participants available for placement, 68 percent were placed in jobs, 28 percent entered or returned to school, and 4 percent entered the armed forces. Since its inception in 1964, over 750,000 young people have enrolled in the Jobs Corps. In 1981 the unit cost for Job Corps operations was approximately $15,000. In constant dollars, however, the per-unit cost was less than the cost in 1968.

Another major program established in the 1960s and placed under the auspices of CETA was the Work Incentive Program (WIN). All applicants for and recipients of Aid to Families with Dependent Children who were 16 years of age or older were required to register for WIN unless legally exempt. In the early years of WIN emphasis was placed upon increasing job readiness through counseling, training, and supportive measures. Later emphasis shifted to immediate job placement, with training and other assistance provided only when placement was not feasible.

In fiscal 1981 there were 1,156,000 current WIN registrants. About one-fifth of these were volunteers who had been legally exempt from registering. Three-fourths of the registrants were women, and more than one-half of the registrants were white. Sixty percent of the WIN registrants had not completed high school. In 1981, many WIN participants, upon completion of their program, entered subsidized employment. Most of them were employed in entry-level jobs at the minimum wage level.

Fiscal 1981 was the first full year of operation of the Private Sector Initiative Program, funded by CETA. It brought together the public and private sectors of the economy in an effort to improve delivery of employment and training programs. During 1981 the program served 117,000 participants, twice as many as the previous year. Of the 86,900 persons leaving the program, 53 percent were placed in unsubsidized employment, 89 percent in the private sector. An additional 10 percent left the program to return to school or enter other employment and training programs. Forty-nine percent of the participants were minorities.

Job Training Partnership Act of 1982. JTPA established a more formal partnership between private industry, the public sector, and vocational training schools to plan, design, and provide federally financed training. Federal resources are targeted to those identified as most in need: economically disadvantaged youth, low-skilled and chronically unemployed adults, and skilled workers who have lost their jobs in declining industries and regions.

Under JTPA training and employment programs are designed both to improve individuals' abilities to obtain and keep jobs by developing job skills and to support services that match individuals with jobs. The major federal activities in this area are financed through grants to states. These grants include a block grant that allows states to design training programs to meet the need of their disadvantaged population and categorical grants for an employment service, public service employment for older workers, summer youth employment and training, and job placement and training for workers displaced by changing economic conditions.

When CETA was terminated in September, 1983, many of its programs were brought under the jurisdiction of the new JTPA. Since then, however, a number of the older programs have been phased out. The 1987 federal budget, for example, reduced the Job Corps from 40,500 to 22,000 slots. Moreover, total outlays for training and employment were reduced from $5.6 billion to $4.5 billion.

Targeted Jobs Tax Credit Program. The federal Tax Equity and Fiscal Responsibility Act of 1982 provides tax credits or wage subsidies to employers who hire youths, welfare recipients, Vietnam veterans, and handicapped persons. Tax credits are available for up to two full years. In the first year the tax credit is equal to 50 percent of the hired person's earnings up to a maximum of $3,000. In the second year the tax credit equivalent is 25 percent up to a maximum of $1,500.

CONCLUSION

The labor force is destined to grow in the future. With this growth and development, however, will come new problems, especially in the absorption of young workers into the employed sector of the labor force and the reabsorption of workers displaced by rapid technological development. In addition, the recurrence of cyclical unemployment, such as that which accompanied the 1980 and 1982 recessions, aggravates the problem of unemployment.

It is expected that the civilian labor force will grow to 127.5 million by 1995. This will be an 8.5 million increase over the 119.0 million labor force of 1987. This means that we will be adding about 1.2 million workers to the labor force annually. During the 1970s, the labor force grew at an annual rate of 2.2 percent. In 1980–1983 it slowed to 1.8 percent and will fall to 1.2 percent in 1985–1990. It will drop below 1 percent in the 1990s. During the ten year period 1985–1995 the number of workers in the 25–34 age group will remain steady.

Fortunately there will be a decline in the number of 16- to 19-year-old workers entering the labor force between 1985 and 1995. This should help reduce unemployment among the youth of the nation, which, on average, is more than double the adult unemployment rate. But there will be about the same number of young blacks in the labor force. Unemployment rates among black teenagers are usually at least double what they are among white teenagers.

The proportion of women entering and reentering the labor force will continue to rise. Especially noticeable will be the reentry of many homemakers into the labor force once their family responsibilities have been reduced with the maturing of their children.

The challenge to maintain full employment is great because nearly one-third of the new entrants to the labor force are high school dropouts who lack training or skills. Unfortunately the demand for industrial, unskilled workers will not increase in proportion to the number of youths coming into the labor force. The largest increase in demand will be for professional, clerical, and skilled workers, and for those occupations which generally require some degree of training and skill or a higher level of education.

Another area of concern is arising as a consequence of increasing life expectancy. There will be a growing number of older persons in the labor force. Many of these persons want to continue in active employment rather than retire completely from the labor force. Probably the greatest difficulty will occur in occupations which will not show substantial growth. Farm workers likewise present a problem since the actual number required by 1995 will be less than the number required today.

Rapid technological development and automation will aggravate the unemployment problem as old skills and occupations disappear and new ones arise in their places. The changing structure of industry from mass production to high tech will also eliminate many jobs, adding to the unemployment problem. At present there are more than 2 million workers displaced each year because of various economic and technological changes taking place. Occupational and geographic mobility will become more essential in maintaining a high level of employment. This is especially so in smokestack industries where current and pending retrenchment and imports have caused widespread job losses. Training and retraining will become more important in the solution to the problem of persistent unemployment.

QUESTIONS FOR DISCUSSION

1. Is 4 percent unemployment a reasonable goal for a full-employment economy?
2. Should formal retraining programs, such as those offered through JTPA, be continued as a solution to the problem of unemployment?
3. Should the federal government make payments to private industry or give them income tax credits for training disadvantaged workers?
4. Should a short work week be promoted as a means of lessening unemployment?
5. Should "discouraged workers" be counted as unemployed?
6. What effect do you think the minimum wage has on teenage employment?
7. Would it be beneficial to promote more vocational education in our high schools?
8. Do you agree with the objective of the Job Training and Partnership Act?
9. Should moonlighting be eliminated or regulated as a means of spreading employment?
10. Should the federal government serve as an "employer of last resort" for the unemployed?

SELECTED READINGS

Bednarzik, Robert W. "Worksharing in the U.S.: Its Prevalence and Duration." *Monthly Labor Review* (July 1980).

Carter, Charlie. "The Outlook for Unemployment." *Economic Review*. Federal Reserve Bank of Atlanta (September–October 1979).

Counting the Labor Force, 1979 (Final Report of the National Commission on Employment and Unemployment Statistics). Washington D.C.: U.S. Government Printing Office (1979).

Finegan, T. Aldrich. "Should Discouraged Workers Be Counted as Unemployed?" *Challenge* (November–December 1978).

Full Employment and Balanced Growth Act of 1978.

Hall, R.E. "Is Unemployment a Macroeconomic Problem?" *American Economic Review* (May 1983).

Krose, S.A. "Give-Back Bargaining: One Answer to Current Labor Problems." *Personnel Journal* (April 1983).

Pearce, James E. "The Use of Employment and Training Programs to Reduce Unemployment." *Voice*. Federal Reserve Bank of Dallas (November 1979).

Rissman, Ellen R. "What is the Natural Rate of Unemployment." *Economic Perspectives*. Federal Reserve Bank of Chicago (September–October 1986).

Rostow, W. W. "Technology and Unemployment in the Western World." *Challenge* (March–April 1983).

Sawhill, Isabel V., and Jean Vanski. *Labor Markets and Stagflation*. Washington D.C.: The Urban Institute, 1982.

Tobin, James. "High Time to Restore the Employment Act of 1946." *Challenge* (May–June 1986).

U.S. Department of Labor. *Employment and Earnings*. Washington D.C.: U.S. Government Printing Office, monthly.

———. *Employment and Training Report of the President*. Washington D.C.: U.S. Government Printing Office, annually.

———. Special Labor Force Report. *New Labor Force Projection to 1995*. (Bulletin 2121).

Weiner, Stuart. "The Natural Rate of Unemployment: Concepts and Issues." *Economic Review*. Federal Reserve Bank of Kansas City (January 1986.)

2

INFLATION AND DEFLATION HOW MUCH AND WHERE DO THEY HURT?

We often hear about the twin evils of unemployment and inflation and what detriments they are to the economy. There are several measures of the cost of unemployment, such as the loss of worker-hours, the loss of wages, and the production gap between actual and potential gross national product, which measures our loss of goods and services. While there seem to be fairly adequate measures of the cost of unemployment, there is much less certainty about the cost, or adverse effects, of inflation in our economy. Interest in and concern about inflation accelerated when it reached double-digit rates in the 1970s, and that concern continues today.

It is said that inflation is the silent thief that robs the consumer of current purchasing power and depreciates the value of savings. It is claimed, too, that inflation distorts profits, forcing businesses to pay higher prices for current assets and higher replacement costs for machinery, equipment, and buildings. Inflation also weakens our international balance of payments position. Moreover, both consumers and investors are forced to pay higher taxes since inflation increases the cost of government services. Seldom, however, do we see a measure of just how much inflation decreases the current purchasing power of the consumer, depreciates the value of savings, adds to the cost of capital, or increases the cost of government.

The Phillips curve, which we shall examine later in the chapter, endeavors to give some estimate of how prices will react to a reduction in unemployment at various levels of unemployment. Figures provided by the Departments of Commerce and Labor and the President's Council of Economic Advisers indicate that the closer we are to full employment the more difficult it is to reduce unemployment further without substantially affecting the price level. But we have no easy or specific measure of the total cost of inflation. Consequently, we have no formula which accurately measures the disadvantages of a certain level of inflation. Such a guide, of course, would be of great use in the evaluation of proposed monetary, fiscal, and psychological measures when the economy is on the brink of inflation.

TYPES OF INFLATION

There are many definitions of inflation. In the simplest sense, inflation is merely a persistent rise in the price level. Inflation, however, may be one of four types: (1) demand-pull inflation, (2) cost-push inflation, (3) structural inflation, or (4) social inflation.

Demand-Pull Inflation

The type of inflation that we have heard about most frequently is known as *demand-pull inflation*. Demand-pull inflation occurs when the total demand for goods and services exceeds the available supply in the short run. This demand-pull inflation, sometimes referred to as excess-demand inflation, is much more likely to occur in a fully employed economy because of the difficulty of producing additional goods and services to satisfy the demand. Competitive bidding for the relatively scarce goods and services forces prices upward. The excess demand, or excess spending, may result from several causes. Consumers may decide to use past savings, consumer credit may be too liberal, government deficits may be too large, commercial and bank credit may be excessive, or the money supply otherwise may be increasing too rapidly. Generally, when the money supply or other forms of purchasing power increase faster than the productivity of our economy, demand-pull inflation results.

Cost-Push Inflation

The second type of inflation is known as *cost-push inflation*. Cost-push inflation may occur in a fully employed or an underemployed economy.

Whether it starts with increased wages, higher material costs, or increased prices of consumer goods is difficult to say. If wages or material costs do increase for some reason, however, producers are likely to increase the prices of their finished goods and services to protect their profit margins. Rising prices will decrease the purchasing power of wages. As a result, wage earners, especially through their unions, may apply pressure for further wage increases. This in turn may lead to further increases in the price of materials and finished products, which in turn leads to further wage increases and develops into what we generally call the *wage-price spiral.*

Cost-push inflation became more pronounced in the 1960s and 1970s with the growth and increased strength of labor unions. It also was aggravated by the use of administered pricing by large and powerful producers. *Administered pricing* is a situation in which a seller can exert an undue influence on the price charged for a product because of the absence of competition.

Structural Inflation

Another type of inflation that may occur in either a fully employed or less-than-fully employed economy is *structural inflation*. This arises when there is a substantial shift in demand to the products of one industry away from other industries. It is assumed that there is a certain amount of both flexibility and immobility among the factors of production. More specifically it is assumed that wages and prices tend to have downward rigidity and upward flexibility due to administered pricing and labor union pressures.

If there is a heavy shift in demand to the products of Industry X and away from the products of Industry Y, for example, it could push production in Industry X to, or near, full capacity. Because of the immobility and scarcity of labor and resources, Industry X may have to pay higher wage and material costs as it tries to increase production. Under such circumstances the increased demand could cause prices to rise in Industry X as a result of demand-pull inflation. This will cause the general price level to rise, since it is assumed that prices in Industry Y will not decline because of inflexibility.

The situation is aggravated when the inflationary effects spill over into other industries. The general increase in the price level could instigate wage increases and subsequently price increases in Industry Y. Although production and employment may be lessened in Industry Y as a result of demand shifts away from it, employers may be forced to pay higher wages to offset the higher living costs in an effort to hold on to experienced and skilled workers. In effect, structural inflation contains elements of both demand-pull and cost-push inflation.

One answer to each of these three types of inflation is increased productivity. In demand-pull inflation, if productivity can be increased to provide the additional goods and services demanded, the inflationary pressure will be removed. On the other hand, the demand for goods and services can be lowered by reducing the money supply or by reducing spendable income. Cost-push and structural inflation can be modified if wage increases are kept in line with increases in productivity. Since wage raises would increase in proportion to the increase in productivity, incomes would stay in balance with the amount of goods produced. Goods and services would be available when wage earners spent their higher incomes.

Social Inflation

In recent years we have witnessed the development of a fourth type of inflation known as *social inflation*. It results from the increasing demand for more government services in the form of higher Social Security payments, improved unemployment benefits, the distribution of more welfare, wider health care coverage, better rent subsidies, and a host of other social services. Social inflation is further encouraged by the rising costs to private enterprise of greater fringe benefits, such as longer vacations, more holidays, shorter hours, better pensions, and broader hospital and insurance coverage for employees. Moreover, the cost of helping to preserve the natural environment through the use of expensive antipollution equipment, either by the government or by private enterprise, exerts increased pressure on the price level. Likewise do the financial requirements of the Occupational Safety and Health Act (OSHA) and the Equal Employment Opportunity Act (EEOC). Social inflation may occur at full employment, adding to demand-pull inflationary pressures, or at other times it may augment cost-push inflationary pressures. Moderation in such demands and measures in recent years has been partially responsible for the disinflation that is now occurring in the economy.

Demand and Supply

Any type of inflation can be aggravated by changes in demand and supply. With a rising demand and/or a shortage of supply, whether real or fabricated, prices will rise. This has been very much in evidence in regard to food, materials, fuel, and energy in the past decade or more. These shortages are sometimes referred to as "supply shocks." Whether price increases due to shifts in supply and demand should be labeled inflation is a debatable issue. Some contend that it is merely a reflection of the market's attempt to allocate scarce resources among users. It is suggested also that higher prices for some goods could cause lower prices elsewhere due to shifts in demand. Nevertheless, such changes can supplement rising price levels and certainly dovetail with the concept of structural inflation.

THE RECENT AMERICAN EXPERIENCE: UNEMPLOYMENT AND INFLATION

During the past 15–20 years, the economy has moved through several phases of economic activity including recession, nagging unemployment, stable prices, full employment, inflation, stagflation, disinflation, and even some months of deflation. During this time various doses of monetary and fiscal measures have been applied in an effort to stabilize the level of economic activity, reduce unemployment, promote a healthy rate of economic growth, and stabilize the price level. A review of past developments will give us a better insight into our current problems.

Economy in Transition

After seven years of *nagging unemployment* and relatively stable prices following the 1958 recession, the economy entered a transition phase in the winter of 1965–1966. By the middle of 1965 unemployment had fallen below 5 percent and by the end of the year it had dropped to 4.1 percent, approaching the full-employment level for the first time in seven years or more. It appeared that we were moving from an economy of nagging unemployment and idle capacity to a condition of full employment and high utilization of capacity characterized by shortages of skilled labor, scarcity of certain materials, and inflationary pressures.

Early in December, 1965, the Federal Reserve raised the discount rate as a protective step against the clouds of inflation it foresaw on the economic horizon. Its action was both praised and criticized. During most of 1966 the Federal Reserve continued to apply some brake against inflation by tightening the money supply. This action contributed to the "money crunch" of 1966, which had a substantial impact on the construction industry.

By January, 1966, unemployment fell to 4 percent and in February it was down to 3.9 percent, the first time the full-employment level had been reached in more than seven years. The Johnson Administration faced a serious problem of deciding whether it should continue its expansionary fiscal measures or whether it should shift to anti-inflationary measures. It was concerned, of course, that anti-inflationary devices could slow economic expansion.

Inflation: 1966–1968

After much discussion and analysis, the Johnson Administration took only limited precautionary measures against inflation. In large part it rejected the idea of increasing taxes or reducing federal spending. Wage and price controls appeared to be an extreme. It did little to encourage or supplement the Federal Reserve's tighter money policy. On the other hand, the Administration did not sit by idly and do nothing. Early in 1966 it rescinded the excise tax cuts on automobile sales and telephone service. It also provided for an accelerated method of corporate tax collection and other minor measures to help avert inflation. But the Administration put its primary emphasis on "jawbone" tactics through its get-tougher policy in regard to implementation of the voluntary wage-price guideposts. Not until it became apparent that the price level was increasing at a 4 percent annual rate did the Administration take further action by suspending in September, 1966, the 7 percent tax credit on new investment and the accelerated depreciation measures.

Although the price increases slowed down a bit toward the end of the year, the Consumer Price Index (CPI) showed an annual increase of 3.4 percent. In December, 1966, unemployment measured 3.8 percent of the civilian labor force compared to 4 percent 12 months earlier.

Mini-Recession—Early 1967. The price level (CPI) stabilized in the latter part of 1966 and the first quarter of 1967. It was evident that the economy was slowing down since the GNP during the first quarter of 1967, in terms of constant dollars, actually declined. This mini-recession of 1967, as it has frequently been labeled, was caused primarily by a decline in the rate of private investment.

Although the President in his 1967 *Economic Report*, issued in January, mentioned the need for a 6 percent surcharge on personal and corporate income taxes to combat the inflationary tendency in the economy, the measure was not pushed to any degree in Congress. Many government officials, economists, and others naturally felt that inflation had subsided and cited the fact that it was accomplished without any sizeable tax increase, without any drastic cut in government spending, or without the imposition of compulsory wage and price controls. In fact, by the end of 1966 and in the first quarter of 1967 the Fed was again displaying a more liberal attitude toward the creation of credit. The discount rate, for example, was decreased from 4.5 percent to 4 percent in April, 1967. As a stimulant to the sluggish economy, the 7 percent tax credit on new investment was restored 6 months ahead of schedule.

Inflation Resumes—Mid-1967. The joys of stable prices, however, were short-lived because by the second quarter of 1967 the CPI resumed its upward movement. By the end of 1967, the CPI had risen sufficiently to show a 3 percent increase for the year and unemployment was down to 3.5 percent. As a result, members of the Administration and others were again talking about the need for restraint. Some government officials were even suggesting that direct controls of various kinds might be needed to cool the economy if management and labor did not hold wages and prices in check. In the latter part of 1967 and early 1968, the Federal Reserve moved toward a tighter money position by raising reserve requirements and moving the discount rate back to 4.5 percent.

The Income Tax Surcharge—1968. In his Economic Report of 1968, President Johnson called for the imposition of a 10 percent surcharge on personal and corporate income taxes as a means of combating inflation. The size of the proposed surcharge was increased from 6 to 10 percent because signs of an overheated economy were more in evidence. The proposed bill to institute the tax became embroiled in a Congressional hassle as to whether it was better to increase taxes or reduce federal spending. As a result of prolonged hearings and debate, final Congressional action on the bill was delayed until June, 1968. At that time, Congress imposed a 10 percent surcharge on personal and corporate income taxes to be effective until June, 1969. The surcharge was made retroactive to April 1, 1968, for individual income and January 1, 1968, for corporate income.

The impact of the surtax fell more heavily on savings than expected, however, as consumers continued spending for goods and services, especially for new cars, and the rate of savings fell sharply. Fixed investment for plant equipment, which increased moderately in the first half of 1968, accelerated in the second half. Capital spending was no doubt spurred by the feeling that the 7 percent investment credit might be suspended as an anti-inflationary measure.

By mid-1968 a number of national wage negotiations had taken place in the economy, adding to the cost-push inflationary element. The shortage of skilled labor, and even unskilled labor, was evident in the economy. Average hourly wage gains of 7 percent in manufacturing industries during the year plus a reduction in savings offset the impact of the income tax surcharge. Consequently, labor unit costs rose sharply.

All this, of course, caused prices to continue their upward momentum, and by the end of 1968, the CPI had risen 4.7 percent. Thus, in spite of the addition of a strong fiscal measure to accompany somewhat restrictive monetary measures, little success was achieved in arresting the upward movement of prices in 1968.

The New Economic Policy

With prices and wages continuing to rise, in 1969 the Federal Reserve adopted a more restrictive policy. As a result of various monetary measures, the rate of growth of the money stock declined. The Federal Reserve, through a series of changes, increased the discount rate from 5.5 percent in December, 1968, to 6 percent in April, 1969. By the summer of 1969 the prime rate for commercial loans offered by banks reached 8.5 percent, and the interest rate on federal funds approached the 10 percent level. The CPI in the first half of 1969 rose at an annual rate of 6.3 percent.

Gradualism Policy. With the inauguration of the Nixon Administration, inflation was cited as the nation's number one domestic issue. President Nixon adopted a policy of *gradualism* to bring inflation under control. He wanted to achieve stable prices without seriously disrupting the growth in economic activity. Among other measures, he asked Congress to retain the 10 percent tax surcharge that was due to expire in June, 1969. The budget was balanced and, in fact, ran a slight surplus in fiscal 1969. Defense and other government spending was cut and the Fed tightened money supply. Although there was some discussion about the need for wage and price restrictions of some type, the President shied away from either formal or informal wage-price measures. By the end of 1969, it was apparent that the measures employed to "cool off" the economy were effective in slowing down production. But they were not as effective in slowing down the rate of inflation. Although the real GNP declined in the fourth quarter of 1969, the price level was increasing at a rate of 6 percent.

With certain reservations the forecasts for 1970 were favorable. It was the hope of the President's Council of Economic Advisers that the inflationary rate, which had been in the vicinity of 6 percent, would recede to 3.5 percent by the end of the year. Economic measures designed to cool the economy and bring about a reduction in inflation were expected to result in a slightly higher level of unemployment, up from 3.5 percent to, perhaps, 4.5 percent of the civilian labor force.

The economy did cool off in the first half of the year for a number of reasons, including a decline in business investment, a cutback in defense spending, a tightness of money, and a slowdown in housing starts. The price level, however, continued to rise at the undesirable rate of more than 5 percent annually. Measures designed primarily to arrest demand-pull inflation failed to contain cost-push price pressures. Unemployment increased more than anticipated and reached a rate of 5.5 percent by midyear. This left the Nixon Administration in the position of deciding whether to continue anti-inflationary measures and risk the possibility of higher unemployment or shift to expansionary measures and risk the resurgence of inflationary pressures.

Unfortunately, the economy did not rebound in the second half of the year. Unemployment for 1970 averaged 4.9 percent and by the end of the year had reached 6 percent. Although 1970 was officially labeled a recession year, the CPI still managed to increase by 5.9 percent.

In the late months of 1970, the Administration shifted its emphasis to expansionary measures. A number of steps were taken to increase effective demand. Earlier in the year, the personal and corporate income tax surcharge was allowed to lapse. Subsequently the administration encouraged the Federal Reserve to liberalize the money supply. In addition, Congress enacted an accelerated depreciation schedule to spur business investment. Discount rates were lowered several times, moving down to 4.75 percent within three months, and the Administration announced that the federal deficit for fiscal 1971 would be in excess of $18 billion and that the projected budget for fiscal 1972 showed an $11.6 billion deficit. The Administration was trusting that any resulting increase in effective demand would not evoke demand-pull inflationary pressures, since the economy was in a state of less than full employment. The administration, however, was concerned about cost-push price pressures. Consequently, in early 1971 it began "jawboning" as a means of holding the price line, and more was heard about the possibility of wage and price guideposts and an incomes policy.

The Game Plan. Economic forecasts for 1971 were good, but not spectacular. The Council of Economic Advisers set year-end goals of 4.5 percent unemployment and a 3.5 percent rate of inflation. The economic "game plan" was to restore full employment and stable prices by mid-1972. Measures designed to attain that growth rate, remove the production gap, and eliminate nagging unemployment could very well add to price pressures and cause the rate of inflation to accelerate. The task of reaching stable prices was aggravated, too, by the fact that wage increases of 30 percent or more spread over a three-year period had been negotiated in the construction, auto, railroad, and tin can industries. Similar wage concessions in the steel industry in the summer of 1971 led to an immediate 8 percent average increase in steel prices. Some members of Congress, who in early 1971 recommended that President Nixon use the authority given to him in 1970 to impose wage and price restrictions, renewed their efforts after the steel settlement.

In July, 1971, in Congressional hearings on the state of the economy, an Administrative spokesman indicated that the Administration was not going to reach its year-end goals of disinflation and reduction of unemployment. It was stated that prices were stickier and unemployment more stubborn than anticipated. A month later, in response to an inquiry of what the Administration was going to do about inflation and unemployment, the Secretary of the Treasury stated that the Administration was not going to impose compulsory wage and price controls, it was not going to adopt voluntary wage-price guideposts, it was not going to increase government spending, and it was not going to reduce taxes.

Wage and Price Controls

Economic pressures regarding prices, wages, and the balance of payments brought about a change of attitude in the White House by mid-1971.

Phase I: The 90-Day Freeze. With the knowledge that progress on his economic game plan was being stifled by substantial wage and price increases, President Nixon in August, 1971, made drastic and sweeping changes of domestic and international economic policies. Among other measures, he declared a 90-day freeze on all prices, wages, and rents, temporarily suspended convertibility of dollars into gold, imposed a 10 percent surcharge on imports, and froze a scheduled pay increase for government employees. He also sought to reinstate tax credits as a means of stimulating investment and jobs and asked Congress to reduce personal income taxes and repeal the 7 percent excise tax on automobiles. The President established a Cost of Living Council to work out details for restoring free markets without inflation during a transition period following the freeze.

Phase II: Control Period. The 90-day freeze was followed by a Phase II control period. President Nixon established a Pay Board and a Price Commission. The Commissions were composed of representatives of labor, management, and the general public. Each was to work out what it considered permissible noninflationary wage and price increases, respectively. The Pay Board subsequently established a 5.5 percent annual wage increase as a maximum. It did allow that certain exceptions could be made to the 5.5 percent figure.

The Price Commission, on the other hand, indicated that it was going to attempt to hold overall price increases to 2.5 percent annually. Since the President did not desire to set up an elaborate formal structure of wage and price controls, such as existed during World War II and the Korean Conflict, much of the stabilization program had to depend on voluntary compliance. The Cost of Living Council exempted most business firms and most workers from any reporting requirements. Others, however, were required to report changes in prices and wages. Larger firms, moreover, had to give prenotification of changes to the Price Commission and/or the Pay Board.

The effectiveness of the wage and price controls in combating inflation can be gauged somewhat by the fact that, during the six months prior to the freeze, prices increased at an annual rate of 4.5 percent. In the five months following the freeze, they increased at an annual rate of 2.2 percent. The price level for 1972, during which price controls existed for the entire year, increased 3.3 percent.

The inflationary period starting in 1966 can be seen in Table 2-1 and Figure 2-1. From 1960 to 1965, for example, prices rose at an average annual rate of 1.3 percent. From 1965 to August, 1971, prices rose at an average annual rate of more than 4.7 percent. Figure 2-2 shows price

Figure 2-1

Consumer Prices

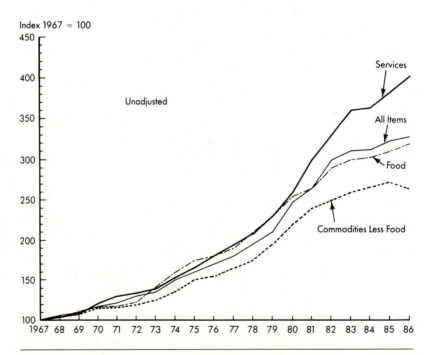

SOURCE: *Economic Report of the President,* 1986.

Table 2-1

Employment, Unemployment, and Prices

Year	Total Civilian Employment[a]	Unemployment[a]	Unemployment Rate	CPI (1967 = 100)	Inflation Rate[b]	Discomfort Index
1965	71,088	3,366	4.5%	94.5	1.9%	6.4
1966	72,895	2,875	3.8	97.2	3.4	7.2
1967	74,372	2,975	3.8	100.0	3.0	6.8
1968	75,920	2,817	3.6	104.2	4.7	8.3
1969	77,902	2,832	3.5	109.8	6.1	9.6
1970	78,678	4,093	4.9	116.3	5.5	10.4
1971	79,367	5,016	5.9	121.3	3.4	9.3
1972	81,153	4,882	5.6	125.3	3.4	9.0
1973	85,064	4,365	4.9	133.1	8.8	13.7
1974	86,794	5,156	5.6	147.7	12.2	17.8
1975	85,846	7,929	8.5	161.2	7.0	15.5
1976	88,752	7,406	7.7	170.5	4.8	12.5
1977	92,017	6,991	7.1	181.5	6.8	13.8
1978	96,048	6,202	6.1	195.4	9.0	15.1
1979	98,824	6,137	5.8	217.4	13.3	19.1
1980	99,303	7,637	7.1	246.8	12.4	19.5
1981	100,397	8,273	7.6	272.4	8.9	16.5
1982	99,526	10,678	9.7	289.1	3.9	13.6
1983	100,834	10,717	9.6	298.4	3.8	13.4
1984	105,005	8,539	7.5	311.1	4.0	11.5
1985	107,150	8,312	7.2	322.2	3.8	11.0
1986	109,597	8,237	7.0	328.4	1.1	8.1

[a]In thousands.
[b]Changes are from December to December.

SOURCE: *Economic Report of the President, 1987* and *Economic Indicators* (February 1987).

increases in average annual percentage increments, categorized as creeping, jogging, and galloping inflation.

Although employment increased substantially during the 1960s, the amount and rate of unemployment, after dwindling early in the decade, rose in the next few years. Note the net increase of over 2 million in unemployment between 1969 and 1971 in Table 2-1.

Phase III: Voluntary Guideposts. In January, 1973, the President removed compulsory controls and announced Phase III, which in effect reestablished voluntary guideposts for price and wage increases. The guidepost figures used at that time were 2.5 percent and 5.5 percent annually for prices and wages, respectively.

Phase IV: Controls Reinstituted. The removal of compulsory Phase II controls proved to be premature, however. During the first five months after decontrol, the CPI rose at an annual rate of nearly 9 percent. Consequently, in June, 1973, President Nixon declared a 60-day freeze on prices. Wages were not affected at this time. Instead of ending the freeze on all goods at the end of the 60-day period, prices were unfrozen selectively, and Phase IV controls were imposed on various categories of goods and services at different times before and after the 60-day period.

Again large firms were required to give a 30-day prenotification of price increases. Unlike Phase II, however, firms did not have to wait for approval by the Cost of Living Council before putting such increases into effect. But the Council had authority to delay any price increases indefinitely, and it reserved the right to reexamine prices at any time.

Figure 2-2

Annual (Year to Year) Inflation Rate

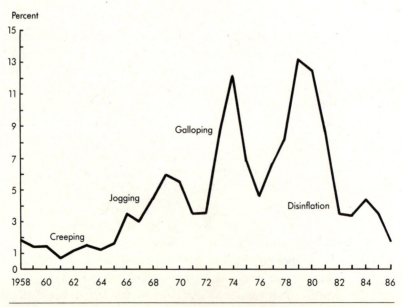

SOURCE: *Economic Report of the President,* 1987.

The new base period established was the fiscal quarter prior to January 12, 1973, the date of decontrol of Phase II. Price increases equal to dollar cost increases subsequent to the base period were to be permitted by Phase IV. No allowance was to be made for a profit markup on these cost increases. Controls were imposed on an industry-by-industry basis, thus providing more flexibility than was available under Phase II. At the time of the imposition of Phase IV, several high Administration officials indicated that they hoped controls could be removed by the end of 1973. Most of the controls were removed by early 1974. Unfortunately the CPI rose 8.8 percent in 1973. Inflation was aggravated by higher fuel prices and an increase of almost 20 percent in food prices.

The Recession of 1974–1975

In spite of efforts to temper inflationary pressures, the price level continued to rise during 1974, reaching double-digit figures within a few months.

Upon taking office in the summer of 1974, President Ford declared that inflation, among other things, was to be a prime target of economic policy. He held a summit conference composed of legislators, cabinet members, business people, labor leaders, and economists. The suggestions to combat inflation were many and varied, including a tax surcharge, tighter money, a balanced budget, higher interest rates, wage-price guideposts, reductions in federal spending, and a return to wage-price controls. Before they had an opportunity to implement any of these measures, however, the recession of 1974 deepened sufficiently to cause the President to deemphasize the problem of inflation in order to concentrate on expansionary measures to offset the adverse effects of the recession. Particularly distressing was the growing rate of unemployment, which reached 7.9 percent in January, 1975. But in spite of the recession, the CPI rose 12.2 percent during 1974.

As the state of the economy deteriorated over the next few months and it became evident to President Ford and his economic advisers that the economy actually was in a recession, the emphasis of economic policy began to shift toward expansionary measures. Demand-pull price pressure had abated. But with a number of sizable wage settlements and administered prices being set, double-digit inflation still prevailed due to cost-push, structural, and social inflationary pressures in the economy.

It was apparent to all that the state of the economy had slipped downward from *stagflation* to *slumpflation*. That is, the economy went from a condition of inflation with very low or no growth to a condition of inflation even with declining economic activity. Consequently, President Ford abandoned his plans to reduce federal spending and balance the budget. Early in 1975 he accelerated a public employment program and recommended an extension of unemployment compensation. He presented a $349.4 billion federal budget, including a huge $52 billion deficit. Included in the budget were a $16 billion tax reduction and other measures to offset growing unemployment. By that time the real GNP had declined for its fifth consecutive quarter. Income taxes were reduced by $23 billion and other expansionary measures were suggested by economists, business people, labor leaders, and Congress.

This threw economists into a new ball game. Previously they thought they knew what measures to use to expand the economy during a recession and, on the other hand, what measures to recommend to combat inflationary pressures. However, when recession and inflation occurred simultaneously, a condition with which they had very little experience, a real dilemma was presented. Should the Administration, for example, emphasize anti-inflationary measures and risk aggravating unemployment, or should it emphasize expansionary measures and risk higher prices? Not only was there an absence of foolproof economic measures to deal with slumpflation, but also prudent decisions had to be made regarding which was the more serious problem—unemployment or inflation.

Fortunately the increase in the price level did drop in 1976 to 4.8 percent, the first time in the 1970s that it had been under 5 percent except for the freeze/control period in 1971–1972.

Double-Digit Inflation Again, 1977–1980

Shortly after he took office in 1977, President Carter proposed a $30 billion economic package, including tax rebates, to stimulate the economy and reduce the existing 7 percent unemployment rate. Before the package received Congressional approval and could be implemented, the employment picture brightened. With the improvement in the economy, the inflation rate unfortunately began to edge upward and reached 6.8 percent in 1977. When the inflation rate continued to rise early in 1978, President Carter sought voluntary cooperation from business, labor, and consumers to exercise restraint. At the same time, the Federal Reserve tightened the money supply. When these measures failed to restrain price increases, the President late in 1978 announced a set of voluntary wage-price standards. These standards sought to hold wage increases to 7 percent annually and keep price increases one-half of one percent under the average annual increase in the base period 1976–1977. Increased prices for OPEC-controlled oil, food shortages, and climbing home prices and mortgage rates caused the CPI to increase by 13.3 percent in 1979.

Double-digit inflation continued into 1980, and the prime interest rate approached 20 percent. On March 14 the President, exercising his authority under the Credit Control Act of 1969, invoked compulsory credit restraints of various types on consumers, banks, and businesses. Most of these controls were implemented through the Federal Reserve System. Many of these restrictions were removed by mid-1980 when the 18 plus percentage inflation rate of the 1980 first quarter began to ease off.

Disinflation of the 1980s

By the time President Reagan took office in January, 1981, inflation was back to a double-digit level. The prime interest rate was 20 plus percent. As had other Presidents, he declared inflation to be a major economic problem and vowed to bring about its demise. Within a few days President Reagan dismantled the wage-price standards of the previous administration and disbanded the Council on Wage and Price Stabilization.

In the next several months the new administration invoked certain supply-side measures to encourage savings, stimulate investment, and motivate worker effort. Many of these measures were contained in the Economic Recovery Tax Act of 1981. In addition some of the more burdensome government regulations of business were lessened.

In the meantime the Federal Reserve maintained tight control over the money supply, eventually pushing the discount rate to 14 percent by June, 1981, in an effort to combat inflation. It maintained the discount rate at double-digit levels until October, 1982. This helped keep the prime rate charged by banks above 15 percent, which discouraged borrowing. At the same time the Administration tried to reduce federal spending. The severe recession of 1982 saw unemployment exceed 10 percent from September, 1982, to April, 1983. The result was wage demand moderation and even give-backs by union membership.

The combination of tight money and credit, high interest rates, decline in loan demand, wage concessions, and other adverse factors associated with the recession of 1982 broke the back of inflation. The inflation rate declined to 3.9 percent for 1982 and through 1986 remained under 4 percent.

Early in 1987 with the federal deficit still in the vicinity of $200 billion annually, the large Third World debt still on the scene, a number of bank failures occurring, sizeable recent increases in the U.S. money supply, and the decline in oil prices expected to bottom out, there was some apprehension in the financial markets about the probability of renewed inflation.

THE PROBLEM

This brings us to the heart of the inflationary problem. It is obvious from an analysis of the previous data that we have had an inflationary trend in the economy since 1966. The initial impact on prices was brought about by demand-pull pressures resulting from exuberant consumer spending, record-level private investment, and large federal deficits during a period of full employment. Since that time, cost-push, structural, and social elements have added to inflationary pressures.

Perhaps it would be simple enough to arrest the inflation by applying conventional but stringent monetary and fiscal measures. But there is always the danger of applying the brakes too strongly and causing the economy to turn downward. Furthermore, any time anti-inflationary measures have an adverse effect on employment, it is frequently the poverty-level wage earner, the teenager, and the nonwhite workers who are thrown out of work. Consequently, anti-inflationary measures begin to take on political overtones. Thus, the problem or task is to find the proper amount and mixture of monetary, fiscal, and psychological measures to arrest the inflation but not cause unemployment to rise above the 4–5 percent level consistent with full employment. In this regard it should be remembered that prices are considered to be stable if the CPI does not move more than 2 percent annually.

THE TRADE-OFF

Once the economy is at full employment, it is difficult to ride the crest of the economy at the point where unemployment is minimized and the price level stabilized. The President's Council of Economic Advisers, the Departments of Labor and Commerce, and others in 1966 thought that we could lower the level of unemployment below the then current goal of 4 percent to 3.5 percent without substantially affecting prices.

The debate about the trade-off between unemployment and an increase in prices renewed interest in the concept known as the *Phillips curve*. This curve was developed by the British economist, A. W. Phillips, who studied the relationship between the level of unemployment and annual wage increases for the United Kingdom for the years 1861–1913. From his studies he found that when unemployment was high, money wage increases were smaller; and when a low level of unemployment existed, wage increases were larger. Phillips concluded that the money wage level would stabilize with a 5 percent unemployment rate.

In the 1960s American economists began to apply a Phillips-type curve to changes in prices in relation to unemployment. This became feasible especially when the level of unemployment began to fall consistently below the 4 percent level. Subsequently a number of Phillips curves were constructed showing the relationship between price changes and the level of unemployment. Any interpretation of these curves must be made cautiously since they are constructed with various assumptions. There is absolutely no certainty, for example, that because a given relationship occurred in the past, it will hold precisely in the future. Figure 2-3 depicts a Phillips curve developed by Paul Samuelson in the mid-1960s. It shows price stability at about 5.5 percent unemployment and a 2 percent price rise associated with 4 percent unemployment.

Several other Phillips curves were developed about that time. One shows a price rise of about 2.5 percent associated with unemployment of 4 percent. Another shows a price increase of around 4 percent associated with a 4 percent unemployment rate. Still others, which have been developed

Figure 2-3

Modified Phillips Curve for the United States

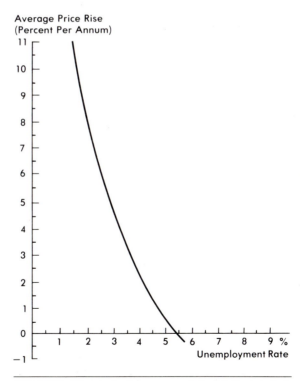

SOURCE: "Analytical Aspects of Anti-Inflation Policy," *The Collected Papers of Paul A. Samuelson,* Vol. II, MIT Press, ©1966 by the Massachusetts Institute of Technology.

Inflation and Deflation—How Much and Where Do They Hurt?   **47**

more recently, depict a rightward shifting of the Phillips curve, indicating a higher price level-unemployment relationship in which both the price level and unemployment rates are higher as one is traded off against the other.

The Phillips curve, of course, has some validity when inflation is demand-pull in nature. But the relationship between unemployment and price changes does not hold so well when inflation is of a cost-push or structural variety. A Phillips curve for the years 1970–1983, for example, would definitely move the price-unemployment line of best fit upward and to the right. Shown in Figure 2-4 is a more recent set of Phillips curves reflecting this movement. The 1983 *Economic Report of the President* noted that the inflation threshold unemployment rate probably lies between 6 and 7 percent. Some analysts challenge the Phillips curve and its relationship between unemployment and price level changes. It should be remembered, however, that the Phillips curve was originally designed to measure the

Figure 2-4

Unemployment-Inflation Experience, 1960–1986

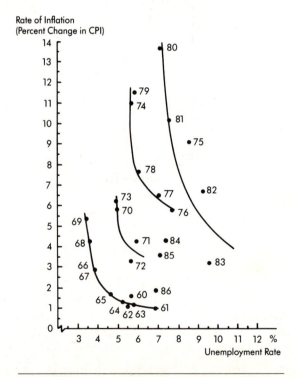

SOURCE: Data from *Monthly Labor Review*.

relationship between demand-pull inflation and unemployment, not cost-push, structural, or other varieties of inflation.

A side effect of unemployment-price change watching has been the development of the *Discomfort Index*, which measures the level of unemployment and the rate of inflation combined. Table 2-i shows the rising level of discomfort as both unemployment and inflation continued to increase in the 1970s and into the 1980s.

COST ESTIMATE OF INFLATION

Whether because of normal expansionary forces in the economy, an insufficient tightening of the money supply, the lack of voluntary restraint on the part of consumers and investors, or the failure of the Administration and Congress to adopt more stringent fiscal measures sooner, the price level rose more in each of the years between 1966 and 1986 than it did in any of the preceding 15 years at a cost to American consumers of several billion dollars. Some idea of the cost of this inflation for one year, such as 1985, may be estimated as shown in the following sections.

Higher Prices for Goods and Services

The consumer price index was 3.8 percent higher in 1985 than it was in 1984. This means that consumers paid out $95 billion more in 1985 in the form of higher prices for goods and services, as spending for consumer goods was $2,601 billion in 1985: ($2,601 billion ÷ 1.038) × 0.038 = $95 billion. The cost of business investment in machinery, equipment, and buildings figures to almost $24 billion more if a 3.8 percent price markup is applied to the $661 billion gross private domestic investment in 1985. Government outlays—federal, state, and local—rose by $30 billion due to the higher price tags on the goods and services purchased by all levels of government. The inflationary impact on net imports was about $2 billion. In short, the 1985 GNP of $3,998 billion was $147 billion more than it would have been had prices not risen in 1985.

At first glance we may want to conclude that the cost of inflation in 1985 was $147 billion, but it must be kept in mind that we consider prices to be stable when the CPI or the implicit price deflator moves 2 percent or less on an annual basis. Thus, if a 2 percent annual increase was considered tolerable, this would mean that excess inflationary pressures in 1985 increased the price level by 1.8 percent (3.8 − 2.0 = 1.8). In this case it could be considered that the total cost of inflation was only $71 billion, instead of the $147 billion cited in the previous paragraph.

Depreciation of Savings

There is more to consider, however, since inflation affects not only current income but also savings held by individuals, businesses, and the government. For example, in 1985, households held total savings of more than $7,000 billion in the form of time and savings deposits in banks, savings and loans, and credit unions, federal, state, and local government securities, and corporate bonds and notes.[1] Consequently, the price rise of 3.8 percent for 1985 depreciated the value of these savings by $256 billion: ($7,000 billion ÷ 1.038) × 0.38 = $256 billion. If only 1.8 percent excess inflationary pressures were applied, the value of these savings still eroded by $124 billion. This adverse effect of inflation on individual savings, however, was offset to some extent by the fact that the nearly $3,000 billion combined mortgage, installment credit, and other debt burden of individuals was eased somewhat as a result of inflation.

Loss of Insurance Protection

The effect of inflation on the protection rendered by life insurance policies must also be considered. At the end of 1985, for example, Americans held more than $5,500 billion worth of life insurance protection.[2] Thus, the 3.8 percent depreciation in the purchasing power of the dollar lessened the protection of these policies by a total of $201 billion, or $97 billion using the 1.8 percent inflation factor.

For those items above, the approximate total cost, in the form of higher prices or loss of value resulting from the 1985 inflation, was somewhere between $604 billion and $292 billion, as indicated in Table 2-2.

Fortunately, by 1985 inflation had abated from its double-digit and near double-digit levels. Otherwise the costs of inflation would have been much greater. In 1981, for example, the inflation rate, as measured by the consumer price index, was 9.7 percent. Using a 9.7 percent inflation rate and measuring the same items as in Table 2-2, the cost of inflation that year would have been a staggering $1,500 billion.

Balance of Trade

Inflation can also have an adverse effect on the economy through the balance of international trade. This is especially so when domestic prices are rising faster than are those abroad. In such a case the relatively greater

---

1. *Statistical Abstract of the United States: 1986.*
2. *Statistical Abstract of the United States: 1986.*

Table 2-2

Total Costs of Inflation, 1985

Category	Cost or Loss (in Billions) at 3.8 percent		Cost or Loss (in Billions) at 1.8 percent	
Current goods and services		$147		$71
Consumer goods	$95		$46	
Private investment	24		12	
Government	30		14	
Imports	−2		−1	
Savings		256		124
Insurance protection		201		97
Total cost		$604		$292

increase in prices in America deters foreigners from purchasing our goods and services, thus causing our exports to decline. On the other hand, the relatively lower price of foreign goods and services encourages Americans to purchase foreign goods, causing imports to increase. A change in the export-import ratio can affect jobs, profits, and incomes. Any attempt to measure the net effect of such changes is difficult, however. Any disadvantage to particular groups has to be offset by any gains to the total economy through the utilization of the law of comparative advantage. Table 2-3 gives some indication that the United States compares favorably vis-à-vis other major nations on the matter of inflation.[3] It can be observed also that there was some slowing down of inflation rates as we moved into the 1980s.

Other Effects of Inflation

Offset against the cost of inflation, of course, must be the gains in the economy obtained as a result of those measures which incidentally brought on the inflationary conditions. Since the current inflationary binge started in the late 1960s, both the number and percent of unemployed have been on an upward trend, as shown in Table 2-4. Consequently, inflation in recent years has not been offset by reductions in unemployment, but a gain in

---

3. For some insight into the effects of hyper-inflation elsewhere in the world see pages 362–365.

Table 2-3

Consumer Price Indexes — Selected Countries

	1984 Index (1967 = 100)	Average Annual Percentage Change in Food Prices	
		1975–1982	1980–1984
United States	311.1	7.2	4.4
Canada	334.9	9.8	6.9
Japan	314.4	4.9	2.9
Austria	249.9	4.7	4.6
France	439.2	9.8	10.9
Italy	702.7	15.8	13.5
Netherlands	284.9	4.9	3.8
Australia	412.4	10.9	8.1
Sweden	287.5	11.6	12.6
United Kingdom	565.1	12.2	6.2
West Germany	207.9	4.0	3.8

SOURCE: *Statistical Abstract of the United States: 1986.*

employment has occurred. In the 10-year period 1975–1985, for example, unemployment rose by 400,000 but total employment rose by 21.4 million.

It might be said that as a result of some measures contributing toward inflation, such as high-level investment, large budget outlays for social programs, and the occurrence of substantial federal deficits, wages in general were higher and profits greater. In fact, it can be claimed that higher prices have to be offset in large part by higher incomes since the total cost of production is equal to our national income.

It can be brought out also that inflation made it easier for individuals, banks, government agencies, and insurance companies to meet their fixed obligations. But many of these gains came at the expense of the consumer.

Looking at American consumers only, the inflation of 1985 cost them somewhere between $46 billion and $95 billion, depending on whether one uses a 1.8 percent inflation factor or a 3.8 percent factor. Even if the minimum cost for consumer outlay, depreciation of savings, and loss of protection on life insurance policies is used, higher prices still cost American consumers, directly or indirectly, $267 billion in 1985 compared to what they would have paid if we had had stable prices.

It might still be argued that if prices rise, current incomes have to rise by a similar amount because our total factor cost is equal to our national income. But granting that the 1985 cost of inflation was completely offset

Table 2-4

Employment, Unemployment, Prices, and Wages, 1965–1986

Year	Total Employment (Thousands)	Unemployment (Thousands)	Unemployment rate	CPI (1967 = 100)	Rate of Inflation[a]	Weekly Money Wage[b]	Real Wage (1977 $)	Percentage Increase in Real Wage
1965	71,088	3,366	4.5%	94.5	1.7%	95.45	183.21	2.7%
1966	72,895	2,875	3.8	97.2	2.9	98.82	184.37	0.6
1967	74,372	2,975	3.8	100.0	2.9	101.84	184.83	0.2
1968	75,920	2,817	3.6	104.2	4.2	107.73	187.68	1.5
1969	77,902	2,832	3.5	109.8	5.4	114.61	189.44	0.9
1970	78,678	4,093	4.9	116.3	5.9	119.83	133.33	−1.3
1971	79,367	5,016	5.9	121.3	4.3	127.31	190.58	1.9
1972	82,153	4,882	5.6	125.3	3.3	136.90	198.41	4.1
1973	85,064	4,365	4.9	133.1	6.2	145.39	198.35	0.9
1974	86,794	5,156	5.6	147.7	11.0	154.76	190.12	−4.1
1975	85,846	7,929	8.5	161.2	9.1	163.53	184.16	−3.1
1976	88,752	7,406	7.7	170.5	5.8	175.45	186.85	1.5
1977	92,017	6,991	7.1	181.5	6.5	189.00	189.00	1.2
1978	96,048	6,202	6.1	195.4	7.7	203.70	189.31	0.2
1979	98,824	6,137	5.8	217.4	11.3	219.91	183.41	−3.1
1980	99,303	7,637	7.1	246.8	13.5	236.10	172.74	−5.8
1981	100,397	8,273	7.6	272.4	10.4	255.20	170.13	−1.5
1982	99,526	10,678	9.5	289.1	6.1	266.92	168.09	−1.2
1983	102,510	10,717	9.5	298.4	3.2	280.70	171.26	1.9
1984	106,702	8,539	7.4	311.1	4.3	292.86	172.78	1.3
1985	108,856	8,312	7.1	322.2	3.6	299.09	170.42	−1.1
1986	111,303	8,237	6.9	328.4	1.9	304.50	170.88	.3

[a]Year to year changes as opposed to December to December changes.
[b]Average weekly earnings in nonagricultural industries.

SOURCE: *Economic Report of the President, 1987* and *Economic Indicators* (January 1987).

by the yearly increase in income, there was still the loss of $221 billion resulting from depreciation of our savings and the loss of protection from our life insurance and annuity policies, using the 1.8 percent inflation factor.

Regardless of how the pie is sliced or the cost allocated, American consumers have paid dearly for the excessive inflation of recent years. From most indications, it would have been far better for the economy to have taken more stringent monetary and fiscal measures in the early stages of price pressures in an effort to combat the pending inflation. In the absence of voluntary restraint by consumers and investors or a reduction in spending by the government, an earlier increase in personal and corporate income taxes would have been beneficial and may have been less expensive for consumers in the long run. Our total payout for goods and services would have been less; excessive inflation might have been avoided; and there would have been no long-term loss in the form of depreciated savings, no decline in protection from insurance policies and annuities, and no inflated replacement costs of assets.

Another bit of evidence of how inflation is hitting the pocketbook of the average American can be seen in what is happening to average weekly earnings. Between 1976 and 1986, for example, the wages of workers in nonagricultural industries increased $129.05 per week or 75 percent. But the purchasing power of the wages declined by $15.97 or 9 percent. Moreover, since the higher money wage moved many workers into a higher income tax bracket, their real wage gains were further obliterated by higher income tax payments. Workers in many categories have had smaller money wage increases than the average. Consequently, it becomes obvious that the average worker made little economic gain during this inflationary period.

Since there is a certain nebulousness about the cost of inflation, and the measures necessary to combat it are seldom politically popular, there is generally a hesitancy to take the vigorous steps necessary to prevent rising prices. It is easy enough to arouse public and political sympathy and support for measures to bolster the economy when an excessive number of people are suffering from unemployment. It is difficult, however, to generate enough public sentiment and political concern to do more to combat inflation, even though the small loss it may involve to each of 241 million consumers may add up to a greater cost than the loss suffered by the excess unemployment (above the 4–5 percent unemployment level) of a million or more workers.

Some Problems of Disinflation

Although a slowing of the rate of inflation, known as disinflation, is generally good for most members of the economy, it can cause hardships for certain individuals or segments of the economy. It especially affects debtors who had anticipated paying off their debt with inflated dollars.

Four broad segments of the economy have been particularly hurt by the disinflation of 4 percent or less during the 1980s: (1) several Third World countries, (2) the energy industry, (3) farmers, and (4) their respective creditors. Third World countries borrowed heavily in the 1970s at high interest rates. They anticipated that their expanding economies would provide funds to pay off their loans. The softening of prices, along with lower sales, however, brought less income from their exports, especially those to the United States. This meant less foreign exchange to meet payments of their debt, including interest, to commercial banks, foreign government agencies and international organizations. This forced them to borrow additional funds to meet interest payments, and it caused their creditors to restructure many loans and reschedule principal repayments, since a number of these nations were on the verge of default. It also forced many of these nations to impose austerity measures to improve their domestic economies. Particularly hurt were nations, such as Mexico, that relied heavily on the sale of oil for foreign exchange. Their creditors, especially commercial banks, were hurt by the need to reschedule loans, allocate more funds to contingency reserves for loan losses, and write off or write down some of these loans as assets on their balance sheets.

The dramatic drop in oil prices, which contributed so much to the disinflation, hurt the domestic energy sector of the U.S. economy. Many companies in the petroleum business were forced out of business as oil prices declined. Widespread layoffs and wage reductions had a dampening effect on the entire oil-producing area of the United States, especially in Texas, Oklahoma, and Louisiana. Numerous bankruptcies and the inability of oil producers to repay loans (originated in the high oil price era of the 1970s) had a negative effect on commercial banks and other creditors of energy companies. As a result several bank failures occurred when banks wrote off or wrote down loans.

Another sector hard hit by disinflation, and to some extent actual deflation, has been U.S. agriculture. During the 1970s and early 1980s, because of strong domestic and foreign sales, high land values, and encouragement by creditors, farmers (like the oil producers and Third World nations) borrowed heavily at high interest rates to expand output and purchase new land and equipment. When domestic and foreign sales slackened and agricultural prices dropped in the 1980s, many farmers found it difficult to meet interest and principal payments on their loans. Moreover, declining land values made creditors reluctant to renew farm loans. As a result more farmers were forced out of business in the 1980s than at any time since the Great Depression of the 1930s. In addition, a large number of farm banks and commercial banks in farm areas failed. Even the elaborate federal farm banking and credit system had to ask for special funding help from the federal government in 1986.

In addition, millions of U.S. workers have experienced layoffs and wage reductions, in part, because of disinflation. Lower prices and the squeeze on profits in the steel, auto, petroleum, shoe, and airline industries forced unions into "wage give-backs" in order to protect jobs and to help companies continue to operate. In spite of wage concessions by both managerial and hourly employees, widespread layoffs still took place in many of these industries.

The problems cited in this section are associated with disinflation or deflation. However, in large part they originated with the overspending, overborrowing, and overconfidence associated with a prior inflationary period.

CONCLUSION

Until 1981 our attempts to arrest or slow inflation had not been successful. Although compulsory wage and price controls in 1971 and 1972 did contain price increases within the 3–4 percent range, shortly thereafter inflation erupted again. Even though inflation did taper off a bit in 1976 to 5.8 percent, it moved up to the double-digit level by 1977 and remained there until 1981. The rise in the price level was abetted by the OPEC price hikes, food shortages, hefty wage increases, widespread indexation of wages and other income, government spending, and a decline in productivity per labor hour. This happened in spite of Presidential jawboning; voluntary, but formal, wage-price standards; short-term credit restraints; and high interest rates. Even the recession of 1980 did not slow the rate of inflation.

Abnormally high interest rates, some reduction in government spending, imposition of supply-side measures, and a severe recession in 1982 caused price pressures to abate. The disinflation that occurred was aided by moderation in wage demands, salary reductions, and even give-backs by some union members. The CPI showed a 10.4 increase in 1981, as measured on a year-to-year basis, but it increased only 8.9 on a December-to-December basis. In 1982, the CPI rose 3.8 percent on a December-to-December basis, remained between 3 and 4 percent through 1985, then dropped to 1.1 percent in 1986.

Although we have had some disinflation for the past few years, and even a few isolated months of deflation, the price level has not yet settled to the low rates of the late 1950s and early 1960s before the current inflationary surge began. We still have some way to go before we have overcome inflation and restored price stability. In fact, there is still much concern by the chairman of the Federal Reserve Board and others about reinflation occurring as the economy recovers, idle capacity declines, unemployment decreases, and exceptionally large federal deficits continue.

QUESTIONS FOR DISCUSSION

1. Should some type of wage and price controls or guidelines be reimposed as a means of preventing future inflation?
2. Do higher prices paid by consumers for goods and services represent a real economic cost to the economy, or are they merely a redistribution cost?
3. Are losses from depreciated savings and reduced insurance protection that result from inflation real economic costs?
4. In your judgment, which is the more serious problem, inflation or unemployment?
5. Do you agree with the provisions in the Economic Recovery Tax Act of 1981 that indexed federal income tax payments to the price level (CPI) to prevent the taxpayer from paying higher tax rates on inflated (nonreal) income?
6. Under the circumstances at the time, do you think the Carter Administration should have taken stronger anti-inflationary measures in the late 1970s? Why or why not?
7. What additional anti-inflationary measures, if any, would you suggest for today's economy?
8. Do you think that a 2 percent rise in the Consumer Price Index is consistent with price stability?
9. Do you think that our unemployment goal today should be 4 percent or something closer to the inflation threshold rate of unemployment?
10. Do you think that all wage rates should be tied to the CPI, as they are in certain industries, to protect the purchasing power of workers?

SELECTED READINGS

Brimmer, A. F. "Monetary Policy and Inflationary Expectations." *Challenge* (May 1983).

Essays on Inflation. Federal Reserve Bank of Richmond, 1986.

Fisher, Stanley. *Indexing, Inflation, and Economic Policy*. Cambridge: MIT Press, 1986.

Galbraith, John Kenneth. *A Theory of Price Control*. Cambridge: Harvard University Press, 1982.

Genetski, R. J. "How the U.S. has Beaten Inflation." *Euromoney* (May 1983).

Harrigan, Brian. "Indexation: A Reasonable Response to Inflation." *Business Review*. Federal Reserve Bank of Philadelphia (September–October 1981).

Humphrey, Thomas. "Early History of the Phillips Curve." *Economic Review*. Federal Reserve Bank of Richmond (September–October 1985).

"Is Inflation Licked—Or Just Resting?" *Inflation Watch*. American Enterprise Institute (May–June 1982).

Kahley, William J. "Adjusting to Disinflation." *Economic Review*. Federal Reserve Bank of Atlanta (June 1983).

"Wage Inflation: Speed-Up Ahead." *Morgan Guaranty Survey* (September 1983).

Weiner, Stuart. "Union COLAs on the Decline." *Economic Review*. Federal Reserve Bank of Kansas City (June 1986).

3
PRODUCTIVITY
DO WE
NEED AN
INDUSTRIAL POLICY?

One of the most discussed and debated topics of the 1980s is productivity. For decades the U.S. economy, because of its high productivity, was considered to be the eighth wonder of the world, and our standard of living was the envy of many people throughout the world. Prior to the mid-1970s America's high rate of productivity was taken for granted. But its decline moved it to the forefront of concern by 1980. The subject was debated in the 1980 Presidential campaign. The Joint Economic Committee of Congress described productivity as the "economic linchpin of the 1980s." Productivity along with capital formation was a major topic of the 1983 *Economic Report of the President*. It was again debated during the 1984 Presidential campaign. Much was heard about the need for an industrial policy. In more recent years numerous proposals for import protection measures have been introduced into Congress. Why? What caused the change? What is the problem? How can it be remedied?

The economist often defines production as the addition to or creation of utility. *Utility* is the want-satisfying ability or usefulness of a good or service. There are several types of utility. *Form utility* arises by changing the shape or composition of a good. This occurs, for example, when manufacturers change iron ore, glass, rubber, plastic, and other materials into an automobile, or when processors convert fats and chemicals into soap, thereby enhancing the usefulness of the materials.

Since goods are more useful in some places than in others, *place utility* is the process of getting goods from the original producers to the ultimate consumer. This production involves merchandising, transportation, and selling. The possession of a good or service may be more valuable to a person at one time than at another. A home is more useful to a couple during the family-formation stage, when the children are young and growing, than it is when they are first married or retired and they can be satisfied with an apartment. Thus, the finance company or bank that provides the mortgage money for them to purchase the home creates *time utility*. It is producing, as is the investment firm that underwrites a bond issue for a manufacturer to obtain funds to construct a new plant.

Production, in the form of services, is also undertaken by doctors, lawyers, teachers, pilots, and entertainers. They, too, are producing and contributing to the GNP, just as the workers on the assembly line or the operators at their word processors are. Government workers are also part of the productive process.

In fact, today 51 percent of our total production for consumption is in the form of services. Moreover, 75 percent of the civilian labor force is engaged in the production of services, and it is anticipated that this will increase to 80 percent before the current decade is over. One can gather from this that services contribute much to our GNP and standard of living and provide many employment opportunities.

HOW PRODUCTIVITY IS MEASURED

Technically, productivity is a measure of the relationship between output (production) and one or more of the units of input, such as land, labor, or capital, used to generate the output. Productivity may be stated in terms of bushels per acre, Btu's of energy per investment dollar, or output per machine. More commonly, however, it is measured in terms of productivity per worker-hour or, more properly, productivity per labor-hour.

Contrary to popular belief, productivity per labor-hour does not necessarily measure changes in the efficiency of labor in the productive process. The source of increased productivity may be additional capital investment, technological development, or improved labor efficiency. Regardless of the source of the increase (or decrease) in production, it is convenient to measure it in terms of productivity per labor-hour. One reason for this is the ease with which labor input can be quantified and related to output. Productivity per labor-hour may be stated in units of physical output, such as tons per labor-hour, or in the dollar value of output, such as $22.30 per labor-hour. Many times, productivity per labor-hour, whether measured in units of output or dollars, will be shown in terms of annual percentage changes. It may also be shown as an index number, which measures the output per labor-hour of every year compared to a base year percentage value of 100.

Regardless of how it is measured, productivity can be increased by spending more investment dollars on improved machinery and equipment, by applying better management techniques, by eliminating waste, by using labor and other resources more efficiently, by improving the skills and training of workers, and, lastly, by developing pride in one's work.

PRODUCTIVITY DETERMINES INCOME

Income flows from production as payments are made in the form of wages, rent, interest, and profits in exchange for services rendered. Consequently, our production was responsible for providing us with the highest income per capita in the world for many years. The growth in total production (GNP) and our real income is generated by the growth in labor-hours plus growth in productivity. It can be seen in Figure 3-1 that in the past 30 years nearly 80 percent of our real GNP growth resulted from increases in productivity. It has contributed much less to our growth since the late 1970s, however.

Productivity and GNP Growth

Most productivity studies indicate that productivity is the engine, or driving force, of economic growth. Usually they designate productivity as the independent variable on a regression analysis chart showing the relationship between productivity growth and GNP growth, as shown in Figure 3-2. This relationship continues to the present day.

Some critics object to this, however, and maintain that in such a relationship GNP growth should be the independent or causal variable. They contend that if there is sufficient demand for output (GNP), productivity will increase via innovation, investment, and other ways to satisfy the demand. These critics feel demand and GNP growth stimulate productivity growth. Whichever way one chooses to argue regarding cause and effect,

Figure 3-1

Growth in Output, Productivity, and Worker-Hours
(Private business Sector)

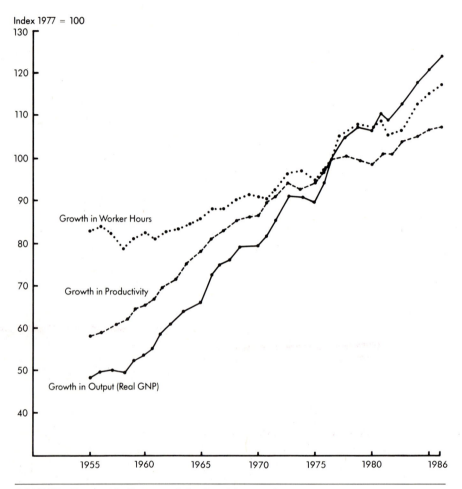

Index 1977 = 100

SOURCE: *Economic Report of the President,* 1987.

there is little doubt that there is a positive relationship between productivity growth in manufacturing and GNP growth.

Investment and GNP Growth

It stands to reason that output and productivity can be increased as more is spent on investment in new and more efficient machinery, equipment, and production facilities. In this regard we often relate investment as a

Figure 3-2

Productivity Growth and GNP Growth

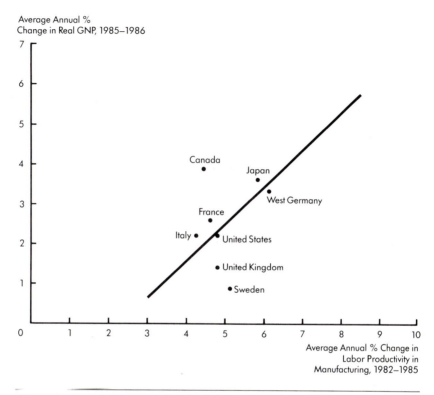

Average Annual %
Change in Real GNP, 1985–1986

Average Annual % Change in
Labor Productivity in
Manufacturing, 1982–1985

SOURCE: Data taken from *Productivity Perspectives,* 1987.

==percentage of GNP in a positive fashion to the growth of the GNP.== There is little doubt that the use of earth movers and cranes in construction work, the application of lasers in scientific experiments, the development of electronics in communications, the use of computer centers in steel production, and the use of machine operations and robots for assembly and other routine work have enhanced productivity and reduced cost.

Another method is to relate fixed capital investment to productivity, as shown in Figure 3-3, and then relate productivity growth to GNP growth, as was shown earlier in Figure 3-2. This relationship still applied into 1987.

Productivity and Business Cycles

Productivity is affected by the business cycle. During the first half of the expansion stage of a business cycle, productivity tends to be high. During

Figure 3-3

Investment in Fixed Capital and Productivity Growth

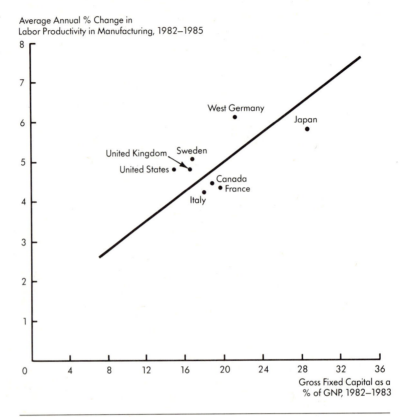

SOURCE: Data taken from *Productivity Perspectives,* 1987 and *Economic Handbook of the Machine Tool Industry,* 1985-86.

this period profits rise, investment and capacity increase, materials and labor are readily available, and input prices are moderate. Productivity tapers off, however, during the latter half of the expansion when businesses become complacent as they experience record profits. They add readily to their work forces, labor turnover rises, and material shortages appear. During this period, managers push toward maximum capacity, press marginal equipment into operation, and become less careful about cost and output efficiencies. Productivity continues to fall during the early stages of the contraction (recession). During this time, final production demand begins to weaken, work forces are maintained, total output declines, and costs continue to rise.

Productivity tends to improve, however, during the latter stage of a recession. This reversal in productivity occurs as profits dwindle and managers become more efficient and cost conscious. As workers are laid off and an attempt is made to maintain the same total output with a smaller work force, productivity per labor-hour rises. During this time capital is substituted for labor, new techniques are tried, and more attention is paid to productivity to prevent profits from dwindling or losses occurring. The change in the pattern of labor productivity over the business cycle as evidenced by the recession of 1982 can be seen in Table 3-1.

Although productivity changes tend to peak, decline, and then rebound over the phases of the business cycle, the annual growth rate in productivity between business cycle peaks has been declining over the past few decades. As a result, some of the gains made during the cycle are lost between cycles. The long-run decline in the growth in productivity per labor-hour is shown in Figure 3-4; shaded areas are recession periods.

Table 3-1

Quarterly Changes in Output per Hour of All Persons for the U.S. Private Business Sector, 1979–1986

1979	1	−1.9%	1983	1	2.0%
	2	−2.6		2	5.9
	3	−2.1		3	1.2
	4	−0.6		4	2.8
Annual		−1.2%	Annual		2.7%
1980	1	1.5	1984	1	4.4
	2	−2.9		2	2.6
	3	1.3		3	−0.3
	4	1.0		4	−0.1
Annual		−0.3	Annual		2.3
1981	1	5.9	1985	1	0.9
	2	2.2		2	2.7
	3	4.7		3	3.4
	4	−4.1		4	−3.2
Annual		1.4	Annual		1.0
1982	1	−0.4	1986	1	3.3
	2	−1.6		2	0.5
	3	1.7		3	−0.4
	4	3.0		4	−2.3
Annual		−0.4	Annual		0.7

SOURCE: *Economic Indicators* (January 1987).

Figure 3-4

Change in Output per Hour, Private Business Sector

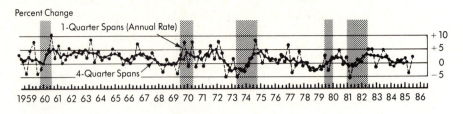

SOURCE: *Business Conditions Digest* (July 1986).

WHAT HAS HAPPENED TO PRODUCTIVITY?

Something happened to production in the 1970s. After decades of high-level production and world dominance in per capita income, the rate of increase in productivity in the U.S. economy slowed markedly. This change in the increase in productivity is very obvious in Figure 3-5, which shows that since the mid-1960s actual productivity has fallen substantially below the trend established in previous decades.

Between 1948 and 1965 productivity in the private business sector of the U.S. economy increased 2.6 percent annually. In the period 1965–1973 the rate declined to 2 percent annually. Since 1973, productivity growth in the private business sector of the economy has averaged less than 1 percent per year, as shown in Table 3-2. In several quarters it showed negative growth rates.

This slowdown has had a pronounced effect on our total GNP, our per capita incomes, and our standard of living. According to testimony before the Joint Economic Committee, a study estimated that if the productivity trend was not reversed, the average American household would lose $8,500

Table 3-2

U.S. Labor Productivity Growth, Private Business Sector

1948–1955	3.4%	1978	0.8%	1982	− 0.4%	1986	0.7%
1955–1965	3.1	1979	− 1.2	1983	2.7		
1965–1973	2.3	1980	− 0.3	1984	2.3		
1973–1977	1.0	1981	1.4	1985	1.0		

SOURCE: *Economic Report of the President, 1979,* and Economic Indicators *(January 1987).*

by 1988,[1] as shown in Table 3-3. Various estimates indicate that if the U.S. had maintained its 1947–1967 growth rate, its GNP in 1988 would have been $400 billion higher. One nation, Switzerland, has surpassed the United States in per capita GNP and income, and in recent years the gaps between the United States and some other nations have narrowed, as shown in Table 3-4.

The productivity slowdown also aggravated our higher than usual rate of inflation by contributing to the cost of production during the 1970s and early 1980s. Lower productivity increases the per unit cost of output, causing producers to raise final product prices or suffer reduced profits.

Figure 3-5

Output per Hour in the Private Business Sector
(1947–1985 Actual Levels and 1947–1967 Trend Extrapolated)

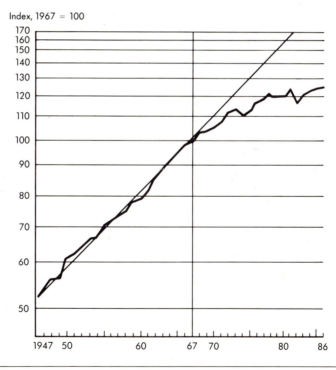

SOURCE: *Economic Report of the President,* 1987.

1. Notes from the Joint Economic Committee, March 2, 1979, by C. Jackson Grayson, Chairman, American Productivity Center, Inc.

Table 3-3

Loss of Household Income from Slower Growth
During Period 1968–1978 and Forecast for Period 1978–1988

Average productivity growth 1948–1968	3.3%
Average productivity growth 1968–78 and forecast for 1978–88	1.5%
Loss in productivity growth	1.8%
Resulting loss of household income in 1978 because of cumulative lower productivity	$3,700
Total loss if slow growth continues to 1988 because of cumulative lower productivity	$8,500

SOURCE: *Joint Economic Committee* (March 2, 1979).

Table 3-4

Per Capita GNP, Selected Countries, 1983

	Per Capita GNP 1983	Per Capita GNP as a % of U.S. Per Capita GNP, 1983	Per Capita GNP as a % of U.S. Per Capita GNP, 1970
Switzerland	$15,633	111%	66%
United States	14,093	100	100
Norway	12,930	92	62
Canada	12,622	90	78
Sweden	10,744	76	86
West Germany	10,672	76	64
Denmark	10,684	76	67
Australia	9,272	69	57
Japan	9,697	69	40
France	9,473	67	61
Netherlands	9,164	65	50
Belgium	8,245	59	56
United Kingdom	7,999	57	45
Spain	6,990	50	20
Italy	6,149	44	36

SOURCE: *Statistical Abstract of the United States: 1986.*

According to Senator Bentsen, Chairman of the Joint Economic Committee, "In the absence of dramatic gains in productivity, our efforts to stem inflation cannot succeed." The inflationary impact of productivity on prices is observable in Figure 3-6, which shows higher percentage changes in output prices for those industries with the lowest rates of productivity.

From another point of view, inflation itself adds to the replacement cost of depreciated assets. Depreciation reserves based on the original costs of assets are usually inadequate to cover the higher replacement costs of

Figure 3-6

Prices and Productivity: Average Rates of Change, 1960–1975

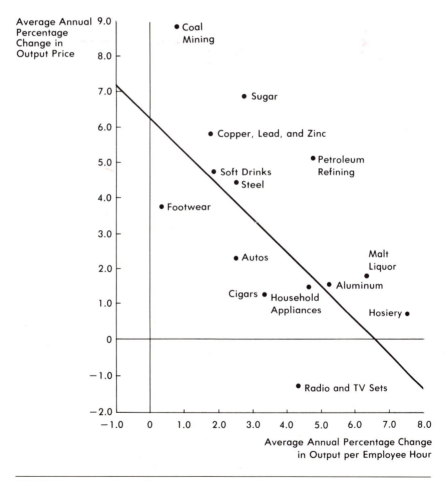

SOURCE: *Productivity Perspectives,* 1983. Reprinted with permission from the American Productivity Center.

the assets. This forces the producer to seek other funds to make up the difference between the reserves and the cost of the replacement asset. The higher cost of the replacement assets, unless these assets are much more efficient, adds to the cost of future production.

WHY THE DECLINE IN PRODUCTIVITY?

Many causes are cited for the decrease in the rate of productivity growth in the United States. The causes include those discussed in the following paragraphs.

Decline in the Capital-to-Labor Ratio

The capital-to-labor ratio is the ratio of nonresidential capital stock to total hours worked in the private nonfarm sector. From 1948 to 1973, a high rate of private investment led to a growth in the capital-to-labor ratio amounting to almost 3 percent per year. Since 1973, because of lower rates of investment and large additions to the labor force, the capital-to-labor ratio growth rate has declined to 1.75 percent per year. This slower growth of capital stock reduced productivity growth by one-half of a percentage point per year. Consequently, the trend rate of labor productivity growth declined from 3 percent annually in the 1948–1973 period to 1.5 percent from 1973 to 1978, and to negative levels in 1979, 1980, and 1982. In spite of economic recovery, the productivity gain in 1985 was only 0.3 percent.

Change in Composition of the Labor Force

Productivity has been reduced somewhat by the shift in the age-sex composition of the labor force. The addition of many young workers to the labor force, plus the addition of many women who lacked work experience or skills, slowed down the growth of productivity. It is estimated that such demographic shifts in the labor force accounted for a 0.4 percentage point reduction in the actual productivity growth rate between 1965 and 1973. As those workers became older and more experienced, that reduction in productivity during the late 1970s and early 1980s narrowed to a 0.33 percentage point decline.

Impact of Economic and Social Regulations

Economic and social regulations have contributed in many ways to the productivity slowdown. Economic regulations, such as those covering pricing and entry in transportation, precluded labor and capital from flowing into those uses that had a relatively high value. Socioeconomic regulations that attempt to reduce pollution, provide greater worker safety, and change income distribution tend to slow down conventional productivity growth. Improvements in the form of purer air, cleaner water, and noise abatement are generally not included in measured output. Thus, when an increased fraction of labor and capital resources are diverted to these uses, measured productivity is reduced. The petroleum and primary metals industries were particularly hard hit by regulations and have suffered severe reductions in productivity, as shown in Table 3-5. It is estimated that economic regulations have reduced total annual productivity growth by a 0.3 percentage point since 1973.

In addition, there are substantial indirect costs associated with social regulation, such as litigation expenses, reduced innovation because of uncertainties of meeting standards, and the cost of keeping records and filling out forms. A federal report on regulatory reform estimates that regulation compliance costs business somewhere between $50 billion and $150 billion annually. Federal environmental regulations alone imposed a direct cost to business of more than $25 billion annually in the mid-1980s. Professor Murray Weidenbaum, Director of the Center for the Study of American Business, concluded that federal regulations cost business more than $100 billion annually in the early 1980s.

Table 3-5

Total Factor Productivity Trends for Manufacturing Industries

	Average Annual Rates of Change	
	1973–1979	1979–1985
Manufacturing	0.7%	2.1%
Food	0.1	2.9
Tobacco	0.3	−8.0
Textiles	3.7	2.6
Printing and publishing	−1.2	−0.9
Chemicals	1.2	2.0
Petroleum	−3.3	−4.0
Rubber	−0.5	4.0
Leather	1.2	−0.6
Primary metals	−2.0	−0.9
Machinery, except electrical	0.8	1.6
Miscellaneous manufacturing	1.9	1.5

SOURCE: *Productivity Perspectives, 1987.*

Some large firms have indicated that they must employ hundreds, and in some cases thousands, of persons in order to comply with the implementation of economic and social regulations. One company, for example, reported that it spends $20 million per year to fill out governmental forms.

Decline in Research and Development Spending

The decline in the amount of research and development (R&D) in the United States also has contributed to the slowdown in productivity. R&D often leads to innovation and new technology, which result in new products, higher productivity, or lower costs. All of these increase our real GNP. Although estimated total spending for research and development in 1983 of $86.5 billion was greater than the $20 billion spent for that purpose in 1965, R&D spending dropped from 2.9 percent of our GNP in 1965 to 2.6 percent in 1985. If we devoted the same percentage of our GNP to research and development today as we did in the mid-1960s, however, R&D expenditures would be about $13 billion higher.

Use of Antiquated Facilities

Insufficient and tardy replacement of depreciated equipment and the lack of an ample supply of additional equipment have resulted in the aging of much of our manufacturing capital stock. This has forced manufacturers to use older, less productive, and more costly marginal equipment and capacity. Studies published in the early 1980s indicated that nearly 40 percent of the plants and equipment in the U.S. economy were over 10 years old, and 16 percent were 20 years of age or older.

From another point of view, 16 percent of all our manufacturing equipment at that time was technologically outmoded, and this ran as high as 50 percent for transportation equipment, 30 percent for steel making, and 25 percent in the auto industry. A survey by the National Machine Tool Builder's Association published in 1985 showed that 67 percent of the machine tools used in metalworking industries were at least 10 years old and 33 percent were over 20 years old. Another study indicated that American business would have to spend $435 billion (in 1986 dollars) within the next five years just to modernize manufacturing facilities.

PRODUCTIVITY AND IMPORTS

Not all industries experienced the same decline in productivity growth. Whereas agriculture, communication, and railroads held their own or showed some improvement in productivity, substantial declines occurred in mining, construction, non-rail transportation, and services, especially finance and insurance. But the fact still remains that there has been a general deterioration in the growth of productivity in the U.S. economy in the past decade or more while productivity growth and product quality of competing foreign producers have been increasing.

It was startling to many Americans to learn early in 1980 that our major domestic steel producer, United States Steel Corporation, had entered into a five-year contract to purchase technology from a Japanese steel firm. Today it is common knowledge among experts from industrial nations that the Japanese are the most efficient steel producers in the world. Our economic pride was dented when a survey of American automobile engineers indicated that a large number of them felt the quality of foreign cars was superior in many respects to that of American-made cars and that routine assembly work on U.S. cars was sloppy compared to that of most foreign imports. Moreover, many cars or parts sold by U.S. auto firms are now being manufactured abroad.

Who would have dreamed twenty or thirty years ago that by the 1980s most of our baseball gloves would be imported from Asia? Or that one of our largest and strongest labor unions, the United Auto Workers of America, would be seeking legislation to force foreign auto producers to locate plants in America? Today in the motorcycle market we hear of Suzuki, Kawasaki, Yamaha, and Honda. Seldom do we see American-made Harley-Davidson bikes. Productivity problems and costs have been in large part responsible for the requests by American manufacturers for restrictions on imports of autos, steel, shoes, TV sets, and other commodities.

In 1986 imports accounted for 22 percent of U.S. steel deliveries, 28 percent of auto sales, 52 percent of motorcycles and bicycles, 40 percent of color TV sets, 50 percent of radios, 70 percent of the stereo equipment, substantial portions of the shoes and apparel worn by Americans, and 46 percent of textile products sold in the United States. Because of cost and productivity advantages, many U.S. producers of shoes, radios, TVs, and stereo equipment operate production facilities in foreign lands or purchase final products abroad for resale in the United States.

The flood of imports caused the closure of many of our steel-making facilities and pressure for the restriction of steel imports from Japan, the Common Market, and even Third World nations. Imported autos continue to compete strongly in the U.S. market in spite of Japan's voluntary restrictions on imports to the United States. Strong reaction to the competition from imports has led to the establishment of Japanese auto plants in the United States, Japanese-American joint production ventures, and Congressional consideration of a management-backed and union-backed U.S. domestic contents bill for autos sold in the United States. Cost disadvantage also brought about the passage of a special tariff on certain motorcycles at the request of that industry. Moreover, in 1987 the steel industry was seeking legislation to limit steel imports to 16 percent of U.S. sales.

HOW DO WE GET BACK ON THE PRODUCTIVITY TRACK?

To arrest the slowdown in productivity, reduce inflation, promote economic growth, and raise our standard of living, it will be necessary to adopt some positive measures. These measures must come from several sources: government, business, consumers, and labor.

Increase Investment

Among other changes, a substantial increase in business fixed investment is essential to sustain economic growth and reduce unemployment. To regain our growth in productivity it will be necessary to devote a significantly larger share of current production to replace, modernize, and expand the capital stock of the U.S. economy. Growth is needed particularly in those sectors of the economy producing intermediate goods that are of critical importance to other industries. This is particularly true of the industries producing basic materials and energy, where substitutes exclusive of imports may be difficult to find.

Although it may seem that we invest heavily, investment in the U.S. economy is meager compared to that in some other major industrial nations. In 1984, for example, fixed investment in the U.S. economy amounted to 16.0 percent of our gross national product. But in West Germany the figure was 20.5 percent, and 28.0 percent of the Japanese output went into capital formation, as indicated in Figure 3-7. The relationship of investment to GNP in various nations is likewise shown in Figure 3-8. Over the past two decades those nations with the highest rate of fixed investment have had the largest percentage increases in productivity.

Restoring the earlier 3 percent annual economic growth rate in the capital-to-labor ratio would contribute greatly to our productive growth. But to do this would require a commitment to devote a larger share of our total output to capital formation.

In addition to the expansion of production, employment, and income, society would benefit in other ways from a higher rate of investment. Of particular importance are the expansion of domestic energy production, especially from new sources; the reduction of environmental pollution; the elimination of occupation hazards; and improvement in product safety. Benefits from investment for social goals, such as clean air or water, are not included in measured production. Currently 5 to 6 percent of capital

Figure 3-7

International Investment Patterns
(Fixed Capital Formation as a Percent of GNP)

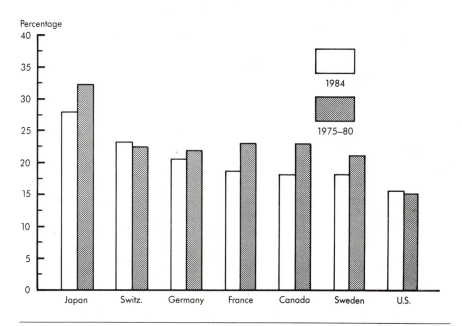

SOURCE: *Economic Handbook of the Machine Tool Industry 1985-86.*

expenditures by business is for pollution abatement. This means that a higher total investment is necessary to meet both our output and social goals for the U.S. economy and that productivity changes are underestimated.

Establish More Favorable Tax Policies

There is a need for government to create a more favorable tax policy to encourage investment. This can be accomplished by lowering the cost of capital or by raising its after-tax rate of return. The corporate and personal income tax reductions provided in the Economic Recovery Tax Act of 1981 were helpful in meeting the goal of increased investment. Especially encouraging were the Accelerated Cost Recovery System, which permitted faster depreciation of assets, and the Investment Tax Credit, which provided specific tax reductions for new investments. These two measures were designed to improve cash flow and lower taxes over the depreciable life of assets. In this regard, however, it must be pointed out that British laws have permitted the instant depreciation of assets (100 percent) since 1973 and the rate of investment is no better than that in the United States.

Figure 3-8

Investment in Fixed Capital and GNP Growth

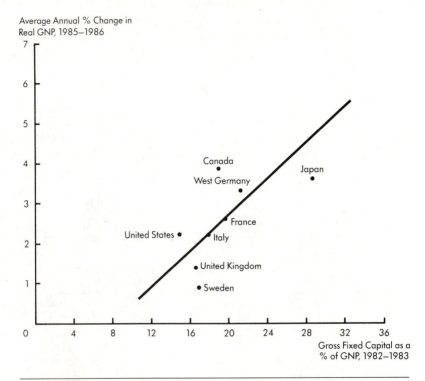

SOURCE: Data taken from *Productivity Perspectives, 1987* and *Economic Handbook of the Machine Tool Industry, 1985-86.*

Others have suggested the use of direct incentives to supplement accelerated depreciation in stimulating investment. Direct investment incentives may take the form of governmental grants or low-cost loans for certain industries or even direct governmental investment in some ailing industries. Such measures, however, are highly controversial.

Encourage Savings

Savings of Americans have averaged 6 to 8 percent in the previous two decades. This compares unfavorably with the double-digit levels of savings in many Western European countries and the nearly 20 percent savings rate of the Japanese as shown in Figure 3-9. The adoption of more employer-sponsored savings plans for employees would be helpful as a means of encouraging savings. Removal or reduction of the double tax on corporate profits that are paid out in dividends and the exemption of more interest income from federal and state income taxes would be helpful in encouraging savings. Tax reductions, along with the tax-exempt All Savers Certificates and tax-deferred IRAs provided by the Economic Recovery Act of 1981, were designed to encourage personal savings. Increased savings could be channeled into investment. Most of these tax advantages were lessened, however, by the Tax Reform Act of 1986.

Lessen Regulation

Expansion of deregulation such as that currently in effect in the airline, banking, and now transportation industries can aid productivity and lower costs. Closer scrutiny of both economic and social regulations such as those of the EPA, EEOC, and OSHA, to make sure that their costs and benefits match can be helpful in increasing productivity and lowering costs.

Figure 3-9

Savings Rates, 1985

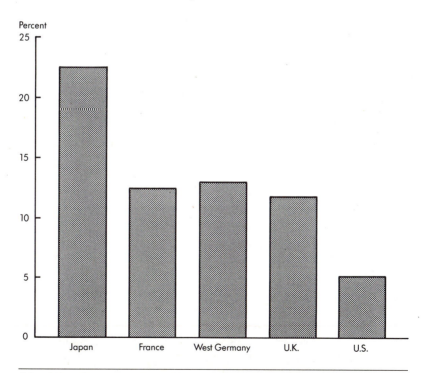

SOURCE: *Economic Review,* Federal Reserve Bank of Kansas City (February 1987).

Productivity—Do We Need an Industrial Policy?   77

Increase Research and Development

More investment is needed from business in the area of research and development. Greater emphasis must be placed on the internal expansion of business than on acquisition and merger. Much needs to be done in the area of product and service innovation in an effort to reduce cost, lower prices, and improve the quality of production.

Improve Managerial Efficiency

There is a crying need for improvement in managerial efficiency if productivity is to be increased. The slowdown in productivity cannot be attributed solely to government interference or worker lethargy. Managers, particularly at the middle and first-line levels, are not maximizing their efficiency. There is a need to apply new and/or better management techniques and to adopt managerial methods designed to improve worker productivity.

Again, it is embarrassing to learn that foreign managers, especially in Japanese-owned plants both in Japan and in America, and even in those plants staffed with American workers, are showing up American managers by insisting on worker discipline and demanding quality work. Perhaps one thing we can learn from Japanese managers is that it is better to insist that work be done right the first time rather than repeating the job or recalling the product.

It is ironic when one remembers the extent to which young potential managers from foreign nations flocked to the United States, especially in the two decades after World War II, to visit our plants and attend our schools to learn the engineering and business techniques developed by U.S. enterprise. Today some of the reverse is taking place. Peter Drucker, the world-renowned management consultant, points out, however, that many of the ideas and techniques being used so successfully by foreigners were actually developed in America. The best thing we can learn from Japanese and German management is how to put these ideas and techniques into practice. In the past few years, we have come to sense that we can learn from others, especially in the areas of quality control, human relations, and technology. Figure 3-10, by showing how far the U.S. lags behind Japan in the use of robots, illustrates that we must learn to take advantage of new technology. In the United States the use of robots has increased considerably. In 1984 alone, for example, we installed more than 5000 robots.

If we are to increase productivity in an effort to improve economic growth, improve product quality, combat inflation, and raise our standard of living, it is necessary that the workers participate in the process. Excessive demands for wages and fringe benefits beyond increases in

Figure 3-10

Robots in Use, 1986

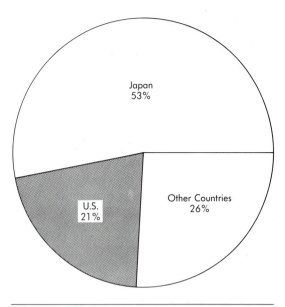

SOURCE: Estimates from the Robotics Industries Association, 1987.

productivity do not solve our productivity problem—they only aggravate it. Consequently, we need moderation in wage demands to keep them more in line with productivity.

What is needed also is better worker training via management-sponsored programs to improve worker skills, more attention to detail, stricter discipline on the job, and the development of pride in one's work. In short, we need a return to the work ethic. Workers themselves, or through management invitation, have to become more involved with the process of developing and implementing productivity improvement. The use of quality circles and quality of work-life groups are good vehicles for this purpose. It may be time also for labor unions and management to modify their adversary roles and cooperate more in the interest of improving productivity.

The Committee for Economic Development (CED) in 1983 released a study on productivity in which it suggested that both government and the private sector can do much to improve productivity.[2] Its major pub-

---

2. *Productivity Policy: Key to the Nation's Economic Future,* Committee for Economic Development, 1983.

Productivity—Do We Need an Industrial Policy? **79**

lic policy recommendations for the government included (1) eliminating unnecessary government constraints on the market system, (2) modifying policies that discourage savings and investment, (3) making basic research an important priority, and (4) continuing economic deregulation in areas in which effective competition is present.

Emphasizing, however, that the critical role in increasing productivity is played by the private sector, the CED study had the following recommendations for business and business managers:

1. Every American business should adopt explicit productivity goals to meet both domestic and foreign competitive challenges.
2. Initiative should be encouraged by top management and promising new ideas nurtured within the organization.
3. Employees (and union leadership in companies with unions) should be substantially involved in designing and implementing productivity-enhancing policies.
4. Financial incentives for improved productivity should be a common element for both labor and management.

The CED added that an appropriate public policy environment, including moderate noninflationary economic growth, is an essential prerequisite for substantial and sustained improvement in productivity.

DOES THE UNITED STATES NEED AN INDUSTRIAL POLICY?

In order to restore U.S. productivity growth, raise per capita GNP, contain inflation, compete better against foreign products, and improve balance of payments, President Carter in 1980 stressed the need to reindustrialize America. Also in 1980, *Business Week* published a special issue on the subject. President Reagan talked about U.S. competitiveness here and abroad in his 1983 *Economic Report of the President* and subsequently established a Commission on Industrial Competitiveness. In that same year he formed a special committee on productivity and competitiveness.

Because of the decline in productivity, competition from imports, and high unemployment (particularly devastating on so-called smokestack industries), there has been a growing call for a more formal industrial policy. An *industrial policy* can be defined as any selective government measures that prevent or promote changes in the structure of an economy. Most nations do have an industrial policy of some type, formal or informal. Even the lack of a formal policy can be considered a policy of letting competition and the market determine the industrial structure of the nation.

Most formal industrial policies involve one or more of the following objectives:

1. Special aid or privileges for growth or "sunrise" industries, such as high tech industries. This may involve low-interest-rate loans, government subsidies for basic research and development, and the allocation of venture capital.
2. Protection for declining or "sunset" industries, such as steel, shoes, and textiles. This may take the form of tax relief, import tariffs, import quotas, and worker retraining and relocation.
3. Targeting of industries. The selection, by an appropriate authority, of certain industries to receive special aid from the government or other private organizations, such as banks.
4. Macroeconomic policy. This involves the development of favorable monetary and fiscal policies that will foster economic development.
5. Nationalization of industry. This is the extreme in industrial policy. The government takes over and operates the basic industries. It is characteristic of socialistic economics.

The United States has always had some type of industrial policy, usually informal. Protectionist tariffs helped early manufacturers in New England, government land grants assisted the construction of railroads, and government subsidies helped build the airlines. The U.S. shipbuilding industry received government aid, and for decades American farmers have received special benefits. On a more formal basis, in the 1930s we had the National Industrial Recovery Administration, which established industrial codes regarding output and prices; the Reconstruction Finance Corporation, which provided long-term, low-interest refinancing; and the Export-Import Bank, which provided loans and loan guarantees for exporters.

In the mid-1970s there was much discussion about national economic planning at the Congressional hearings which preceded the passage of the Full Employment and Balanced Growth Act of 1978 (Humphrey-Hawkins Act). In the 1970s also we had the U.S. Synthetic Fuels Corporation, which provided funds to promote the development of synthetic fuels. More recently the government bailouts of Chrysler and Continental Bank and the automobile domestic contents bill[3] now in Congress are *ad hoc* forms of industrial policy.

Recently several individuals, organizations, and political groups were advocating a formal U.S. industrial policy. The movement was evident in

---

3. This would require 70 percent of the parts in autos sold in America to be made in the United States.

Congress. Early in 1984 an industrial policy bill, the Industrial Competitiveness Act, was approved by the House Banking Committee. Representative Ottinger of New York was leading a "national economic recovery project" to develop a "high production strategy for the United States." Four other House members wrote a master plan called the National Industrial Strategy Act. It would create an Economic Cooperation Council that would solicit the cooperation of government, business, labor, and others to recommend steps to improve U.S. industrial competitiveness. In addition, it would establish a National Industrial Development Bank with authority to lend $12 billion over a four-year period and grant an additional $24 billion in loan guarantees. Senator Edward Kennedy developed a similar type of bill. President Reagan established a Presidential Commission on Industrial Competitiveness, and industrial policy became an issue in the 1984 Presidential campaign. As late as 1986 and 1987, with the flood of imports and loss of U.S. jobs, there was still much discussion about the adoption of an industrial policy for the United States. Moreover, early in 1987 Senator Bentsen, Chairman of the Senate Finance committee, promised to introduce a new trade bill regarding import protection.

Industrial Policy in Japan

Some have suggested that the United States ought to have a national industrial policy for productivity similar to that existing in Japan. In Japan there is much planning and cooperation among the government, financial institutions, business leaders, manufacturers, trading companies, managers, and workers regarding the promotion of productivity. For example, the government may request banks to extend long-term, low-interest-rate loans to certain industries. The Ministry of International Trade and Industry (MITI) may recommend direct government aid for some firms or industries. In short, some analysts maintain that the entire Japanese economy is operated like a huge corporation, sometimes referred to as Japan, Inc. The major objectives of this so-called corporation are economic growth, increased exports, and a rise in the Japanese standard of living.

Several advocates of industrial policy have cited the success of Japan. However, it should be noted that while the Japanese have been very successful rebuilding and developing their steel, auto, and TV industries, they have experienced some failures elsewhere. Their shipbuilding industry, of late, has incurred some problems. Japanese agricultural policies have led to one failure after another. Then too, not all of Japan's success is due to the plans and policies espoused by MITI. For example, MITI's original plan for the auto industry was to restructure it so that only two firms would operate, Nissan and Toyota. Instead the industry blossomed via competition, and now Japan has more auto producers than any nation in the world.

One important reason for the early and current success of MITI policies was the need for Japan to catch up. This they did, of course, with the help of engineering, scientific, economic, and managerial aid from the United States and elsewhere. Now that they have caught up and have gone ahead in many respects, their rate of gain may diminish. Moreover, the experience of Great Britain and France with industrial policies and nationalization of industries has not been successful. In fact, these countries are now in the process of denationalization and are privatizing many of their state-owned and operated enterprises.

Problems of a Formal Industrial Policy

Any formal industrial policy involves the process of picking the winners and losers. Who will select the industries targeted for aid or oblivion? What type of aid will it be? On what basis will the selections be made? These are but a few of the questions.

Of course, there are difficulties in adopting such a policy in the United States: the independence of not only the commercial banks but the Federal Reserve itself, the presence of strong antitrust laws, the adversary character of U.S. labor unions, and in general the separation of the U.S. economic system from the nation's political structure. Also of concern to many Americans regarding the adoption of such a major plan or policy is the probable erosion of many economic freedoms cherished by business firms, banks, labor unions, and individuals.

There is also some disagreement about whether a productivity problem exists. Although there are several nations which are experiencing faster productivity growth, the United States is still ahead of all other nations in terms of total productivity per labor-hour. Figure 3-11 shows gross domestic product (GDP) per hour for several nations compared to the United States. The GDP is measured per labor-hour by Angus Maddison and per employee by the Bureau of Labor Statistics. In spite of its high productivity growth rate, Japan is low on the scale because Japanese workers work substantially longer hours than do workers in the United States and some other nations.

But the major question remains: Should the United States have a formal industrial policy? The arguments pro and con abound. Among economists there is no lack of support for either side. Some economists, such as Robert B. Reich of Harvard, Lester G. Thurow of MIT, and government consultant Pierre A. Rinfret, have spoken in favor of an industrial policy. On the other hand, Charles L. Schultze, Murray L. Weidenbaum, Alan Greenspan, and Herbert Stein, all former Chairmen of the President's Council of Economic Advisers, have spoken out against the adoption of formal industrial policy.

Figure 3-11

International Productivity Levels (United States = 100)

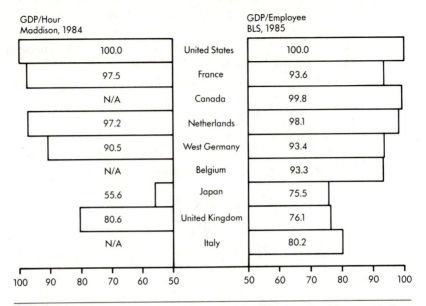

GDP/Hour Maddison, 1984		GDP/Employee BLS, 1985
100.0	United States	100.0
97.5	France	93.6
N/A	Canada	99.8
97.2	Netherlands	98.1
90.5	West Germany	93.4
N/A	Belgium	93.3
55.6	Japan	75.5
80.6	United Kingdom	76.1
N/A	Italy	80.2

SOURCE: *Productivity Perspectives,* 1987. Reprinted with permission from the American Productivity Center.

CONCLUSION

After leading the world in technology, productivity, and economic growth, the United States experienced a noticeable decline in the rate of its productivity growth during the 1970s and early 1980s. Total production and per capita income are determined by productivity growth and growth in labor-hours. Productivity growth accounts for approximately 80 percent of the growth in our real GNP. Among other factors, productivity is affected by investment, business cycles, the capital-to-labor ratio, and the amount of research and development taking place in the economy.

The slowdown in productivity in the U.S. economy has been attributed to an inadequacy of investment, the use of antiquated capital facilities and equipment, the changing composition of the labor force, the impact of economic regulation, the slowdown in R&D spending, management inefficiency, and worker lethargy. The U.S., for example, devotes a smaller share of its GNP to investment than do most other major industrial nations.

In order to improve productivity and our economic rate of growth, it has been suggested that we need a more favorable tax policy to stimulate investment, an increase in our rate of savings, a reduction of the restrictions caused by economic regulation, a modification of our social goals, improved managerial efficiency, increased spending on R&D, more worker involvement in the productive process, and closer cooperation between management and labor. It has also been suggested by some that we should adopt a formal industrial policy similar to that of Japan as a means of restoring productivity and world competitiveness in the U.S. economy.

There is no question that the decline in the rate of growth in productivity in the United States created or aggravated some major economic problems. How we should go about improving productivity and whether or not we should adopt an industrial policy are open questions.

QUESTIONS FOR DISCUSSION

1. Do you think that investment generates economic growth or that increases in GNP induce investment?
2. Should the Congress enact an industrial policy bill?
3. Do you favor the use of direct government aid to business, such as the Chrysler loan guarantee of 1980, to assist U.S. business?
4. Do you think we should modify our national environmental requirements in order to improve productivity?
5. Should the United States restrict foreign imports to aid American businesses?
6. Do you agree with corporate and union leaders in the auto industry that Congress should enact a domestic contents law for the auto industry?
7. Do you believe that the double tax on corporate profits (tax on profits and dividends) should be eliminated?
8. To what extent do you think that the U.S. decline in productivity is due to managerial inefficiency or to worker lethargy?
9. Should the United States adopt a national "Buy American" program to fight off imports?
10. Should the U.S. steel industry be given protection against lower-cost imported steel?

SELECTED READINGS

"America's Human Resources—Keys to Productivity." *Perspectives on National Issues* (April 1982).

Baily, Martin Neil. "Productivity Growth Slowdown by Industry." *Brook-*

ings Papers on Economic Activity. Washington D.C.: The Brookings Institution, 1982.

Denzau, Arthur T. *Will an "Industrial Policy" Work For the United States?* St. Louis: Center for the Study of American Business, 1983.

Economic Handbook of the Machine Tool Industry. National Machine Tool Builders Association, biannually.

Industry Analytical Ratios for the Business Sector. BLS (November 30, 1983).

International Financial Statistics. Washington, D.C.: International Monetary Fund, annually.

Koch, Donald L. "Productivity: Regaining the U.S. Edge." *Economic Review*. Federal Reserve Bank of Atlanta (October 1983).

————. "Productivity: The Micro Solution." *Economic Review*. Federal Reserve Bank of Atlanta (September 1983).

Littman, Daniel. "The Implementation of Industrial Policy." *Economic Review*. Federal Reserve Bank of Cleveland (Summer 1984).

Productivity Perspectives. Houston, TX: American Productivity Center, annually.

Productivity Policy: Key to the Nation's Economic Future. Committee for Economic Development, 1983.

Reich, Robert B. *The Next American Frontier*. New York: Times Books, 1983.

Schultze, Charles L. "Industrial Policy: A Dissent." *The Brookings Review* (Fall 1983).

4

AGRICULTURE
WHY THE
CRISIS?

For the past 60 years or more the United States has grappled with a farm problem that has assumed many forms. For a few years in the early 1970s beef and crop shortages, accompanied by strong domestic and foreign demand, caused a dramatic rise in farm prices and farm incomes. At that time the farm problem appeared to be one of shortages and scarcities. In 1974 and 1975, however, when surpluses began to appear and farm prices and farm incomes started to decline, indications were that the American farm problem was reverting to what it had been—one of abundance. In fact, in 1977 groups of farmers were demonstrating throughout the country and in Washington, D.C., with tractor parades in protest against low crop prices and high operating and living costs. In the mid-1980s the severity of the farm problem reached its most dramatic level since the Great Depression 50 years ago. It culminated in a rash of farm bankruptcies and widespread country bank failures.

The long-range problem of excess supply, low prices, and inadequate farm income has been attacked with a variety of remedies and has been the cause of almost continual controversy at the highest levels of government. But still it remains largely unsolved. Just what is this problem? Is it so complex that it defies clear identification and solution?

THE LONG-RANGE PROBLEM

For decades, many authorities have stated that the American farm problem is essentially one of abundance. Their position holds American agriculture to be so productive that it is capable of supplying the needs of a population vastly larger than our current numbers.

If we accept this view of the farm problem, then its cause is primarily in the economics of the situation. The rapid increase in agricultural productivity simply outstripped the economy's ability to consume farm output. Years ago it was suggested that rising population, new industrial uses, and broadening foreign markets would eventually bring the demand for farm products into balance with supply. Farm prices would then stabilize at some satisfactory level, the government could withdraw its support, and the problem would solve itself, provided farm productivity ceased growing so rapidly. This did not occur, however.

The continual growth in agricultural output, far in excess of current needs, maintained a constant downward pressure on farm prices and on farm incomes. While farm incomes remained low relative to the incomes earned in other sectors of the economy, the prices paid by farmers for goods and services to meet their production and consumption needs continued to rise. In the face of this problem the federal government intervened in the market for farm products in an effort to raise farm incomes to a level that would enable the agricultural sector to remain a viable part of the American economy.

Subsidies to Agriculture

One aspect of the farm problem that disturbs many Americans is the use of their tax dollars to support a particular segment (and a diminishing segment at that) of the national economy. Paying out tax dollars to farmers for not producing seems contrary to American values that have traditionally dictated rewards for the producer and penalties for the laggard. To them, the solution to our farm problem is obvious: Congress need only cancel all programs designed to raise farm incomes. Market forces would then dictate farm prices, and only the most efficient producers would manage to survive the shifts in supply and demand caused by changes in climatic conditions, consumer tastes, the rate of population increase, and the demand of industry for raw materials. Farm efficiency would continue to rise as farmers sought to lower their costs of production and thus gain larger profits. But a flaw is encountered in this solution. What of the farmers who could not compete in the free market, for whatever reason? What would become of them?

They would remain a part of the economy; but they would somehow have to be assimilated into other sectors, and this assimilation would be a difficult and time-consuming process. In 1985 the farm population comprised about 6 million people, or about 2.5 percent of the nation's total population. It has been estimated that America could meet its present and foreseeable needs for farm products with a farm population roughly two-thirds that size. If the other 2 million now living on marginal farms[1] were suddenly displaced and forced to relocate in urban areas, the burden of this nonproductive group would be difficult for urban economies to bear. Massive financial aid would be necessary to sustain these people until they equipped themselves to compete in an urban industrial society. There would also be a hardship on the displaced farmers and their families. Uprooted from the only way of life many of them have known, they would undoubtedly be sorely pressed to conform to urban standards and values. It is easy to foresee antagonisms developing between the original urban dweller and the interloper. It is not necessary to dwell on the consequence of such a move, however, except to state that it would be serious for urban society.

Social Aspects of the Farm Problem

In considering the consequences of final and complete removal of government agricultural subsidies, we are forced to come face to face with the essence of the farm problem. The trouble is found not only in the economics of the farm industry, but also in the social problems that arise therefrom. The American people have long praised the efficacy of a free enterprise economy, but they have never hesitated to interfere with the workings of such an economy when it began to exact social costs detrimental to the welfare of certain groups of citizens. They recognize that competitive pressures have placed American farmers in a situation that they cannot face unaided. If we, as a people, were willing to accept the displacement of marginal farmers because they could no longer compete with their more efficient counterparts, and if we were willing to accept the fact that a considerable number of these people would be forced to exist in substandard

---

1. We shall define *marginal farmers* as farm owners or operators (tenants) who would be unable to support themselves and their families in the absence of government crop support payments or other aids that supplement their income. The causes of their marginal position can be many and varied. The essential characteristic, however, remains their inability to survive, as farmers, in the absence of government intervention in the market of agricultural products.

conditions, then the solution to the farm problem would be easier. The humanistic values of the American people, however, are strong enough to prevent such a social calamity, and we live with what we term a "farm problem" even though the problem developed because of our efforts to solve a problem of much wider dimensions.

A Paradox

There have been and are many farm problems and several ramifications to each one. The farm problems in the United States, except for the shortages that occurred in the mid-1970s, were different from the problems in many other parts of the world.

The Real Problem. Since the dawn of recorded history, the human race has struggled with a much more realistic and serious farm problem. A majority of the world's population still faces this problem—not enough food. Historically, famine has been coupled with disease and war as the most terrible ravagers of the human race. Outright starvation has taken a heavy toll, but far more insidious have been the effects of inadequate diets. The life expectancies of people in countries with sufficient food of the proper type inevitably are higher than those of peoples existing in countries where food supplies are just sufficient to keep them at the margin of subsistence. It is only recently, in the last 200 years, that some fortunate countries, including the United States, have managed to solve this problem.

The countries of North America and Western Europe have managed to provide diets for their people that have been well above subsistence levels. The application of scientific methods to farming, the development of more efficient farm implements, and the substitution of mechanical energy for animal energy in the heavier farm tasks aided in the solution of the farm problem for these fortunate countries. It is now possible to produce, in a given year, a surplus of farm products to tide the country over in the event of a poor production year. No longer must the peoples of the agriculturally advanced countries live in constant fear of uncontrollable changes in the weather. In short, the peoples of these countries, particularly the United States, need no longer fear the ravages of starvation. For them the real farm problem has been solved. Herein lies the paradox.

The American Problem. In considering the cause of the farm problem in the United States, we found that it lies in the social costs caused by rapidly increasing agricultural productivity and the resulting abundance of farm products. We need only consider our farm problem in light of the real farm problem to see the paradox. We are faced with an embarrassment of agricultural riches because we refuse to subject a relatively small portion of our population to inordinate economic and social pressures, while much of the rest of the world does not have enough to eat. When viewing it in this context, one begins to wonder whether the United States really has a farm problem.

Having placed our farm problem in its proper dimension vis-à-vis the world food situation, we can now consider it somewhat more objectively. In the first place, failure to solve the American farm problem will not precipitate starvation for the American people. Thus, a solution predicated over a period of years is practicable. Second, in seeking a solution we are not motivated by a lack of economic efficiency or insufficient resources; rather we are motivated by a concern for human dignity and values. We want our marginal farmers to assume productive places in our society. To gain a better understanding of the complexities of the American farm problem, we must turn to a consideration of its historical evolution.

HISTORICAL DEVELOPMENT OF FARM PROBLEMS IN THE UNITED STATES

It is easy to forget that for more than the first 100 years of this country's existence agriculture was the most important sector in the total economy. In fact, as late as 1910 the majority of Americans still resided in rural areas. Although 25 percent of the U.S. population still live in *rural areas* (defined by the Census Bureau as places of less than 2,500 population) only a limited number actually live on farms or directly depend on the agricultural sector of the economy for their livelihood.

Early Importance of Agriculture

The root cause of the farm problem can be traced to the dominance of agriculture during the developmental years of our country. The vastly rich, undeveloped areas of the United States were an open invitation to settlement and cultivation during the nineteenth century. The federal government, through a series of public land acts, made available to its rapidly growing population immense tracts of fertile land on a free or nominal cost basis. The great westward shift in population began in earnest in the years immediately after the Civil War, and by 1910 distinct areas of the country were specializing in the production of crops best suited to their soil and climate characteristics. The middle section of the country had three broad belts, the northern and southern belts raising spring and winter wheat, and the middle belt specializing in corn and hog production. To the west, the vast grasslands of the Great Plains encouraged beef cattle operations on an unprecedented scale. Cotton began a westward movement, and by 1900 Texas ranked first in cotton production. Tobacco then became the major crop of the deep South.

The net result of the shifting of crop and livestock production to areas best suited to their requirements was a vast increase in farm output, although this fact alone was not sufficient to explain the increase. Not only were the land and climate better suited to the crop raised thereon, but also much larger amounts of land came under cultivation. In short, the major cause of increasing farm output during the latter years of the nineteenth century and early years of the twentieth was that both more and better land was being cultivated.

Rising Level of Agricultural Technology

Throughout this period the development of improved farm machinery played a major role. Threshing machines, steel plows, seed drills, mowers, rakes, cultivators, and reapers significantly increased the farmer's efficiency. With these machines the farmer could cultivate much larger acreages than was possible with the primitive hand tools that had historically been available. The mechanization of the American farm was widespread by 1915 and, in addition, the development of the gasoline engine had solved the age-old problem of the limitations imposed by animal power. Some of the steam-operated farm machines of the late nineteenth century had become so large that 40 horses were necessary to move them. The advent of the gasoline-powered farm tractor in the early years of the twentieth century cut the last bond that tied farmers to the relatively puny resources of animal power, and the farmer's capability literally soared.

The period from World War I to the present has witnessed a somewhat different form of improvement in agricultural technology. Prior to World War I, practically all of the increases in farm output could be ascribed to more extensive cultivation of the land. After the war, advances in the pure sciences began to contribute to the increase in output per acre, and intensive cultivation became important. Genetics (the development of improved strains of crops and livestock), soil chemistry, highly efficient fertilizers, and the development of fungicides and insecticides enabled the farmer to raise per-acre output to unprecedented levels. Scientific farming, a general term used to describe the application of the latest mechanical, scientific, and methodological improvements to farm operations, became a popular term; and the Department of Agriculture became very active in disseminating the latest of these developments to every portion of the national farm community. The Department has maintained its efforts to increase farm productivity through the years. In addition, the agriculture colleges in various states have played a significant role in the basic research needed to raise farm productivity.

Increasing Productivity

In summary, we can conclude that the rise in real farm output during the past 100 years has been phenomenal. More recently, the index of farm output compiled by the Department of Agriculture indicates that real output increased 5.5 percent in the period 1960–1981. But even more important for our purposes, farm output per worker-hour during that period increased 250 percent and the worker-hours required on farms declined by 57 percent. Of course, these trends are more meaningful when placed in a wage-price context. That shall be our task in the next section. The increase in the farmer's individual productivity, more intensive cultivation of the land and minimal acreage reduction have been major elements of our farm problem. These trends continue today unabated, and unless they are kept constantly in mind, a meaningful analysis of the future of our agricultural community is improbable.

THE ECONOMIC ASPECTS OF THE FARM PROBLEM

A family farm is a producing unit, and the farmers who operate them have as their goal an income sufficient to cover all costs of production plus a residual amount that will enable them to support their family in a reasonably comfortable manner. This is a basic fact that must be remembered in discussing the farm problem.

During the pioneering years of American agriculture, the family farm was almost a self-sufficient unit. Located far from centers of manufacture and trade, and hampered by a still-developing transportation system, American farmers were forced to rely on their own ingenuity for survival. With the construction of regional and national railroad networks and the building of a farm-to-market system of highways, their lot improved. Farmers could begin to specialize in the crop for which their land was best suited, and they could sell this output in ever-broadening national and international markets. The money income that they earned allowed them to purchase the goods of other specialists at prices much lower than the value of their own efforts used in producing similar goods. Thus, American farmers entered the market economy and began to face the rigors of competition.

The Market for Farm Products

The farm economy has always been a prime example of the freely competitive market. The producers of farm products fit well the assumptions of pure competition in that price and output decisions in the market have never resulted from the decisions of one or a small group of producers. American agriculture has historically been characterized by the relative insignificance of the output of the single farm when compared to the entire supply forthcoming from all farms. In addition, since farm products, by their nature, are homogeneous, it is very difficult to distinguish the corn produced on Farm A from that produced on Farm B. As a consequence, it is practically impossible for any farmer to gain any substantial degree of pricing power. Thus, both the larger number of relatively small producers of farm products and the homogeneity of the products themselves tend to negate the chance of monopolistic power developing in the agricultural community. The possibility of controlling the amount of farm goods produced in a given year is likewise largely precluded because of the wide dispersal of farm ownership.

Price Determination

The absence of price power on the supply side of the market for farm products has been one of the most important economic aspects of the farm problem. Since it seems clear that individual farmers are powerless to set the price for their products, they must be considered price-takers. They can only sell their products at the going price. If they choose to charge more than the going price, they can sell little or nothing. If they charge less than this price, they are acting irrationally since the market will take all of their output at a higher price. But if farmers are price-takers, who is the price-maker?

The answer to this question must necessarily involve some discussion of economic principles. Probably the first idea that confronts most beginning students in economics is the interaction of the forces of supply and demand to form a market price. Given a schedule of the units of a good that sellers will sell at various prices, and a similar schedule of what buyers will purchase at various prices, it is possible to determine what the market price will be. Table 4-1 illustrates such a schedule. It is apparent from this table that the market for commodity X will be cleared at a price of $5. Five units of the good will be offered for sale at this price, and all five units will be taken off the market. Thus, we state that the forces of supply and demand determine both market price and also the number of units of the good that changes hands.

While this simple example preserves the essentials of supply and demand interaction, it at the same time is a heroic oversimplification of reality. It does not tell us why buyers and sellers will exchange five units at a price of $5 each, nor does it tell us over what period of time this exchange will take place. To answer these questions, we must probe much

Table 4-1

Schedule of Supply and Demand for Commodity *X*

Price	Number of Units Sellers Will Sell	Number of Units Buyers Will Buy
$10.00	10	0
9.00	9	1
8.00	8	2
7.00	7	3
6.00	6	4
5.00	5	5
4.00	4	6
3.00	3	7
2.00	2	8
1.00	1	9

deeper into the nature of the supply and demand schedules. We have seen something of the supply side of the market for farm commodities, but until we consider the nature of the demand side we can draw no meaningful conclusions.

Inelasticity of Demand. The demand for most farm products is price inelastic. Moreover, the supply is relatively price inelastic. Although economists may be familiar with such terms, their meaning may not be immediately evident. When the economist speaks of price inelasticity for a product, this simply means that the buyer's desire for the good is not significantly affected by changes in the price of the good. Conversely, when the economist states that the demand for a product is price elastic, this means that the buyer's desire for the good will be significantly affected by changes in its price. Within this definitional framework we can now discuss the nature of the supply of and demand for farm products.

Probably the most significant limitation faced by the American farm is the size of the human stomach. The consumer can eat just so much food within a defined time span. Whether the price of the food is high or low, it will not significantly affect the amount consumed. The same limitation is true of other farm products that cannot be classified as edibles. Thus, wool and cotton are limited by the population's ability to wear out its clothing, and tobacco growers can hope to sell no more of their output than smokers can possibly smoke. The producers of hides are constrained in the amount of leather they can supply by the ability of people to wear out shoes and other leather goods. What we are really saying is that the demand for farm products is relatively stable over long periods of time, and it is largely determined by population growth and consumption patterns rather than price levels.

We must be careful, however, to keep clearly in mind that we are speaking of the aggregate demand for farm products, not the demand for a particular product. It is entirely possible that falling meat and dairy prices may cause people to shift their demand away from the cheaper products, such as bread and potatoes, and toward the more expensive high protein foods.

Rising incomes may have a similar effect. The overall demand for farm products is income inelastic as well as price inelastic, which simply means that higher incomes will not cause people to consume proportionally more food. Higher incomes, however, will probably cause people to purchase more of the higher priced foods, such as meat and dairy products, and less of the cheaper products.

Variability of Supply. If the demand for farm products is stable over long periods of time, then what of supply? Have American farmers been able to regulate their output so that it just meets the stable but growing demand and thus balances farm prices and maintains incomes at some normal level? Unfortunately, no. The variables that determine the level of crop and livestock production are too numerous and unpredictable to submit to the control of humans. Rainfall, days of sunshine, mean temperature, insects, plant diseases, floods, blizzards, and hail all combine to make farming an unpredictable business at best. Even today, after years of intensive research directed toward controlling at least some of these factors, the average farmer can still lose an entire crop because of the vicissitudes of nature. About the only variables that the farmer can regulate are the number of acres planted, the heads of livestock raised, and the amount of growth that can be induced through the application of labor and scientific farming methods.

The Parity Ratio

What then has been the result of this situation where demand is relatively stable and predictable, and supply is highly variable? The best indicator is the movement of farm prices. In Figure 4-1 this movement is traced for the years 1978–1986. In addition, an index of prices paid by farmers is plotted together with a parity ratio. This last measure indicates the change in the purchasing power of farmers. A rise in the parity ratio above 100 shows that the index of prices received by farmers (the numerator) is rising faster than the index of prices they pay (the denominator), and they are enjoying a relative increase in the purchasing power of their

Figure 4-1

Prices Received and Prices Paid by Farmers

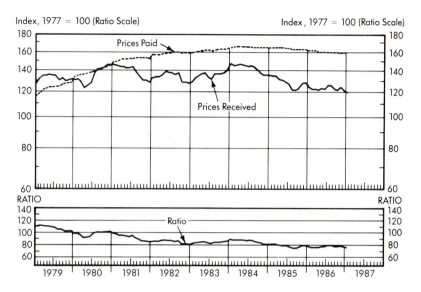

Index, 1977 = 100 (Ratio Scale)   Index, 1977 = 100 (Ratio Scale)

[1]Ratio of index of prices received to index of prices paid, interest, taxes, and wage rates, on 1910–1914 = 100 base.

SOURCE: *Economic Indicators* (February 1987).

income. A drop in the parity ratio indicates that their income is relatively less valuable and its purchasing power is decreasing.

An examination of parity ratios gives a fairly clear indication of the plight of American farmers for the 60 years prior to 1987. In only 17 years (1916–1920, 1941–1952, and 1978–1979) did the parity ratio stand in favor of the farmers. Since 1980 it has moved downward.

The record traces a rather dismal picture of the financial plight of the farmer, but it also yields an insight into the nature of the supply of farm products. It indicates that the producers of goods and services sold to farmers—the manufacturer, the retailer, the banker, and the professional person—were better able to match their supply of goods and services to a shifting demand, and the prices they could charge did not vary as much as the prices of farm products. Unfortunately the farmer did not enjoy a similar degree of control over the supply of farm products. Given favorable growing conditions, the farm community would produce a bountiful crop. In the face of stable demand, this increase in supply would cause a fall in farm prices. The farmer, witnessing a drop in income caused by falling prices, could react only by planting more acreage or by raising more livestock the next year to counteract the decline in income. But this additional increase in farm output would only worsen the price situation and cause a further decrease in income. The only way out of this vicious circle was a decline in farm output caused by adverse growing conditions, or an artificial increase in demand stemming from war. Favorable parity years coincide almost exactly with the years of World Wars I and II and a brief period of general high inflation in the 1970s.

The only other solution, voluntary control of supply by the farm community, has never been successfully used. Many attempts at voluntary control of farm output have been made, but most were ineffective, largely because of the independent nature of American farmers. They have simply refused to accept the fact that by increasing their output they may well be acting to their own detriment.

Economic Effects of Rising Farm Productivity

The crucial point in this entire discussion centers around the farmer's ability to increase output in an effort to offset falling income. We discussed previously the technological development that has taken place in American agriculture. Figure 4-2 traces the results of this development for the years 1929–1985, as it has affected the productivity of the farm worker and farm

Figure 4-2

Farm Employment and Worker Productivity, 1929–1984

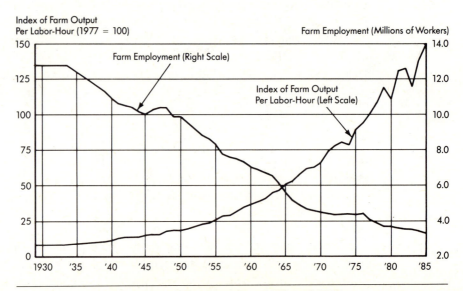

SOURCE: *Economic Report of the President,* 1978 and 1985.

employment. The effects of soaring farm labor productivity are evident. As farmers added successive increments of capital equipment and fertilizers, the need for farm labor declined, as did the farm labor force. In the period between 1929 and 1985, the farm labor force was reduced by more than two-thirds, from 12.8 million to 3.5 million. The significant fact about this decline is that it occurred in the face of a farm output that more than doubled. Over the same period, 1929–1985, the volume of total farm production (1977 = 100) rose from 44 to 117.

The revolution in agricultural technology also had a direct effect on the size of the average American farm. As equipment became larger and more specialized, as well as more efficient, farmers were forced to use it on larger acreage. The days when the general purpose farm of 100 or so acres would support the farmer and the farmer's family are fast fading. Table 4-2 shows clearly the accelerating trend toward larger farms in the United States. This trend parallels the surge in farm labor productivity that is traced in Figure 4-2. As the average size of the American farm increases, it takes fewer and fewer farm workers to handle it efficiently. As available farm jobs decline, agricultural workers must find other types of employment or face, at best, a marginal existence on less than marginal farms. You will recall that earlier in this chapter we stated that estimates by agricultural experts indicate that the American agricultural sector could meet the country's present and foreseeable needs with a farm population roughly two-thirds its present size. If these estimates are correct, then it follows that the remaining one-third of the American farm population is existing on marginal farms. But what sustains them? If their farms

Table 4-2

Number of Farms, Classified by Acreage, 1959–1982

Size of Farm	1959	1969	1974	1978	1982
Total	3,711	2,730	2,314	2,258	2,241
Under 10 acres	244	162	128	151	188
10–49 acres	813	473	380	392	449
50–99 acres	658	460	385	356	344
100–179 acres	773	542	443	403	368
180–259	414	307	253	234	211
260–499 acres	472	419	363	348	315
500–999 acres	200	216	207	213	204
1,000–1,999 acres	70	91	93	98	97
2,000 acres and over	57	60	62	63	65

SOURCE: *Statistical Abstract of the United States: 1986.*

are not efficient enough to compete with large, capital-intensive farming operations, in the absence of nonfarm income, how do they manage to survive?

COMMODITY PRICE SUPPORTS AND THE CONCEPT OF PARITY

During this century American farmers witnessed a variety of changes in the agricultural sector of the economy in regard to production, employment, and especially income.

The Golden Age

The first two decades of the twentieth century were the golden age for the American farmer. Markets grew along with industrial production, the work force increased rapidly, and on the farms prices and incomes continued to rise. When war broke out in 1914, the production of agricultural commodities virtually ceased in Europe. The American farm sector rapidly was transformed into the "breadbasket" for the entire allied group.

Farm prices rose sharply and farm output jumped almost 20 percent in the period 1916–1918. Additional acreage was pressed into production to meet the surging demand, much of it originating in Europe. With the end of hostilities, however, the American farmer's golden age came to an abrupt end. European farms began to return to production. Markets became saturated, prices dropped, and farm income fell rapidly due to the greatly enlarged farm output flowing from war-enhanced farm capacity. The plight of the American farmer became very serious by 1920 and began to attract government attention.

Farm Legislation of the Twenties

In 1920 and 1921 President Harding signed into law legislation that raised tariffs on those agricultural commodities that were imported into this country. The law worked well with respect to commodities such as wool and sugar where foreign producers were an important element in the market. Unfortunately this action ignored the fact that the United States was almost completely self-sufficient in agriculture. The producers of cotton, tobacco, feed grains, and pork received virtually no help from the tariff since imports of these commodities were limited.

The 1920s were good times for most Americans. Industries prospered and the age of mass production and mass consumption arrived. Unfortunately the farmer did not participate in this prosperity. While the cost-price relationship (parity) of farm products was not excessively out of balance, a new factor had been injected. During the war years many farmers had borrowed heavily to finance the cultivation of new acreage, or to bid away available acreage from their counterparts. The high level of farm incomes had been capitalized in farm land values that rose to record heights. With the arrival of depressed farm prices and incomes in the 1920s, the fixed payments for interest and principal repayments on loans contracted during the war became excessively burdensome. As a consequence, farm foreclosures became common, particularly in the Middle West, where many small banks were forced into bankruptcy because of the decline in value of their prime assets, mortgages on farm lands.

The Great Agricultural Depression

By 1929 the situation in agricultural income had reached a stage serious enough to prompt the Hoover Administration to attempt corrective measures. The Agricultural Marketing Act of 1929 created the Federal Farm Board and gave the board power to support the stabilization activities of farmers' cooperative marketing associations. The board promptly entered the market for cotton and wheat, and attempted to support prices at 16 cents a pound for the former and $1.15 a bushel for the latter.

By the summer of 1931, the domestic price of cotton had fallen to 6 cents a pound and that of wheat to 39 cents a bushel, and the board held some 3.5 million bales of cotton and 257 million bushels of wheat. The failure of this program was highly predictable. It is possible to subsidize farm prices domestically through purchase and storage if sufficient funds are available and purchased surpluses are to be followed by periods of shortage; but the board was limited in the funds available to it, and it was trying to deal in farm output that was perennially in surplus. A feasible solution would have been the introduction of acreage production controls. The board, however, was forced to plead for voluntary reductions in farm output. Its plea was largely ignored and the program failed.

The Introduction of Parity. With the election of the Roosevelt Administration in 1932, the philosophy of government support payments to augment agricultural income was made explicit. Stripped of all embellishment, this philosophy was simply "economic equality for agriculture." The Administration was committed to the idea that the purchasing power of farm income was to be the criterion for judging the adequacy of farm policies. The concept of parity was introduced as a standard of the relative well-being of the farm sector. In the first New Deal farm legislation, the Agricultural Act of 1933 redefined support to mean parity; that is, farm prices were to be established and supported at a level that would give farmers an income from agricultural commodities that would result in purchasing power comparable to that of a base period, August, 1909 to July, 1914. Remember that this base period occurred in the middle of the golden age of American agriculture.

Parity is perhaps best understood by means of a simple example. In Table 4-3 the prices of a bushel of wheat (price received) and a shirt (price paid) are listed for the base period (1909–1914) and some assumed year such as 1935. Price indexes of prices paid and received by farmers are calculated with the base period equal to 100. A parity ratio is computed using the index of price paid as the denominator and the index of prices received as the numerator. The parity ratio of .43 that results shows that while the prices of both goods have fallen, the price of a bushel of wheat has declined much further. The bushel of wheat commanded less than half as much purchasing power, relative to the shirt, as it did in the base period. In the 1909–1914 period, the sale of 1.33 bushels of wheat ($2.00)/($1.50) would have yielded the farmer sufficient income to buy the shirt. In 1935 the farmer had to sell 3.12 bushels of wheat ($1.25/$.40) to purchase the shirt.

To maintain equality of purchasing power for the farmer, the actual price of wheat in 1935, $.40 per bushel, must be adjusted upward by dividing it by the parity ratio. Thus, wheat must be supported at $.93 per bushel, a price that would still allow the farmer to sell 1.33 bushels ($1.25/$.93) and use the proceeds to purchase the shirt.

With the introduction of the parity concept, the farm problem took on explicit social dimensions that had previously been only implicitly present. The federal government committed itself to the philosophy that the farm

Table 4-3

Computation of Parity Price

	Good Sold		Good Purchased	
	1909–1914	1935	1909–1914	1935
Wheat, 1 bu.	$1.50	$.40	Shirt $2.00	$1.25
Price index	100	27	100	63

$$\frac{\text{Index of prices received}}{\text{Index of prices paid}} = \frac{27}{63} = .43 = \text{parity ratio}$$

$$\frac{\text{1935 price of wheat}}{\text{Parity ratio}} = \frac{\$.40}{.43} = \$.93 \text{ Per bushel of wheat, parity price}$$

population should not be subject to an economic situation which placed it in a significantly inequitable position regarding income and purchasing power relative to the rest of the population.

Agricultural Adjustment Act of 1933. The mechanisms for preserving this equity have changed in detail over the years, but they have all been ordered to the same end and have employed essentially the same means. The Agricultural Adjustment Act of 1933 established the Commodity Credit Corporation (CCC). This corporation was similar to the Federal Farm Board under the 1929 Act in that its purpose was to engage in loans, purchases, and storage operations of agricultural commodities. The loans were of the nonrecourse type and stipulated that if the value of crops held as collateral went above the CCC loan, the farmer could repossess the crop, sell it, pay off the loan, and pocket the difference. If prices fell below the loan support price, the farmer would be permitted to default on the loan payment, keep the money loaned, and allow the government to take permanent possession of the crop that was serving as collateral.

The program differed from the 1929 Act in that it did provide for controlled production. The Secretary of Agriculture was empowered to enter into voluntary contracts with farmers who would agree to restrict crop acreage and livestock breeding to a specified percentage of a base period. Another provision of the 1933 Act permitted payments to be made in the form of rentals for acreage taken out of production. The funds to implement this program were to come from a tax on the processors of the supported commodities, such as meat packers, wheat millers, cotton ginners, and canners.

In January, 1936, the Agricultural Adjustment Act of 1933 was held unconstitutional by the Supreme Court. The Court ruled that the tax on processors was being employed in the interest of a particular group (the farmers) rather than in the general welfare. Furthermore, the court stated that the benefit payments it financed were being used to purchase conformity with a program which Congress, under the Constitution, had no power to enact.

The Soil Conservation and Domestic Allotment Act passed in 1936 had as its purpose the conservation and improvement of soils. Direct federal payments were made for planting "soil-conserving" crops in place of "soil-depleting" crops. Since the crops that historically had been supported were defined as *soil depleters,* the result was another method of payment by the federal government for acreage reductions in the principal cash crops.

Agricultural Adjustment Act of 1938. The Agricultural Adjustment Act of 1933 was followed by the Agricultural Adjustment Act of 1938. The new AAA was essentially the same as the Act of 1933 with one significant difference. In the new Act, Congress specifically directed the Secretary of Agriculture to intervene with nonrecourse loans whenever the prices of basic commodities fell below defined levels, or supplies rose above certain levels. In addition, the loans were to be at rates between 52 and 75 percent of parity. Since that time, Congress has specified support prices at a certain percentage of parity prices.

World War II and After

The war years, with the tremendous increase in demand for American farm products, resulted in the use of a substantial part of the pre-World War II agricultural hoard. In 1945, the ratio of farm to nonfarm per capita income, on the 1909–1914 base, stood at 151. Not until 1948 did world agricultural supply catch up with demand. When it did, the consequences for the American farmer were quickly felt. Prices of farm commodities fell drastically and the cry for government supports was quickly taken up by various farm organizations.

New farm legislation was passed in 1948 and 1949, but it was new only in the sense that its date of passage was more recent than the acts of the 1930s. There was little appreciable change from past practices. The CCC was placed on a permanent basis, and it began support operations in a manner substantially unchanged from its prewar practices.

Since 1948 there has been little basic or essential change in the government's efforts to support farm incomes. New-sounding phrases have been heard, but none involve any real departure from the programs we have discussed. Flexible price supports were simply supports that were not rigidly defined by Congress, but which could fluctuate within set limits depending on potential supply. Parity was redefined in the 1949 Agricultural Act as taking into consideration prices received and prices paid during a fairly recent period (previous ten years according to this specific act) instead of the relationships existing in the 1909–1914 period.

An innovative plan introduced in 1949 by the Secretary of Agriculture, but never enacted by Congress, would have substituted direct income payments to farmers in place of parity price supports. Food prices to consumers would then be determined by the marketplace. Lower crop prices would be offset by higher direct income payments to farmers.

A soil bank program similar to the soil conservation program of the 1930s was legislated in the 1950s. It provided diversionary payments to farmers who shifted production acreage into a nonproducing soil bank (idled land). Flexible price supports were included in the Agricultural Act of 1954 and the Secretary of Agriculture tried to shift farm policy toward a freer market. The Food for Peace program sought to reduce surplus crops through food aid programs to other nations.

The Food and Agriculture Act of 1965 removed many of the mandatory production controls and substituted voluntary acreage limitations, lower price supports, and higher direct payments to farmers in an effort to regulate farm supply. The Agricultural Acts of 1970 and 1973 liberalized output restrictions on individual farms and certain crops in an effort to benefit family farms. The Federal Food Stamp program was expanded to give assistance to low income families and help the sale of foodstuffs. It exceeded $12 billion annually by the mid-1980s.

The Agriculture and Consumer Protection Act of 1973 established the concept of *target prices* and deficiency payments. Participating farmers were to make production decisions on the basis of a target price, but the entire crop was to be sold at the market price. The difference between the target price and the average market price in the first five months of the marketing year was to be paid to the farmer in the form of a deficiency payment per unit of production. Using this plan, the government hoped to avoid accumulating stockpiles.

A major disadvantage of target pricing and deficiency payments was its potentially large cost to the government when there is a wide difference between target price and the loan rate. To overcome this disadvantage the program required the farmer to reduce the amount of acreage planted.

The Food and Agriculture Act of 1977 established a three-year CCC loan program, known as the Farmer Owned Reserve (FOR). In exchange for a higher loan rate, a farmer who satisfies an acreage reduction requirement can place commodities in the FOR for three years. After the first year, the loan interest is free. The Department of Agriculture pays for the storage cost of all years of the loan. In return the farmer agrees not to sell grain until the market price rises to a specified release price.

The Agriculture and Food Act of 1981 continued the target price/deficiency payment program and the farmer owned reserve program. In addition it introduced the acreage reduction program (ARP). The ARP, similar to past programs, required participating farmers to limit a crop to only a portion of its previously established base acreage and to devote the remainder of the base acreage to conservation usage.

As a result of continued large crop surpluses and accumulated government stockpiles, the Payment-in-Kind Program (PIK) was initiated in 1983. PIK provided compensation (payments) to farmers in the form of commodities, from government storage, for diverting a specific amount of acreage into conserving uses. Participating farmers, in turn, could sell the crops on the open market.

Finally, at the end of 1985 Congress passed a five-year farm bill that took a stab at curbing the rise in outlays for farm subsidies, which by that time exceeded $16 billion annually. It provided for cuts in federal price supports for many farm programs in an effort to reduce the government's role in agriculture. As a result, early in 1986, the Secretary of Agriculture announced reduced price supports for crops along with acreage cutbacks.

Tremendous stocks of surplus farm commodities continued to pile up throughout the 1950s and early 1960s in the face of government efforts to contract supply. Government outlays to support farm prices continued to mount through the period. Table 4-4 gives an indication of the growth in value of price support inventories and loans between 1955 and 1985. This value represents total CCC holdings of loans and crops. Government investment in farm price support activity grew rapidly in the mid-1950s and then leveled off around the $6 billion mark. There was a noticeable and substantial decline in surplus stocks in the latter half of the 1960s as a result of our greater aid to foreign nations. In the early 1970s there was a dramatic decline in CCC holdings as a result of crop shortages in the United States and elsewhere throughout the world. There was, however, a substantial increase in CCC holdings in the early 1980s, with some decline in 1984 as a result of the PIK program.

PROBLEMS OF PRODUCTIVITY

What was the cause of this seemingly intractable situation? What maintained the gigantic government stockpiles of farm commodities in the face of almost constant efforts to contract supply through the use of acreage allotments, soil banks, and land retirement? The answer to this question should, by now, be apparent. Rising farm productivity was the source of the difficulty. Even though farm acreage was reduced through acreage quotas or land retirement, rising production per acre more than offset the yield lost through land taken out of production.

Table 4-5 shows this trend for selected farm commodities and gives some idea of the dimension of the growth rate in output per acre. Such yield increases can only be explained by the tremendous advances in soil chemistry, insecticide chemistry, farm machinery technology, and scientific farming practices, such as crop rotation, contour plowing, and modern farm management. As government price support programs set smaller and smaller acreage allotments in an effort to bring supply into balance with demand, and as the modern farmer obtained more and more production from this shrinking acreage through the application of increasingly sophisticated materials, machinery, and techniques, eventually something had to give. Actually something had been giving all along.

As acreage bases shrank, more and more marginal farmers found it

Table 4-4
CCC Loans and Commodities Owned, 1955–1985
(Millions of Dollars)

Year	Loans Outstanding	Commodities Owned	Wheat	Sorghum	Corn	Cotton
1955	$2,137	$4,572	$2,297	$133	$934	$266
1960	1,347	6,021	2,452	646	1,700	880
1965	2,534	3,892	1,297	648	595	1,123
1970	2,952	1,853	405	173	293	225
1975	335	415	2	Z	Z	Z
1980	5,119	2,737	705	93	596	1
1983	15,084	10,227	1,533	520	3,392	174
1984	8,571	7,358	1,510	290	1,045	74
1985	12,631	6,921	1,951	359	757	46

Z = Less than $500,000

SOURCE: *Statistical Abstract of the United States: 1969* and *1987.*

Table 4-5

Output per Acre Comparison for Selected Crops

Crop	1956–1960 Yield per Acre	1985 Yield per Acre	Percentage Increase
Wheat	23.4 bu.	37.5 bu.	60
Corn	51.2 bu.	118.0 bu.	130
Sorghum	32.4 bu.	66.7 bu.	106
Soybeans	23.2 bu.	34.1 lbs.	47
Tobacco	1,591.0 lbs.	2,196.0 lbs.	38
Cotton	434.0 lbs.	630.0 lbs.	45

SOURCE: *Statistical Abstract of the United States, 1978 and 1987.*

impossible to support their families through farm work alone, even with price supports supplementing their incomes. Their acreage allotment had become too small to support efficient machinery, and they had turned to outside sources of income, primarily jobs in the city. It can be seen from Table 4-6 that more than half of the income of farm families comes from off-farm sources.

Farmers in increasing numbers sold their land to larger operators and moved to urban communities. As a result the corporation or commercial farm was rapidly becoming the major source of farm output. This sort of farming operation is usually large enough and well enough financed to

Table 4-6

Personal Income Received by Total Farm Population,
1980–1986 (Billions of Dollars)

Year	Net Farm Income	Off-Farm Income
1980	$21.5	$37.7
1981	30.1	34.7
1982	22.1	36.3
1983	15.0	43.4
1984	34.5	39.0
1985	30.5	41.0
1986	28.7	—

SOURCE: *Economic Report of the President, 1987* and *Economic Indicators (January 1987).*

take advantage of expensive, but highly efficient, innovations in farming methods. The question then arose whether the government should subsidize marginal farmers and large farming corporations equally, even though the latter did not need support to survive and prosper. This was a dilemma faced by the federal agricultural policy makers as late as the mid-1980s.

FARM PROBLEMS IN THE 1980s

The farm problem reached another critical stage in the 1980s. Encouraged by the inflationary environment of the 1970s that saw farmland values rise at a compound annual rate of 12 percent, farmers often made investment and production decisions on the assumption that good times and inflation would continue.

Financial Stress

With the rapid growth in both earnings and equity (farmland values) outstripping inflation, farmers invested freely in machinery, in equipment and buildings, and in such land improvements as clearing, irrigation and terracing. Substantial acreage previously used for pasture or conservation under government-sponsored programs was converted to grain and crop production in order to capitalize on booming export markets.

Lulled by the same feeling of optimism, farm lenders, including commercial banks, federal land banks and government agencies, were willing, and often eager, to finance farm expansion and production. As a result farm debt, in spite of very high interest rates, grew rapidly in the 1970s and early 1980s, virtually matching the three-fold increase in farm asset values.

Unfortunately for many farmers, the realities of the 1980s have not matched the expectations of the 1970s. Widespread drought in 1980 and 1983 had a dampening effect on earnings. Export markets shrank instead of growing. The heavy external debt of many foreign countries and the rising strength of the U.S. dollar made it costly for foreigners to buy U.S. food products. From their peak in 1981 farm exports dropped 28 percent by 1985. Likewise, growth in U.S. per capita meat production and consumption declined. Inflation, which had been at double-digit levels in the late 1970s and early 1980s, has been at 4 percent or less since 1982 and down to 1.1 in 1986.

Farm income has dropped. Net real cash income in the first half of the 1980s was down 22 percent from the 1970s, and net farm income fell in some years. By 1985 the value of farmland, which accounts for 75 percent of total farm assets, had declined by 33 percent from its 1980 peak. Total farm assets, in the meantime, had declined 20 percent. On the other hand, farm debt continued to rise in the 1980s.

Large loans, originally collateralized by high land values at high interest rates, left thousands and thousands of farmers in a bind as they struggled to service their debt (meet principal and interest payments) out of reduced earnings. Loan payment deficiencies became widespread throughout the country. Especially hard-hit was the Midwest. With their asset values less than their debts many farmers became insolvent and could not pay off their loans even by selling their farms.

A U.S. Department of Agriculture study released in 1985 indicated that 17 percent (370,000) of the nation's 2.2 million farm operators were financially vulnerable. Ten percent of the operators had serious financial problems (meaning that they could become insolvent in about four years), 3.3 percent had extreme financial problems and could become insolvent within two years, and 3.3 percent were already insolvent. The study indicated further that 25 percent of family-size operators fell into the financial vulnerability categories.

Under the circumstances many farm operators voluntarily sold out or were liquidated via bankruptcy and foreclosure by their lenders. As a result many farmers were forced off their land. In the Kansas City Federal Reserve district in 1985, for example, it was estimated that 6.7 percent of all farms and ranches were fully liquidated (2½ times normal) and another 6.7 percent (six times normal) were partially liquidated.

Aggravating the farm problem was the fact that U.S. farmers depressed prices by harvesting record and near-record crops in 1985. Planted acreage was very large despite farm programs that had reduced acreage as much as 30 percent in some cases. Carryover stock and CCC holdings were still at high levels in spite of the two-year-old PIK program. Early in 1986 it was apparent that U.S. agriculture was facing another difficult year. Farm income declined further due to weak crop prices and sluggish exports. Land values, too, continued their decline. Many farm lenders who had postponed foreclosures in previous years were themselves in poor financial condition. Therefore, farm liquidations, both full and partial, again ran well above normal in 1986.

Lender Problems

In addition to commercial banks there are other sources of farm credit. The largest percentage of farm loans come from the elaborate Farm Credit

System (FCS). The FCS is composed of 12 Federal Land Banks (FLB) and more than 400 Federal Land Bank Associations (FLBA); 12 Federal Intermediate Credit Banks (FICB) and 370 Production Credit Associations (PCA); and a Bank for Cooperatives (BC). The banks, once owned by the government, are now wholly owned by the federally chartered affiliated FLBAs, which in turn are owned by the borrowers. FCS obtains capital through retained earnings, through the requirement that borrowers buy stock in the associations from which they borrow, and through the sale of bonds. Its bonds are sold to individuals, banks, insurance companies, pension funds and others.

Government agencies such as the Farmers Home Administration (FmHA), the Commodity Credit Corporation (CCC) and the Small Business Administration (SBA) also provide financing for farmers. FmHA makes both real estate and non-real estate loans to farmers, CCC provides crop loans and SBA specializes in business loans.

Other farm lenders include insurance companies and a category of lenders composed of "individuals and others." This last category includes merchants, S&Ls and local credit organizations. Table 4-7 shows the percent of farm debt held by each of these lenders.

With farm debt exceeding $210 billion, farm lenders faced severe problems in the mid-1980s. In 1986, for example, the FCS was holding $74 billion in farm loans of which at least 15 percent were uncollectable. Consequently, some FCS officials were seeking help through an infusion of capital from the U.S. Treasury. In addition to widespread farm bankruptcies, more than 100 U.S. banks failed, many of them country banks servicing the farm community. The picture was no brighter in early 1987, with more delinquencies, foreclosures, and bank failures anticipated.

Table 4-7

Distribution of Farm Debt by Lender, 1985

Lender	Percent of all Farm Debt ($213 Billion)
Banks	23.4%
Farm Credit System	31.9
Federal gov't. agencies	17.3
Life insurance companies	5.8
Individuals/others	21.6
	100.0

SOURCE: *Economic Perspectives,* Federal Reserve Bank of Chicago (November/December 1985).

RECOMMENDATIONS

Numerous recommendations for alleviating current farm stress have been made by various individuals, agencies, and organizations. Among these recommendations are:

1. Moderate bankruptcy laws as they apply to the farmer that would require the government, through some type of transfer payment, to share the financial loss of foreclosure along with the creditor and debtor.
2. Enact a debt moratorium law similar to that enacted in the 1930s that would restrict or limit the use of foreclosure proceedings against farmers who can not meet their debt obligations.
3. Provide loan guarantees by federal or state agencies to protect lending institutions from default on the part of the borrower. Thus, in effect, the debt collection or debt obligation would be transferred to the government.
4. Restructure farm debt and stretch out, or write down, principal and interest payments to give the debtor more time to meet credit obligations.
5. Establish a program of asset leasebacks in which the creditor would take title to real property in lieu of farm debt repayment and then lease the property back to the original debtor. This would keep the property off the foreclosure market and thereby reduce the pressures on declining land and other asset values.
6. Develop a program of debt-equity exchange. This would encourage the farm debtor to seek outside equity from individuals, current debt holders and equity institutions in exchange for debt. In fact, a special bank, as used in several Third World countries, might be established by the government to give the farm sector an infusion of capital by purchasing equity in exchange for farm debt.[2]
7. Form a government-financed agency to absorb farm assets—loans, land, and farm machinery—out of farm lenders' portfolios.
8. Increase price and income supports for farmers.
9. Provide a direct infusion of government funds or encourage reliance on the Federal Reserve as a lender of last resort.
10. Establish an insurance fund similar to the Federal Deposit Insurance Corporation for the Farm Credit System.

---

2. A good review of some of these and other recommendations is contained in an article "Policy Options for Agriculture," *Economic Perspectives,* Federal Reserve Bank of Chicago (November/December, 1985). The article was written by Michael D. Boehje, Assistant Dean of the College of Agriculture at Iowa State University.

11. Weaken the foreign exchange value of the U.S. dollar to encourage exports.
12. Lower U.S. interest rates to lessen farm loan repayment pressures.

THE 1985 LEGISLATION

In the 1980s debates on both overall farm policy and a solution to the current crisis were widespread among farmers, bankers, government officials and members of Congress. Suggestions for general long-range farm policy were that: (1) the government should extradite itself from agriculture and let the free market decide the long-term farm problem; (2) the government should continue its current and past measures of price supports, acreage allotments, and other financial support to farmers; and (3) government should move gradually toward the establishment of free markets in agriculture and set target dates for reaching certain levels of reduced government intervention.

Finally, in December of 1985 Congress passed a five-year Food Security Act following the third suggestion. The Act tries to maintain farm income while allowing market forces to influence farmers' decisions to a greater extent. Four important provisions of the act are as follows:

1. The reduction of crop loan rates toward the level of world market prices in order to make U.S. farm products more competitive in world markets.
2. Freezing of target prices for two years followed by a decline in target prices thereafter.
3. The expansion of exports through the use of export credit guarantees over the next several years and the extension of the Food for Peace program to help the needy abroad.
4. The establishment of a long-term Conservation Reserve to idle as many as 45 million acres of marginal land.

The legislation assumes that the FmHA loan programs will remain a key source of credit to financially stressed farmers and made $4 billion available for these programs in 1986. It further proposes legislation to grant help to the FCS through the establishment of a back-up line of credit with the U.S. Treasury.

There is not much that is new in the 1985 legislation. Crop loans have been lowered before and target prices have been adjusted. Export expansion over the years has been encouraged, but will depend on the economic status of foreign nations and the foreign exchange value of the U.S. dollar. The Food for Peace program is a few decades old and the Conservation Reserve idea is reminiscent of past programs, especially the Soil Bank of the 1950s.

CONCLUSION

Provisions of the 1985 farm legislation were expected to cost the federal government $52 billion during 1986–1988 and keep it very much involved in agriculture. By the end of 1986, however, the cost had exceeded $25 billion and was expected to exceed $70 billion for the full three-year period. In the meantime problems will continue for both farmers and farm creditors as they attempt to cope with the most severe farm crisis since the Great Depression of the 1930s. Although the number of farmers has been decreasing over the decades, the United States still has more resources devoted to farming than its productivity warrants.

At the economic summit meeting of the seven leading industrial nations in Venice during June of 1987, President Reagan suggested that all nations begin reducing farm subsidies and move toward completely free-market farming by the year 2000. Regardless of the direction farm policy takes in the coming years, the voice of the farmers will still be heard. Despite their shrinking numbers, farmers are still a significant economic and political force. They are the prime producers of those goods that Americans need most to sustain life. American farmers and their problems will not be ignored, just as they have not been ignored in the past.

QUESTIONS FOR DISCUSSION

1. Are the social implications of the American farm problem more serious than its economic implications?
2. Is it possible to reconcile the American ideals of free enterprise and individual dignity with federal support programs as these programs have evolved over the years?
3. Do you think that the United States should move more toward agribusiness and stay away from the family-farm concept?
4. Does the concept of parity seem to be a usable device in the farm price support programs?
5. Should there be a ceiling imposed on the size of government subsidy that can be paid to any individual farmer or farm corporation?
6. Does there appear to be a conflict between the welfare of consumers and the desire of the government to improve the income of farmers?
7. Do you think we should concern ourselves with farm bankruptcies any more than we do with any other business bankruptcies?

8. Would a forecast of greater or lesser federal involvement in American agriculture seem more likely to you at this time?

9. Do you think the most recent farm bill (1985) was a step in the right direction? Why?

SELECTED READINGS

Benjamin, Gary L. "The Financial Stress in Agriculture." *Economic Perspectives*. Federal Reserve Bank of Chicago (November/December 1985).

Boehje, Michael D. "Policy Options For Agriculture." *Economic Perspectives*. Federal Reserve Bank of Chicago (November/December 1985).

Brenton, C. Robert. "An Ag Banker's View." *Economic Perspectives*. Federal Reserve Bank of Kansas City (November/December 1985).

Drabenstott, Mark. "U.S. Agriculture: The Difficult Adjustment Continues." *Economic Review*. Federal Reserve Bank of Kansas City (November 1985).

————. "U.S. Agriculture: The International Dimension." *Economic Review*. Federal Reserve Bank of Kansas City (November 1985).

"Financial Characteristics of U.S. Farms, January 1984." *USDA Agricultural Information Bulletin No. 495*, July 1985.

Furlong, Frederick T. and Randall J. Pozdena. "Agricultural Credit Conditions." *Weekly Letter*. Federal Reserve Bank of San Francisco (August 12, 1985).

Gregorash, George M. and James Morrison. "Lean Years in Agricultural Banking." *Economic Perspectives*. Federal Reserve Bank of Chicago (November/December 1985).

Henry, Mark, Mark Drabenstott and Lynn Gibson. "A Changing Rural America." *Economic Review*. Federal Reserve Bank of Kansas City (July/August 1986).

Irwin, George D. "The Farm Credit System: Looking for the Proper Balance." *Economic Perspectives*. Federal Reserve Bank of Chicago (November/December 1985).

Norris, Kim. "The Farmers Home Administration: Where is it Headed?" *Economic Review*. Federal Reserve Bank of Kansas City (November 1986).

Owens, Raymond E. "The Agriculture Outlook for 1986. . .Continued Financial Weakness Seen." *Economic Review*. Federal Reserve Bank of Richmond (January/February 1986).

5

BANKING DEREGULATION SUCCESS OR FAILURE?

Without question, the many changes that have occurred for banks and other depository institutions are massive and revolutionary.[1] Not only has the scope of change been enormous, but the pace of change has been very swift.

Why have these profound and far-reaching changes occurred? What factors could have allowed these institutions to break so cleanly from the past? In general, the answer is found in some basic laws of economics involving reactions to competitive forces of the marketplace. To a large degree, market forces pressured bankers into developing financial innovations to circumvent a body of regulation originally enacted in response to the traumatic banking crisis of the 1930s. Moreover, sweeping federal legislation of the early 1980s and numerous changes in state laws covering depository institutions have had great impact. Both federal and state legislation not only eliminated a number of regulatory constraints of the

---

1. Depository institutions are four in number: commercial banks, savings and loans, mutual savings banks, and credit unions. Savings and loans, mutual savings banks, and credit unions are collectively referred to as thrifts.

past, but also allowed depository institutions to alter the nature of many of their assets and liabilities. Add to these factors the explosion in technological progress in the computer and communications fields (which greatly increased the speed of change), and you have the ingredients of a truly remarkable revolution in the financial services industry.

Although the competitive situation for depository institutions became virtually intolerable, especially during the years 1966–1980, a review of the regulatory framework that prevailed prior to that period provides an appreciation of the problems, market forces, and other factors underlying the need for the many recent changes.

OVERVIEW OF BANKING LEGISLATION OF THE 1930s

The most significant force that helped to shape the body of regulation existing prior to the 1980s was reaction to the nationwide banking crisis that accompanied the Great Depression. Many of the critical problems and crisis issues faced by depository institutions during the 1970s and early 1980s were brought about by depression-spawned banking legislation. Even today, some of that legislation is part of our regulatory framework.

Near-Collapse of the Banking System

Between 1930 and 1933 more than 9,000 banks in the United States failed. (See Table 5-1 and Figure 5-1.) During this period "runs on banks" occurred with such regularity that failures averaged about 40 per week.[2] The emergency became so severe that in March of 1933, after the onset of still another major bank panic, President Roosevelt declared a "bank holiday" and closed all banks until the situation calmed down. A wave of bank reform legislation quickly followed.

Banking Legislation of the 1930s

The banking panics of the early 1930s were blamed on (1) speculative activities of the 1920s and apparent conflicts of interest resulting from the

---

2. This dramatic period of bank failures had followed a substantial decline in the number of banks during the economic prosperity of the 1920s. From 1921 to 1929 almost 6,000 banks suspended operations; many of these were small one-office banks located in declining agricultural areas.

Table 5-1

Number of Failed Banks (1900–1986)

Year	Number
1900–1920	1,789
1921–1929	5,712
1930–1933	9,096
1934–1941	467
1942–1951	61
1952–1961	47
1962–1971	60
1972–1981	147
1982–1986[a]	419

[a]As of July, 1986.

SOURCE: *FDIC, Annual Report*, various issues.

participation of banks in investment banking functions and (2) excessive competition for deposits. In order to alleviate the crisis, Congress was faced with the difficult task of drafting legislation designed to quickly restore confidence in the banking system. In addition, lawmakers sought to create a strict body of regulation aimed at insuring long-run stability and freedom from bank failures. These goals were accomplished by the passage of two highly significant acts: the Banking Act of 1933 and the Banking Act of 1935.

Banking Act of 1933. The Banking Act of 1933, known popularly as the Glass-Steagall Act, was revolutionary for its time. In effect it was a "package" of acts, consisting of highly important provisions generally designed to curb excessive competition among banks, to increase federal supervision and regulation, and to establish a system of deposit insurance. The following provisions were of major importance:

1. It established the Federal Deposit Insurance Corporation (FDIC) to insure deposits at commercial banks and mutual savings banks. The obvious intent of this provision was to restore the public's confidence in the banking industry. Because of the rapid achievement of this goal, the FDIC provision is often called one of the most significant pieces of legislation in the history of banking.
2. It prohibited commercial banks from engaging in investment bank-

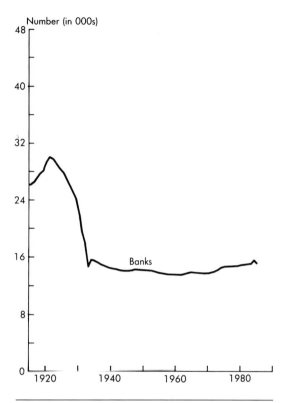

Figure 5-1

Commercial Banks in the United States

SOURCE: Board of Governors of the Federal
Reserve System, 1985 *Historical Chart Book.*

ing activities, thus separating commercial and investment banking.
Investment banking involves the underwriting and trading of securities. Prior to the passage of this act, commercial and investment
banking were virtually integrated. The mixture of these two types
of banking allegedly led to speculative activity and potential conflicts of interest. For example, it was argued, that some banks,
when acting as investment bankers, would place slow-selling securities into their own commercial bank portfolios or into their trust
departments, hence increasing the risk of losses.

3. It prohibited member banks of the Federal Reserve System (Fed)
from paying interest on demand deposits, in an effort to reduce

Banking Deregulation — Success or Failure?   **119**

excessive competition for such funds.[3] Discussion of this and the following provision will be expanded later in the chapter.

4. It gave the Federal Reserve the authority to establish ceiling rates on savings and time deposits of all member banks. The Fed used its Regulation Q to implement this authority.

5. It generally left the branching restrictions of the McFadden Act of 1927 unchanged. It thus supported those provisions of the McFadden Act that (a) prohibited interstate banking and (b) required national banks to conform to the branching restrictions imposed by the laws of the states in which they were located.

The Banking Act of 1935. The Banking Act of 1935 primarily was intended to strengthen the powers of the Fed and to extend the provisions of the Banking Act of 1933 to nonmember banks. Some provisions of the act are as follows:

1. It prohibited nonmember banks from paying interest on demand deposits.

2. It extended the Fed's power to set ceiling rates on savings and time deposits of nonmember banks.

3. It created the Federal Open Market Committee, which is the Fed's powerful decision-making body regarding matters of monetary policy.

4. It expanded the power of the Fed's Board of Governors to set reserve requirements.

5. It placed greater restrictions on entry into the banking industry in an effort to reduce bank failures.

We will return to the long-run ramifications of some of these depression-born regulations when discussing the key problems that were instrumental in creating the need for the passage of the banking legislation of the early 1980s.

Thrifts During the Great Depression

Because the commercial bank was by far the largest and most important depository institution, the great bulk of reform legislation of the 1930s

---

3. A member bank is one with membership in the Federal Reserve System; a bank not belonging to the system is called a nonmember. Federally chartered (national) banks must be members of the Fed, but membership is not obligatory for state-chartered banks.

dealt with that institution. Nevertheless, the turmoil in the financial markets during the Great Depression did result in limited, but significant, legislation designed to aid the much smaller thrift industry.

The Federal Home Loan Bank System (FHLB), established in 1932, consists of 12 regional banks with which thrifts can be affiliated. A Federal Home Loan Bank Board (FHLBB), located in Washington, supervises the activities of the 12 FHLBs. In 1934 the Federal Savings and Loan Insurance Corporation (FSLIC) was created to provide deposit insurance for savings and loan associations, much like the FDIC does for commercial banks and mutual savings banks. Also, like the FDIC, the FSLIC examines and supervises its insured institutions and oversees mergers and liquidations of failing thrifts.

Throughout most of the history of depository institutions, legislation has tended to completely separate commercial banks from thrifts, both in terms of their deposits and other sources of funds (liabilities) and their uses of funds (assets). Regulation gave the commercial banks monopoly over demand deposits. Moreover, the banks could hold a much more diversified portfolio of loans and investments than could the thrift institutions. In contrast, the sources of funds for the thrifts legally consisted of savings and time deposits, and a very large percentage of the loan portfolio of the typical thrift consisted of long-term mortgages.

OVERVIEW: WORLD WAR II–1965

The banking regulations, that had developed largely because of the impact of bank failures on the nation's economy, generally did not interfere with the everyday functioning of the financial system until the mid-1960s. Deposit insurance adopted in the 1930s continued to be a rousing success.[4] Bank failures dropped to fewer than six per year during the post–World War II period. In general, Regulation Q ceilings had little effect on depository institutions because market interest rates remained below ceiling limits. In addition, there was an adequate spread between the average interest rate paid for the sources of funds and the average rate earned from the uses of such funds. Despite this relatively long period of generally profitable conditions, major problems were on the horizon. These problems and other key factors will be addressed in the following sections.

---

4. Credit unions did not have deposit insurance until 1970, when Congress established the National Credit Union Share Insurance Fund.

THE DEPOSITORY INSTITUTIONS DEREGULATION AND MONETARY CONTROL ACT OF 1980

For several years prior to the reform legislation of the early 1980s, much effort was expended by regulators, banking and thrift leaders, and members of Congress to change the body of regulations under which depository institutions had operated since the 1930s. It had become increasingly apparent that some regulations had been made obsolete by significant changes in the everyday functioning of financial services markets, technological advancements, long-term changes in the levels of inflation and interest rates, and general economic conditions. Several groups conducted very thorough private and public studies of the financial marketplace. Most of the studies recommended sweeping reforms, some of which were adopted in the federal legislation of the 1980s.

The most important of these studies were conducted by the President's Commissions on Financial Structure and Regulation of 1971 (Hunt Report) and Financial Institutions and the Nation's Economy of 1975 (Fine Study). Both studies emphasized key problems in the financial markets, and each strongly recommended that major steps be taken to deregulate depository institutions. The proposals offered by the reports were widely supported by many financial industry participants, regulators, and politicians. In fact, some of the recommendations of the Hunt Study were incorporated in a proposed federal financial institutions act in 1973, while the Fine Study led to the introduction of banking legislation in 1975. However, neither attempt at regulatory reform was passed by Congress.

Despite the increasing recognition of the problems and structural changes in the financial services industries, not until the near-crisis conditions of 1979 and early 1980 did Congress act by passing the Depository Institutions Deregulation and Monetary Control Act (DIDMCA) in March, 1980. The Senate Banking Committee Chairman noted that the act would "create a level playing field" to help reduce competitive inequalities among different types of financial institutions. Senator Proxmire said the act was ". . .the most significant legislation. . .since the passage of the Federal Reserve Act of 1913."

The DIDMCA, an omnibus bill often popularly referred to as the Monetary Control Act, contains nine titles, each an act in itself. Each title deals with an aspect of reform of the financial system. We will list only the most important provisions of various titles and then provide overviews of the significant problems and other factors that brought about the need for specific reform measures.

Reserve Requirements and Pricing of Fed Services

Title I of DIDMCA attempts to enhance the ability of the Fed to implement monetary policy. One method chosen to achieve this goal was to subject all depository institutions to uniform reserve requirements.

Membership Problem. For a number of years it had been argued that uniform reserve requirements were necessary both for greater competitive equality among depository institutions and for better control of monetary conditions. In the years prior to the passage of DIDMCA, the proportion of bank deposits subject to the Fed's reserve requirements was rapidly declining. This decrease occurred because member banks of the Fed were required to hold noninterest-earning assets as reserves, while nonmember state banks could often satisfy state reserve requirements by holding interest-bearing assets, such as U.S. Treasury securities.

This membership "tax" caused many banks to withdraw from the system because they were put at a severe disadvantage compared with state nonmember banks. The lost income, or opportunity cost, of holding nonearning reserves greatly increased in the latter 1970s as interest rates rose to record levels. Subsequently, not only did newly formed banks choose to remain nonmembers, but an increasing number of member banks dropped out of the system. Even some national banks (which by law are required to be members) switched to state charters in order to escape membership.

The membership problem was one of the key factors that put pressure on Congress to initiate a legislative remedy. Once fully implemented, the new uniform reserve requirements will reduce the burden on member banks since nonmember banks and other depository institutions will be subject to the same reserve requirements.

Pricing Fed Services. Title I also requires the Fed to impose explicit charges for services that were traditionally provided free to member banks. Moreover, these services must be made available to all nonmember depository institutions on the same terms offered to member banks. In general, the Fed prices must be based on all direct and indirect costs incurred in providing services.

A number of free services provided by the Fed over the years, such as check clearing and wire transfers, were defended as necessary to promote an efficient payments system. Nevertheless, for many years these services have been offered for a fee by correspondent commercial banks for their respondent banks.[5] It was argued that forcing the Fed to price its services

---

5. A correspondent bank is one (usually a larger institution) that provides services and advice for a respondent bank (usually a smaller institution).

would give incentives for private suppliers of similar services to offer more competition, thus enhancing overall efficiency of the payments system.

In past years the reluctance of the Fed to charge for services also was closely related to the membership problem. The provision of free services partially offset the opportunity cost to member banks of holding nonearning reserves, and so helped to slow down both withdrawals from membership and erosion of that portion of total bank reserves subject to the Fed's reserve requirements. Inasmuch as member banks can no longer avoid the cost of holding reserves by leaving the Federal Reserve System, the DIDMCA eliminated the Fed's fear of explicitly pricing services.

Deregulation of Interest Rate Ceilings

Title II of DIDMCA required that all interest rate ceilings on savings and time deposits were to be phased out over a six-year period. The long-standing Regulation Q authority of the Fed to set interest rate limits was turned over to the newly formed Depository Institutions Deregulation Committee (DIDC).[6] The DIDC was charged with implementing this provision to achieve an orderly phaseout of all ceiling rates by March 31, 1986.

Historical Background of Regulation Q Ceilings. Both the prohibition on the payment of interest on demand deposits and the ceilings on interest rates that may be paid on savings and time deposits date from the Banking Acts of 1933 and 1935, but calls for a ban on interest on demand deposits occurred several times during various financial crises in the 1800s.

Demand Deposit Interest Prohibitions. Although the financial sector performed relatively well for several years after passage of the Federal Reserve Act in 1913, large numbers of banks failed in the 1920s, followed by massive bank failures during early years of the Great Depression. The practice of paying interest on demand deposits was alleged to be a major cause of widespread bank failures during the 1920s and 1930s. Critics insisted that such interest payments were highly destabilizing, because they led banks to compete for demand deposits by paying excessive rates to attract greater amounts of funds. To offset these higher costs and maintain profit margins, it was argued, banks were forced to seek out higher yielding but much riskier earning assets than would otherwise be the case. Although

---

6. The voting members of DIDC were the Secretary of the Treasury, the Chairman of the Federal Reserve Board of Governors, and the heads of the Federal Deposit Insurance Corporation, the Federal Home Loan Bank Board, and the National Credit Union Administration. The Comptroller of the Currency was a nonvoting member.

this explanation for bank failures was widely accepted, studies in the 1960s indicate that there is little empirical evidence for these claims.[7]

Time Deposit Interest Ceilings. During the 1930s, people also believed that excessive competition for bank savings deposits led to a situation similar to that caused by the competition for demand deposits. In this case as well, however, subsequent studies have revealed little evidence to support this explanation of bank failures.

Ceiling Rates and Disintermediation

During the 25 years following the Banking Act of 1933, the debate regarding the wisdom of setting interest rate ceilings was a moot point. Until the mid-1950s market interest rates remained below Regulation Q ceiling rates. Once market rates began to exceed ceiling rates, however, the Fed initially responded by raising the ceiling limits. This policy of accommodation prevailed until the credit crunch of 1966.

Faced with large expenditures for the Vietnam War, greater inflationary pressures, and other pressing economic problems, the Fed resorted to a very tight credit policy in 1966. As part of this stance, the Fed refused to raise interest rate ceilings on deposits when market rates exceeded ceilings. In fact, the Fed requested and won approval from Congress to extend ceilings on deposits of savings and loans (S&Ls) and mutual savings banks (MSBs), with such ceilings administered by the Federal Home Loan Bank Board (for S&Ls) and the Federal Deposit Insurance Corporation (for MSBs). Also, in an effort to insulate thrift institutions from too much commercial bank competition, interest rate ceilings on thrift deposits were set slightly higher than those established for banks.

As money market rates moved significantly above Regulation Q ceilings in 1966 and in other tight credit periods, massive but often temporary transfers of deposits from depository institutions into various money market investments (such as U.S. Treasury bills) took place. These transfers, called *disintermediation*, were of course very disruptive and led to occasional but severe restrictions of housing credit.

Beginning in 1977, a sustained rise carried market interest rates well above Regulation Q limits. (See Figure 5-2.) This sustained increase in market rates further aggravated the disintermediation problem, and the soaring opportunity cost of holding non-interest-earning reserves encouraged

---

7. See study by Albert H. Cox, Jr. "Regulation of Interest on Demand Deposits," Michigan Business Studies 17, no. 4 (1966). Also see Byron Higgins, "Interest Payments on Demand Deposits: Historical Evolution and the Current Controversy," Monthy Review, Federal Reserve Bank of Kansas City (July–August,1977).

Figure 5-2

**Regulation Q Ceiling Rates at Thrift Institutions
Compared with 3-Month Treasury Bill Rates**

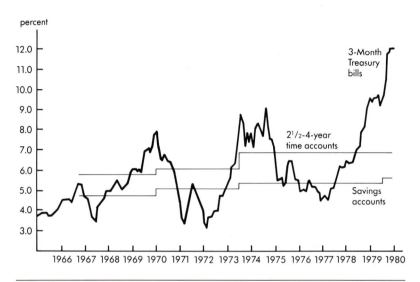

SOURCE: Board of Governors of the Federal Reserve System.

even more member banks to withdrawal from the Federal Reserve System. Moreover, rapid technological advances plus the overall profitability of commercial banking provided strong incentives for nonbank business firms to greatly expand their offerings of bank-like and other financial services.

In addition to the above factors, a major innovation of the 1970s that greatly increased the severity of the disintermediation problem was the money market mutual fund (MMMF). The MMMF is a type of mutual fund that offers its shares for sale to both large and small investors and then uses the funds to acquire money market securities, such as U.S. Treasury bills, commercial paper, negotiable certificates of deposit, repurchase agreements, and so forth. Although MMMF shares were not insured, investors were attracted by the high market rates of interest offered, the limited check writing privileges, the ease of entry and exit without load fees or withdrawal penalty charges, and the relative safety of the high quality, very liquid, short-term investments held by the funds.

In 1978 total assets held by MMMFs were less than $4 billion, but by 1980 the figure rose to more than $60 billion. This expansion then soared to almost unbelievable levels, reaching more than $230 billion by 1982. Clearly, the rapid growth of MMMFs put enormous competitive pressure on depository institutions, since most of the increase in assets came from deposits transferred from these institutions. The loss of deposits, plus other problems, forced hundreds of depository institutions, especially thrifts, to suspend operations or merge with stronger firms.

The astonishing growth of the MMMFs, the rapid expansion of bank-like and other financial services offered by nonbank firms, plus other problems to be discussed later, put enormous competitive pressure on banks and thrifts. These depository institutions, together with the housing and building industries, exerted substantial political pressure on Congress to provide relief for the overwhelming problems caused by deposit losses and other issues. Thus, the disintermediation problem was a very important factor in forcing Congress to move toward a major overhaul of regulations covering depository institutions.

NOW and Other Check-Like Accounts

The provisions of Title III of DIDMCA were designed to contribute to competitive equality among the depository institutions by permitting each of them to offer nationwide interest-bearing "check-like" accounts. These are essentially transactions accounts that are functionally equivalent to demand deposits, but unlike demand deposits, they bear explicit interest. Commercial banks, savings and loans, and mutual savings banks were authorized to offer checking accounts called negotiable order of withdrawal (NOW) accounts. Credit unions were allowed to offer share draft accounts. Both accounts earned interest, subject to uniform interest rate ceilings. The act, however, did not authorize the payment of interest on demand deposit accounts held by corporations.

Also legalized were automatic transfer services (ATS) for shifting funds from savings to demand deposit accounts, and remote service units, which facilitate deposits to and withdrawals from accounts, loan payments, and related transactions. The act also increased deposit insurance from $40,000 to $100,000 at federally insured depository institutions.

Usury Laws

State usury laws were overridden by Title V of DIDMCA. Except for home mortgage loans and mobile home loans, the usury provisions were temporary. Also, states could override DIDMCA, and the act exempted state usury laws on business and agricultural loans only for a period of three years.

A usury law is one that limits the interest rate a lender can charge for certain types of loans. In general, the laws have been aimed at protecting lower-income individuals and small business borrowers from excessive interest charges. Ironically, when market interest rates have moved above government-imposed usury limits, often the effect has been to reduce, and in some cases eliminate, the availability of credit to the very borrowers the law was intending to protect. For example, if the usury rate in a state is 12 percent and the prime rate of interest is 18 percent, many borrowers such as small businesses, home builders and buyers, and farmers are likely to be forced out of the market as lenders seek to ration credit to preferred customers. Lenders have also designed practices, such as charging points on mortgage loans, to override the usury limits. In effect, points require borrowers to pay a portion of the total interest charges up front as a condition of obtaining loans.

Overall, as interest rates rose to record levels in the late 1970s, it became evident that usury limits had become wholly unrealistic as lenders in some areas at times virtually ceased all lending to various borrowers. Some states exerted a great deal of effort to amend their usury laws. In some cases such efforts faced stiff opposition; in others a change in the laws required amendments to state constitutions. Although some states had succeeded in tying usury rates to market rates, others fared poorly in their attempts for changes, with borrowers suffering the consequences. Thus, it was evident that federal legislation was necessary.

The Thrift Problem

Title IV to the DIDMCA focused on the "thrift problem." This problem involves the imbalance between the assets and liabilities of thrifts (savings and loans and mutual savings banks). The act attempted to relieve the maturity imbalance by allowing federally chartered thrifts to make short-term consumer loans and variable-rate mortgage loans.

Over the years thrifts, constrained by regulation, tax incentives, and overall management philosophy, had specialized in long-term, fixed-rate mortgage lending. For these assets to add to the overall profitability of lenders, their earnings had to exceed the costs of customer deposits and other sources of funds.

From the mid-1960s through the 1970s, generally high rates of inflation and rising, volatile market rates of interest changed a reasonable spread between returns on thrift portfolios and the interest costs on sources of funds into serious liquidity and solvency problems. By 1981, many thrifts experienced negative yields because they had to pay more for funds than could be earned on their assets. Even without rate ceilings on deposits, disintermediation, competition from MMMFs, and other problems, the maturity imbalance between assets and liabilities and the steady decline in spreads during the late 1970s and early 1980s would have greatly limited the ability of thrifts to compete and remain viable under the old body of regulations.

Other Titles of DIDMCA

For the purposes of this chapter, our focus has been on some provisions of five of the nine titles of DIDMCA. The emphasis on these provisions provided a look at the most important goals of DIDMCA. Because the remaining provisions are not essential to our discussion, they will not be considered.

PROBLEMS CONTINUE: 1980–1982

For a number of months after the passage of DIDMCA, an increasing number of thrift institutions found themselves in severe difficulty. The previously discussed thrift problem became a thrift crisis. In mid-1980 interest rates started to rise rapidly and did not decline until late summer of 1982. The prime rate of interest reached 20 percent in April, 1980 and stood at 21.5 percent by the year's end. By the summer of 1981 the prime rate was still fluctuating between 20 and 21 percent. At these very high market rates of interest, long-term assets of thrifts were often earning far less than the sources of funds. (See Figure 5-3.)

To add to the overall problems, a great many MMMFs were paying returns well in excess of 15 percent, while passbook savings accounts in banks and thrifts were held to 5.25 and 5.5 percent. The disintermediation problem worsened and continued to deplete the deposits of depository institutions. In 1981, it was reported that about 80 percent of all thrifts were losing money, and losses in the S&L industry soared to $4.6 billion. (See Figures 5-4 and 5-5.) During 1981 the FSLIC provided about $1 billion in aid to troubled thrifts.

Under these adverse conditions, the Depository Institutions Deregulation Committee (DIDC), established under DIDMCA, could not provide the assistance needed. If the DIDC, for example, raised the interest rate ceiling on passbook savings, say from 5.5 to 7 percent, it is clear that this would have simply aggravated the already serious cost problems for thrifts. Thus, DIDC's tinkerings with Regulation Q could eliminate neither disintermediation nor the maturity imbalance of thrifts' balance sheets.

Although many types of new deposits were permitted in the last half of the 1970s and early 1980s, none of these allowed depository institutions to compete effectively with MMMFs. Even new accounts that permitted the payment of money market rates of interest were burdened with sizeable

Figure 5-3

Percent Spread for FSLIC-Insured S&Ls

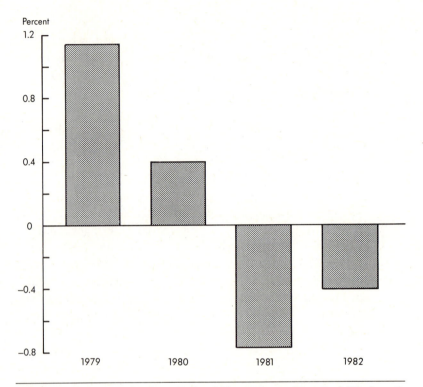

SOURCE: *Quarterly Review,* Federal Home Loan Board of Cincinnati, 1984.

minimum deposits and costly penalties for early withdrawals. The completely liquid MMMFs, on the other hand, had much smaller minimum deposit requirements and no penalty for withdrawals. Hence, it became increasingly apparent that the only realistic way for depository institutions to compete with the MMMF was to receive authority to offer a very similar deposit account.

Despite the fact that the 3.9 percent inflation rate for 1982 was far below the rates of the previous three years, the prime rate during the first half of 1982 remained between 15 and 17 percent and closed the year at 11.5 percent. Although the declining inflation and interest rates were favorable developments, for a large number of thrifts help arrived too late. In fact, 1982 was in some ways a worse year than 1981. During the fall of 1982 thrifts were failing at an average of three per week. (For many years prior

Figure 5-4

Percentage of FSLIC-Insured Institutions Posting Losses

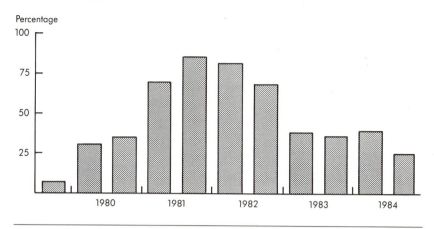

SOURCE: *Federal Reserve Bulletin* (March 1985).

to 1981, failures averaged less than three per year.) Losses for the S&L industry for 1982 totaled almost $4.3 billion. Table 5-2 illustrates the thrift problem in terms of the rapid decline in the number of firms. After much delay and debate by lobbyists, regulators, financial market participants, and politicians, the extreme crisis atmosphere finally forced Congress to take legislative action.

The Garn–St. Germain Act of 1982

Like DIDMCA, the Depository Institutions Act of 1982 (Garn–St. Germain Act) contains a number of titles. The main focus of the act was to prevent the collapse of the thrift industry. One group of provisions emphasized the emergency powers of regulators to facilitate the immediate rescue and support efforts for thrifts until fundamental reforms incorporated into the legislation could become effective. Another major category of provisions attempted to change the nature of the industry to such an extent that, in the long-run, thrifts could exist as viable, competitive institutions under various and rapidly fluctuating economic conditions.

Emergency Powers. The act provides the FDIC and FSLIC a framework to be used for arranging emergency acquisitions for failing depository institutions across both state lines and institutional barriers. Actually, prior to the passage of the act the Fed and the FHLBB had authorized both interstate and interindustry mergers; thus, this provision legalized what regulators had already accomplished. The FDIC was granted powers to authorize the sale

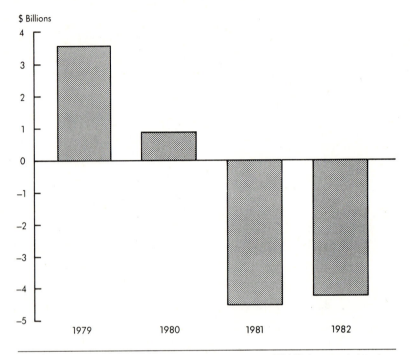

Figure 5-5

Net Income of FSLIC-Insured S&Ls

SOURCE: Combined Financial Statements, FSLIC-Insured Institutions, 1979–1982, Washington, D.C.: Federal Home Loan Board.

Table 5-2

Number of All Savings and Loan Institutions

Year	Federally Chartered	State Chartered	Total Institutions
1960	1,873	4,447	6,320
1970	2,067	3,602	5,669
1979	1,989	2,695	4,684
1980	1,985	2,628	4,613
1981	1,907	2,385	4,292
1982	1,727	2,098	3,825
1983	1,553	1,949	3,502
1984	1,478	1,913	3,391
1985	1,419	1,825	3,244

SOURCE: FHLBB and United States League of Savings Institutions.

of a large, failed commercial bank, or a closed or troubled mutual savings bank to another federally insured institution, whether in-state or out-of-state.[8] In addition, the FSLIC could exercise similar powers, regardless of the asset size of the failing thrift involved.[9]

Other emergency powers relate to the aid provided by the FDIC and FSLIC for closed, insolvent, or troubled banks and thrifts. The insurers may make loans to, put deposits in, and make contributions to troubled institutions or to firms that acquire them. They may also buy or assume an insured institution's assets or liabilities, set up extraordinary mergers and acquisitions, arrange charter conversions, and, when necessary, guarantee against loss the rescuing institution or the company controlling it. Similar powers were given to the National Credit Union Administration (NCUA) in order to provide aid for troubled credit unions.

The act also permits the use of net worth certificates, backed by the FDIC and FSLIC, to aid some seriously troubled institutions. The net worth certificate is a special security issued by a depository institution with severely deficient net worth. The certificate is exchanged for a promissory note of the FDIC or FSLIC. The certificate is then treated as capital for regulatory purposes. When the depository institution returns to a profitable position, it must redeem the net worth certificate.

Long-Run Fundamental Reforms. Long-run reforms likely will prove to be the most important aspects of the Act because they greatly alter the possible sources and uses of funds of thrift institutions. The most important provision of the Act granted depository institutions the authority to offer money market deposit accounts (MMDAs) with no interest rate ceilings. This provided a definite break from past regulatory barriers restricting competition for funds. It also offered an account that could directly compete with the MMMF. Title III of the act noted that the MMDA was an account "directly equivalent and competitive with money market mutual funds."

The new account was offered on December 14, 1982 and had an original minimum balance of $2,500 (changed to $1,000 on January 1, 1985 and zero on January 1, 1986), paid interest similar to market rates, and

---

8. A *large* commercial bank or mutual savings bank was defined as one with assets in excess of $500 million.

9. The FDIC and FSLIC may solicit offers to buy banks and thrifts from any qualified purchaser according to the following priorities: like in-state institutions; like out-of-state institutions; different type of in-state-institutions; different type of out-of-state institutions. Priority is to be given to adjacent state institutions, if out-of-state offers are involved. Also, if the lowest acceptable bid is from an out-of-state institution, then the agencies must accept reoffers from original in-state bidders, if their offers were within 15 percent or $15 million (whichever is less) of the lowest acceptable offer.

allowed six transfers per month—three by check and three by preauthorized, automatic, or telephonic means. There was no reserve requirement on personal MMDAs, but nonpersonal accounts were subject to a 3 percent reserve requirement.[10]

The act broadened the types of customers who could hold various deposits. First, federal, state, and local governments were authorized to hold NOW accounts. Previously, only persons and nongovernment, non-profit organizations could hold these accounts. Second, federally chartered S&Ls were permitted to offer demand deposits to customers who have a business loan relationship with the S&L. The act also required DIDC to eliminate Regulation Q rate differentials between commercial banks and thrifts by January 1, 1984.

On the asset side of the balance sheet, S&Ls were permitted to invest up to 55 percent of their assets in three categories of commercial loans: (1) up to 40 percent of assets in loans secured by commercial real estate; (2) up to 5 percent of assets in secured or unsecured commercial loans; and (3) up to 10 percent of assets devoted to leasing. Under DIDMCA the S&Ls were given authority to invest up to 20 percent of assets in consumer loans. Garn-St. Germain increased this to 30 percent of assets. DIDMCA's authorization for unlimited power to invest in federal government and municipal general obligation securities was extended to include state and local revenue bonds.

EFFECTS OF THE 1980s LEGISLATION

The legislation of 1980 and 1982 provided extensive freedom for depository institutions to innovate. The new lending and deposit powers of thrifts opened the door for competition in a broader financial market. Many larger, more aggressive S&Ls quickly ventured into unfamiliar and more risky types of lending and investing—in some cases with disastrous results. However, most thrifts, some of which were ailing and battered, were not in a position to take advantage of their new powers. While it is true that the new and increased powers may someday lead to an end of the distinction between banks and thrifts, by mid-1987 a great many thrifts had done relatively little to erase the differences.

---

10. Shortly after the October, 1982, passage of Garn–St. Germain, the DIDC acted to authorize the Super-NOW account. This account allowed unlimited checking and unregulated interest rates. But it was subject to a 12 percent reserve requirement because it was classified as a transactions account.

As previously indicated, the major purpose of Garn–St. Germain was to rescue the troubled thrift industry. By 1983 the S&L industry was able to reverse the record losses of the two previous years. It is clear, however, that the legislation had little to do with the improvement in profitability. Rather, the significant drop in interest rates, starting in August, 1982 and continuing irregularly through early 1987, was much more beneficial for the industry than the emergency provisions of Garn–St. Germain. Despite the legislation and the fortuitous decrease in interest rates and the return to profitability for the industry as a whole, a large number of S&Ls continue to experience grave financial difficulties. This ongoing S&L crisis will be covered in a subsequent section of this chapter.

By 1983 the deregulation of interest rate ceilings was virtually complete as DIDC eliminated rate limits on all time deposits of more than 31 days. As scheduled, all Regulation Q ceilings were totally eliminated on March 31, 1986.

The MMDAs authorized by the 1982 act proved to be an immediate success. As many billions of dollars moved into the new deposit accounts, MMMFs experienced massive declines in assets. Only six weeks after their introduction, MMDAs exceeded the $242 billion peak of MMMFs reached in December, 1982. As Figure 5-6 indicates, MMDAs soared to almost

Figure 5-6

Comparison of the Growth of MMDAs with MMMFs

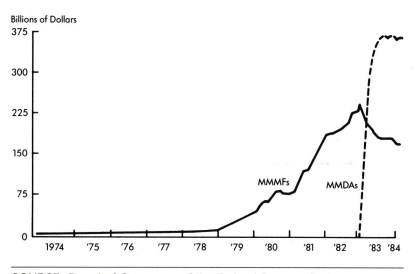

SOURCE: Board of Governors of the Federal Reserve System.

$400 billion by 1984. Most of the funds that went into MMDAs were transferred from other lower-yielding deposits of banks and thrifts, thus increasing the overall cost of funds for these institutions. On the other hand, in the year following the introduction of MMDAs, total assets of MMMFs declined about $100 billion. In early 1987, however, assets of MMMFs totaled $232 billion, as savers found both the funds and deposit accounts to be attractive financial alternatives.

What the Legislation Did Not Consider

Although the legislation of 1980 and 1982 made sweeping, fundamental changes in the U.S. financial system, both acts failed to address certain highly controversial matters. These include:

1. The entry of nonbank business firms into the financial services marketplace.
2. Restrictions on interstate branching.
3. The growth of nonbank banks.
4. The separation of commercial and investment banking.

From 1983 to early 1987, the House and Senate banking committees worked on new legislation, much of which dealt with the above issues. Various comprehensive banking bills have been introduced since 1982 and extensive hearings have taken place, but as of early 1987, there was little indication that passage was imminent. The rest of this chapter will consider some of the current trends, problems, and controversial issues facing the banking and thrift industries.

INTERSTATE BANKING

For several decades following legislation of 1927 and the mid-1930s, banks enjoyed unique protection from competition. Generally, the legislation attempted to ensure the safety and stability of the commercial banking system. Among other restrictions, the laws placed limits on geographic competition by prohibiting interstate banking and forcing national banks to abide by the same regulations imposed on state banks in the host state. During the last few years, however, the geographic constraints on banks have been bypassed and broken by market forces and legislation. Some of the methods and innovations used to circumvent the federal prohibition on interstate banking are described next.

Bank Holding Companies

Prior to the passage of the Bank Holding Company Act (BHCA) of 1956, multibank holding companies (BHC) were used to establish interstate banking networks. A multibank holding company owns or controls two or more banks. A section of the Act, known as the Douglas Amendment, was written to prevent the BHC from circumventing federal restrictions on interstate banking. The amendment closed loopholes in previous federal laws by preventing interstate acquisitions unless the states where the acquired banks were located specifically allowed such purchases. Until 1975, no state permitted these acquisitions. Under the grandfathering provisions of the Douglas Amendment, the BHCs already engaged in interstate branching were allowed to retain those operations.

Nonbank Banks

The 1970 amendment to the BHCA defined a commercial bank as "any institution which (1) accepts deposits that the depositor has a legal right to withdraw upon demand, and (2) engages in the business of making commercial loans." Any firm that owns a bank, as defined, is a bank holding company (BHC) subject to the regulations associated with the act. The key word in this legal definition is *and*. To be a commercial bank an institution must accept demand deposits *and* make commercial loans. If it receives a bank charter and offers demand deposits *or* commercial loans, but not both, then it is not a commercial bank. We call this unique institution a *nonbank bank*.

By exploiting this loophole in the law, nonbanking business firms, such as Sears, J.C. Penney, and E.F. Hutton, found that they could offer banking services by creating nonbank banks. As owners of these institutions, the firms are able to avoid BHC status, thus escaping the requirement to divest themselves of all business activities not permitted to BHCs (e. g., retailing and manufacturing). At the same time, the companies can own institutions that are chartered by banking regulators and therefore can be members of the Fed and obtain FDIC deposit insurance. Hence, nonbanking business firms enjoy the benefits of owning an institution that is almost like a commercial bank, without being subject to the burden of the BHCA regulations.

Although BHCs themselves are subject to the BHCA, they also have reason to exploit the loophole. The nonbank banks they create can be used as vehicles to cross state lines with what are considered legally to be nonbanking subsidiaries. Because the loophole allows BHCs to bypass the Douglas Amendment's prohibition on interstate banking, they can establish nonbank banks in various states without permission from the states involved.[11] The BHC gains major advantages by locating nonbank banks in states where it would not be permitted to operate full service banks.

1. New banking locations offer the opportunity to expand its deposit base. Although a nonbank bank may choose not to accept demand deposits, it can offer NOW accounts, MMDAs, certificates of deposits, and other accounts.
2. New locations permit diversification of business loan customers.
3. New locations can help the BHC position itself as it waits for laws that permit entry of out-of-state full service banks.

By mid-1986 there were about 40 nonbank banks operating in the United States, but about 280 applications to open such banks in 38 states were awaiting final approval.

The Chairman of the Federal Reserve Board of Governors, Paul Volcker, has been a strong foe of the nonbank bank and has told Congress that one of his highest legislative priorities is the closing of the BHCA loophole. Volcker and his allies, however, face some formidable opponents—the Reagan Administration, the Comptroller of the Currency, and most importantly, the U.S. Supreme Court.

Although it is not likely that the nonbank bank will remain in its current form during the next few years, its existence will have a lasting effect on banking. It has helped to focus attention on interstate banking issues and is likely to help speed the final dismantling of interstate restrictions. It has also allowed other nonbanking business firms to offer more banking services, further eroding the separation of banking and commerce. Hence the nonbank bank has played an important role in the continuing movement to deregulate both geographic and product markets in banking.

REGIONAL BANKING

From 1983 to early 1987 one of the hot topics in banking was the state legislative action to establish regional interstate banking agreements. The lack of federal legislation had tossed the interstate banking issue into

---

11. Recall that the Douglas Amendment prohibits a BHC from purchasing a commercial bank in another state, unless that state permits the acquisition.

the laps of state legislatures. In the early months of 1985, about 25 states responded by introducing bills to allow some type of regional banking. By the end of 1986, more than 40 states had enacted laws that allowed interstate expansion as permitted under the Douglas Amendment to the BHCA. The most important legislative action takes one of the following forms:

1. *Reciprocity laws*. Massachusetts, for example, will permit BHCs located in other New England states to acquire Massachusetts banks on a reciprocal basis. New York will reciprocate with any state.
2. *Unrestricted Laws*. Alaska and Maine, for example, will allow out-of-state banks entry, but requires no reciprocal treatment.

The first regional banking compact was established by four New England states. Regional acquisitions and mergers, however, were delayed by lawsuits filed by Citicorp and a small Connecticut bank. These institutions challenged the regional reciprocal limitations of the New England compact. But in June, 1985, the U.S. Supreme Court ruled that regional compacts are constitutional and banks from outside states can be barred from participating. The decision removed the temporary cloud from this form of interstate banking expansion.

Because of the rapid increase in bank stock prices in 1985 and the regulatory pressure to raise capital ratios, the costs of major interstate acquisitions forced large money-center banks to scale back acquisition plans during late 1985 and 1986. Many large banks switched their attention to the acquisition of troubled out-of-state thrifts as their major hope for interstate expansion. Citicorp, for example, gained entry into California, Nevada, Illinois, and Florida by acquiring failing thrifts.

In 1986, *superregional* banks came to the forefront. The superregionals were the result of mergers and acquisitions between relatively large regional institutions. The merger of First Atlanta Corp and Wachovia Corp, for example, gave rise to First Wachovia Corp, a superregional with almost $17 billion in assets.

Texas, California, and Pennsylvania approved some form of interstate banking in 1986. In December, 1986, Chemical Bank of New York announced the acquisition of Texas Commerce Bancshares, Houston's largest bank. The $1.19 billion takeover marked the first out-of-state purchase of a major Texas BHC under the state's new interstate banking law. The acquisition created the nation's fifth largest banking firm, with about $75 billion in assets. It is evident that state legislation has tended to increase the pressure for federally approved interstate banking, and all signs point to continued action in this fast-moving area in the near future.

Garn–St. Germain and Interstate Banking

The large increase in the number of troubled thrifts brought about the need for more merger and capital-raising solutions, even if that assistance had to come from across state lines. The Garn–St. Germain Act provided for interstate acquisitions of failing thrifts by healthy thrifts or, if necessary, by commercial banks. Since the passage of the act in 1982, a large number of interstate acquisitions have occurred. It should be noted, however, that the act does not permit one healthy institution to acquire another sound one.

Other Methods of Achieving Interstate Banking

Actually, an institution does not need a brick-and-mortar presence to engage in certain types of interstate banking business. A New York bank, for example, might sell a negotiable certificate of deposit to an investor in Oregon or Texas, thus raising funds without a physical presence. In most cases, however, banking does require this presence. We will conclude our discussion of interstate banking by reviewing the more traditional methods of accomplishing some degree of interstate business.

Loan Production Offices. These facilities cannot accept deposits, but they do provide the means to solicit local commercial loan business. The home offices of the bank must approve each loan, however. In 1985 about 45 banks operated over 200 offices in nearly 35 states.

Nonbank Subsidiaries of BHCs. For years BHCs have established non-bank subsidiaries interstate. The BHCA of 1970 provides that BHCs may engage in certain nonbank activities that are closely related to banking.[12] Because the subsidiaries are not banks, once established they may open offices in any number of states. A 1982 survey revealed that about 5500 interstate subsidiaries were in operation.

Edge Act Corporations. These are subsidiaries of U.S. banks that are formed to engage in international banking and financial operations. These offices, which date from 1919, can operate interstate. Currently there are about 200 Edge Act offices in the United States.

---

12. Some of the nonbank subsidiaries approved by the Fed include: mortgage banks, finance companies, credit card firms, industrial banks, leasing firms, insurance underwriting companies, bookkeeping and data processing firms, and investment and financial advising firms.

Grandfathered Interstate Banking. Under the Douglas Amendment's grandfathering provision, the twelve BHCs engaged in interstate branching when the BHCA of 1970 was enacted were allowed to retain such operations. Prior to the passage of the International Banking Act of 1978, foreign banks operating in the United States could freely cross state lines. Domestic banks, of course, did not have this authority. Because of this and several other disparities, Congress agreed to correct the competitive inequities. The 1978 act required foreign banks to face the same branching and interstate prohibitions as domestic banks. The foreign banks, however, enjoyed the benefits of the act's grandfather clause, which permitted more than 100 banks to retain interstate banking operations.

THE BANK FAILURE PROBLEM

For many decades prior to World War II, the United States had the dubious distinction of possessing the highest bank failure rate of any major nation. After the FDIC was established in 1934, however, a remarkable decline took place, and post–World War II failures averaged about six per year until 1981. Unfortunately, the number of bank failures has increased steadily over the past several years. The number reached 138 in 1986, more than in any other year since the Great Depression. In addition, the FDIC's list of "problem" banks soared to 1,400 in September, 1986 and was estimated to be growing at an average of one per day. (See Table 5-3.)

Table 5-3

Bank Failures, 1980–1986

Year	Number of Failed Banks	FDIC Problem Banks
1980	10	217
1981	10	223
1982	42	340
1983	48	642
1984	79	848
1985	120	1,140
1986	138	1,400[*]

[*]September, 1986

SOURCE: FDIC, problem bank lists and annual reports.

Failures in Tennessee and other areas in the early 1980s were largely the result of fraud and insider abuse. Fraud and embezzlement were, and still are, major causes of bank failures. (The FDIC reported that between 1949 and 1970, no less than 65 percent of failures resulted from fraud.) In more recent years, the FDIC noted that the failures of large banks in San Diego, New York, Oklahoma City, and Knoxville were caused by "irregular" and "unusual" loan losses.

During 1985 and 1986 failed banks were generally smaller than those that failed in the early 1980s. This decrease in average size mainly reflected the greater concentration of failures within farmbelt states, where banks tend to be relatively small. Depressed farmland and commodity prices are expected to persist well past 1987 and will likely cause additional farm bank failures.

Also in 1985, a substantial number of bank failures occurred in Texas and Oklahoma, two major oil-producing states. The problems worsened in 1986 as much larger banks in these two states suffered massive losses from energy loan defaults. Other recent failures were related to problem loans in the real estate markets. Surprisingly, bank regulators maintain that defaults on foreign loans do not appear to have contributed significantly to recent bank failures, although many billions of dollars in nonperforming international loans are held by some large domestic banks.

Some critics argue that recent deregulation gives banks incentives to take excessive risks. Moreover, it is often suggested that excessive risk taking is fostered by a system of deposit insurance that at least partially insulates banks from potential losses that may result from higher levels of risk. Others discount the theory that deregulation is a direct cause of failures. They do admit, however, that deregulation, plus heightened and new competition from both banking and nonbanking firms, has meant that banks are forced to operate in a much harsher environment where abuses such as fraud, embezzlement, insider abuse, and general mismanagement exact much greater penalties than ever before.

Table 5-4 lists the largest bank failures in U.S. history but does not include the FDIC "bailout" of Continental Illinois in 1984. Without question the nation's seventh largest bank would have collapsed without government help, although the situation technically was not a bank failure. The repercussions of this near collapse and federal rescue are still being felt.

The frailty of Continental became a matter of concern in 1982 when the bank suffered heavy losses on energy loans purchased from the failed Penn Square Bank. Subsequent revelations indicated that Continental was anticipating major problems with loan losses in its foreign, energy, agricultural, and industry loan portfolios. As rumors about its nonperforming

Table 5-4

Largest Bank Failures in U.S. History

Bank	Year	Total Assets (Millions)
1. Franklin National Bank, New York	1974	$3,655
2. United States National Bank, San Diego	1973	1,265
3. United American Bank, Knoxville	1983	760
4. Banco Credito y Ahorro, Ponceno, Puerto Rico	1978	713
5. Penn Square Bank, Oklahoma City	1982	511
6. Hamilton National Bank, Chattanooga	1976	412
7. Drovers National Bank, Chicago	1978	227
8. American Bank and Trust, New York	1976	225

SOURCE: FDIC, *Annual Report,* various issues.

loans spread within the community, the bank lost $4 billion in deposits in only three days. Within a short period, massive runs on the bank by depositors resulted in a total drain of $15 billion.

Bank regulators took prompt action to curb depositor runs. First, the FDIC guaranteed that *all* depositors and creditors would be protected in full, regardless of the dollar amounts involved. Second, the Fed provided the ailing bank with billions of dollars of loans needed for the bank's survival as deposit withdrawals continued. After months of patchwork efforts, the FDIC finally completed a $4.5 billion bailout program for the troubled bank. The regulator did the following:

1. Assumed $3.5 billion of Continental's debt to the Fed in exchange for $4.5 billion of the bank's problem loans.
2. Provided $1 billion of new capital in exchange for preferred stock, in effect, nationalizing the bank.
3. Declared that if, after five years, the FDIC suffered losses in excess of $800 million under the loan package agreement, it would have the right to acquire all of the remaining common stock of the firm for the nominal price of 1/1000 of 1 cent per share. In December, 1986, the FDIC announced that its losses will far exceed the $800 million limit.

The repercussions from the Continental Illinois crisis have been far-reaching, and federal regulators have been strongly criticized on several counts. First, the FDIC had revealed its double standard of deposit insurance coverage. The regulator announced that no large money-center bank would be allowed to fail, but this same guarantee could not be given to smaller institutions. Thus, depositors at large banks would be given full insurance coverage, regardless of the amounts involved, but the statutory limit of $100,000 in protection would hold for deposits held in smaller institutions. This gross inequity in the insurance system was made clear to all. The more sophisticated depositors enjoy the benefits of deposit insurance in excess of the statutory cut-off point, despite the fact that deposit insurance was originally established to protect small and unsophisticated depositors.

A second major criticism concerns the potential attitude that may result from the policies of the federal regulators. If the world is told that large domestic banks will never be allowed to fail, then bank creditors—both insured and uninsured—need not concern themselves with the safety of their funds. This approach could erode market discipline, since investors and depositors have no incentive to appraise the financial strength of the institutions at which they place their funds.

Savings and Loan Failures

As previously indicated, losses for the S&L industry were at record levels in 1981 and 1982, but earnings were positive for the next three years. In 1986, however, the industry slumped badly. The stunning setback reflected huge losses by a relatively few S&Ls, primarily on loans to finance commercial real estate projects. In October, 1986, regulators noted that gross negligence in real estate lending, coupled with an overall recession in commercial real estate, largely caused loan losses of $6.86 billion during the previous five quarters. Regulators and many thrift executives generally agreed that the massive losses had not yet peaked.

The setback in industry earnings followed the good and bad news of 1985. The good news: industry profits more than tripled in 1985. The bad news: despite a favorable interest rate environment, the financial condition of the nation's weakest S&Ls actually worsened in 1985. In fact, by the end of 1985, about 20 percent of the 3,240 FSLIC insured S&Ls were losing money at a collective rate in excess of $10 million per day. In terms of generally accepted accounting practices, about 450 S&Ls were insolvent (liabilities exceeded assets) and another 680 institutions were very close to insolvency with net worths of 0–3 percent of total assets.

The S&L crisis is further complicated by the rapid decline in the reserves of the FSLIC. In 1986 the federal agency revealed that it faced enormous future losses and will have to spend at least $25 billion to merge or close insolvent S&Ls in the next several years. The FSLIC held reserves of only $4.6 billion at the end of 1985, and by January, 1987, only $1.9 billion of reserves were available to insure deposits totaling $890 billion.

Congress considered but failed to pass a plan to provide $15 billion in additional reserves in 1986.

Despite the critical problems caused by the imbalance between the assets and liabilities of thrifts, in 1986 many S&L executives returned to lending policies that shocked many industry analysts. In mid-1986 fixed-rate mortgages soared to 70 percent of total loans originated, compared with only 25 percent two years earlier. This rapid setback for the industry's efforts to restructure its loan portfolios will likely increase its vulnerability to rising interest rates. Many S&L executives argued that they were avoiding interest rate risk by making only those fixed-rate loans that can be packaged into mortgage-backed securities, which are then sold to the secondary mortgage market. The U.S. League of Savings Institutions, however, reported that in 1986 about 80 percent of the total amount of fixed-rate loans originated by S&Ls were retained by the industry, either in the form of loans or mortgage-backed securities. Anthony M. Frank, chairman of the large thrift, First Nationwide Bank, noted that "It's no sin to make a fixed-rate loan, but it's close to one to keep it." Will S&L industry–lending policies of 1986 prove to be the nightmares of 1990? Only time will tell.

The Collapse of Private Deposit Insurance Systems

Prior to 1985, the use of private deposit insurance was recommended as a supplement to federal deposit insurance. With the collapse of private insurance systems in Ohio and Maryland during 1985, however, it is unlikely that the use of similar protection will be seriously considered in the future. The use of private and state deposit insurance for thrifts proved one major point: state regulators had not done their homework! The history of state insurance has clearly demonstrated that while such systems could prevent individual failures, they could not prevent widespread bank panics. The Ohio and Maryland examples underscore problems faced by such insurance systems during emergencies.

The Ohio fiasco began in March, 1985 with the failure of ESM Government Securities, Inc. The small Florida-based firm was closed after regulators discovered over $300 million in losses resulting from fraudulent transactions involving repurchase agreements. It was reported that Home State Savings Bank of Cincinnati, Ohio would suffer about $130 million in losses because of the failure of ESM. Within two days, the S&L lost over $150 million in deposits and was closed. The failed S&L had its deposits insured by the private Ohio Deposit Guarantee Fund (ODGF). A panic quickly followed as depositors of the other 70 thrifts insured by ODGF rushed to withdraw their funds. Within a few days, about 20 percent of the deposits of the ODGF-insured thrifts were withdrawn, leaving many institutions on the brink of failure. Subsequently, the governor of Ohio declared a bank holiday and closed all privately insured thrifts. This was the first action of this kind since the Roosevelt-mandated bank holiday of 1933. Each of the closed thrifts was reopened only after acquiring FSLIC deposit insurance.

The first act of the Ohio drama had barely ended when it was center stage again in Maryland. Two Maryland S&Ls were closed by runs on their deposits after disclosures of problems at these institutions. Within a few days, runs on the additional 100 privately insured thrifts severely depleted their deposits. The governor of Maryland quickly imposed withdrawal restrictions at these institutions. Subsequently, the thrifts had to obtain federal insurance or agree to merge with federally insured institutions before unrestricted operations were permitted.

Shortly after these private insurance crises, other state insurance funds in North Carolina, Pennsylvania, and Massachusetts started making plans to either shut down or drastically reduce their systems. In all three states, the member thrifts were told to obtain federal deposit insurance within a specified period of time.

CONCLUSION

Over the years the U.S. financial system has undergone dramatic change, the pace of which has quickened remarkably in the 1980s. The salient element of the evolving structure of the financial services industry is increased competition in the provision of various services. Banks want to go into other businesses, while other nonbank financial institutions and even nonfinancial firms want to enter the banking business. Because of loopholes in existing legislation, some providers of financial services are regulated as banks, but many others are not.

To a great extent, the market for financial services has become national in scope, creating pressure for the removal of legal barriers to geographic expansion. Moreover, calls for additional reform of the statutory framework have been heard from several sources, including depository institutions, regulators, the Reagan Administration, some members of Congress, and the general public. Unfortunately, while there is a strong consensus that far-reaching reform is needed, there is little consensus as to the specifics of such reform.

Despite the substantial regulatory changes of the early 1980s, the piecemeal approach of those changes failed to address significant long-term problems. By 1985 it was clear that a new, more workable delineation of the roles of depository institutions, regulators, and nonbank suppliers of financial services was needed to correct the prevailing state of confusion and disarray. In early 1987, however, the outlook for legislative action was highly uncertain. Numerous legislative restrictions were still intact despite rapid technological advances, innovation in the financial system, and crisis and near-crisis situations in the S&L industry and, to a lesser degree, in commercial banking. Most observers argued that additional reforms were needed. Also, both the general public and the providers of financial services had experienced the benefits (and some problems) of deregulation, and the vast majority tended to like the net results. Hence, there was little question that further substantive and sweeping legislative reform was both needed and politically popular. The forward motion of the revolution in financial services has been temporarily slowed, but was almost certain to continue.

QUESTIONS FOR DISCUSSION

1. What were the major reasons for bank failures during 1985 and 1986?
2. What were the arguments during the 1930s for (a) Regulation Q interest rate ceilings? (b) the separation of commercial banking and investment banking? and (c) the provision of deposit insurance?
3. Outline the major provisions of the DIDMCA and discuss the problems that brought about the need for this legislation.
4. What was the Fed's "membership problem," and how did DIDMCA help to solve the problem?
5. Compare and contrast the money market mutual fund (MMMF) and the money market deposit account (MMDA).
6. What are nonbank banks, and how do they contribute to the spread of interstate banking?
7. What is regional banking, and why did it expand so rapidly during 1985–1986?
8. Discuss the S&L industry's overall performance in the mid-1980s. Did deregulation help the industry? Explain.
9. What were the major reasons for the passage of the Garn–St. Germain Act?
10. Explain the term disintermediation. Given the major changes that have taken place in regulations, could disintermediation ever be a problem again? Why?

SELECTED READINGS

Arshadi, Nasser. "The Impact of Deregulation on S&Ls: Slow Use of New Opportunities." *Journal of Retail Banking* (Spring 1985).

Fortier, Diana, and Dave Phillis. "Bank and Thrift Performance since DIDMCA." *Economic Perspectives*. Federal Reserve Bank of Chicago (September/October 1985).

Goudreau, Robert E., and Harold D. Ford. "Changing Thrifts: What Makes Them Choose Commercial Lending?" *Economic Review*. Federal Reserve Bank of Atlanta (June/July 1986).

Heaton, Gary G., and Constance R. Dunham. "The Growing Competitiveness of Credit Unions." *New England Economic Review*. Federal Reserve Bank of Boston (May/June 1985).

Holt, Robert N., and Karl S. Walewski. "Why do Some Banks Outperform Others?" *Bank Administration* (April 1985).

Keeton, William R. "Deposit Deregulation, Credit Availability, and Monetary Policy." *Economic Review*. Federal Reserve Bank of Kansas City (June 1986).

Mahoney, Patrick I., and Alice P. White. "The Thrift Industry in Transition." *Federal Reserve Bulletin* (March 1985).

Merris, Randall C., and John Wood. "A Deregulated Rerun: Banking in the Eighties." *Economic Perspectives*. Federal Reserve Bank of Chicago (September/October 1985).

Morris, Charles. "The Competitive Effects of Interstate Banking." *Economic Letter*. Federal Reserve Bank of Kansas City (November 1984).

Pavel, Christine, and Harvey Rosenblum. "Banks and Nonbanks: The Horse Race Continues." *Economic Perspectives*. Federal Reserve Bank of Chicago (May/June 1985).

Wall, Larry D. "Nonbank Activities and Risk." *Economic Review*. Federal Reserve Bank of Atlanta (October 1986).

———, and Robert A. Eisenbeis. "Risk Considerations in Deregulating Bank Activities." *Economic Review*. Federal Reserve Bank of Atlanta (May 1984).

6

GOVERNMENT REGULATION WHY THE CONTROVERSY?

The economic activities of production, distribution, and consumption in the United States are influenced by a large network of federal regulatory programs. Although some types of government control of economic activity date back to colonial times, the extent and form of government regulations today are unprecedented in our nation's history. Increased government regulation has become more necessary as our economy has grown in both size and complexity. But in recent years, the areas of economic activity to be regulated and the extent to which the public sector should intervene have become hotly debated issues. In addition to regulating particular industries, the federal government has used its regulatory functions to assist the private sector in attaining socially desirable ends that cannot be achieved in the marketplace.

Seldom does anyone deny that the goals of federal government regulation are laudable. The controversy focuses on the question of whether the ultimate gains from regulation are exceeded by the costs imposed upon society from such regulation. Critics contend that the "heavy hand" of government has replaced the "invisible hand" of market competition in allocating resources and determining prices and incomes.

With increased regulation, it is feared that losses in economic freedom will eventually lead to losses in personal freedom.

The call for regulatory reform on the part of the Ford, Carter, and Reagan Administrations is seen by some as an encouraging sign that the pendulum is swinging away from increased regulation. In recent years, transportation, communications, energy, and financial industries have undergone extensive deregulation. If regulatory reform applied to other regulated sectors, it is argued that the economy could save billions of dollars by releasing resources for other uses. This would help to control inflation and increase our nation's lagging productivity. However, most critics view such cases of deregulation as isolated ones, and believe that the encroachment by the federal government into the private market sector will continue to increase in the years ahead.

This chapter will limit its scope primarily to regulation by federal agencies and its impact on our economy. It must be recognized, of course, that state and local governments engage in regulatory activities. States and localities impose additional layers of regulation on those already mandated by the federal government as well as regulating some activities the federal government ignores. Consequently, the role of government in the everyday activities of our economy is much greater than the information presented in this chapter would suggest.

JUSTIFYING REGULATION

The reasons for government regulation are many. Historical, political, and social factors, as well as economic ones, loom large in explaining the demand for regulation. Let's review some of the justifications given for government regulation.

Natural Monopoly

A natural monopoly is characterized by large economies of scale necessary for efficient production. Capital requirements are so great that fixed costs are very high. Therefore, the long-run average cost per unit of output falls dramatically with expanded production. The first firm to establish sufficient output to achieve very low production costs could drive out competing firms with higher costs, leaving a monopoly situation. The cost structure of the industry serves as a barrier to entry. The unregulated monopolist may be in a position to charge excessive prices, restrict output, and reap monopoly profits. Thus, the government awards monopoly franchises to such firms, protecting them from competition and, at the same time, regulates services, prices, and profits. The natural monopoly argument is given as justification for regulation of many industries. In fact, however, cases of natural monopoly are few, with public utilities being the major example.

Excessive Competition

A more common justification for industry regulation is the prevention of excessive competition. In this case, government maintains at least several competitors, and often a relatively large number, but again regulates pricing, service, and additional entry. Unlike natural monopolies, these industries are usually characterized by a cost structure comprised largely of variable rather than fixed costs. This allows relatively easy entry and exit into the industry by many small producers. Wide swings in industry output and great instability in prices and profits in both the short and the long run often result. These conditions create a maximum of uncertainty for producers and consumers. The case for government regulation rests on the presence of excessive competition rather than excessive monopoly power. Regulation is justified to provide industry-wide stability. The problems of unregulated competition in the trucking industry, for example, led to regulated competition with the passage of the Motor Carrier Act of 1935.

Discriminatory Pricing

A third justification for government regulation is the prevention of injurious discriminatory pricing. Firms that have a very high ratio of fixed to variable costs can increase profits by charging markedly different prices for the same good in separated markets. A firm possessing strong monopoly power could charge a lower price to those customers who had very elastic demands while charging a higher price to customers with less elastic demands. In order to be classified as discriminatory, the different prices cannot be justified by differences in the cost of service. The result can be a misallocation of resources and can cause serious competitive injury to customers in less elastic markets. The original justification for federal railroad rate regulation was a reaction to the railroads' blatantly discriminatory pricing structure as well as to their monopoly power. By regulating maximum rates for the benefit of customers and minimum rates for the benefit of railroads, the Interstate Commerce Commission (ICC) lessened discriminatory pricing and eliminated "ruinous" competition between railroads.

Cream Skimming

Government regulation has also been justified to prevent destructive cream skimming in industries already regulated. It is argued that once an industry has been regulated on the grounds of being a natural monopoly it must be protected from unrestricted entry. If entry were free, new firms would select only the most lucrative markets to enter (the cream), leaving to the established regulated producers the burden of continuing service to the non-profitable or only marginally profitable markets. The key to understanding the cream-skimming problem rests with a regulated rate structure which "cross-subsidizes" markets. In order to provide an interlocking network of reasonably priced service, some regulated industries charge prices well above costs in markets with heavy volume in order to offset losses in poorer, smaller, or geographically separated markets. The monopoly protection awarded to these industries allows them to engage in a form of discriminatory pricing deemed to be in the public's best interest. But if unrestricted entry into the high-priced cream markets were permitted, prices and profits would fall, and regulated firms would no longer have the financial wherewithal to continue service to lesser markets, resulting in elimination of service to large segments of the country. This has been the justification used to regulate entry into such fields as postal delivery, trucking, airline transportation, telephone service, and many others.

Consumer and Worker Information

A fifth justification for government regulation is based on the need to provide better information to market participants. Government intervention, which is a form of social regulation, is based on the belief that consumers are unable to judge quality in advance of purchase. This has led to direct regulation in the form of minimum product specifications and standards of safety or cleanliness. Because search costs for consumers are expensive and the information gathered by individuals difficult to disseminate, the government intervenes to improve the flow of information for individuals.

Regulations seek to protect consumers from a wide variety of unsafe or ineffective products as well as misinformation stemming from fraudulent advertising. On the factor resource side, the same rationale has been applied to justify intervention in protecting workers from unsafe working conditions in factories, mines, and offices and on construction sites. The information problem has also been used to justify the practice of licensing as a means of determining quality of service before purchase. Examples include barbers, beauticians, plumbers, and realtors, to name just a few. The effect of such licensing, and in many cases the primary reason for its advocacy, is to restrict competition by increasing barriers to entry.

Ill-Defined Property Rights

One remaining justification for government regulation concerns the problem of ill-defined property rights. The workings of a market economy are predicated on recognizable private property rights. But where property rights are unassigned or indefinite, anyone can use the property without paying for it as long as no one else is using it. The result may be that social costs may differ from private costs, since users will not bear the full costs and are likely to overuse the property. In the communications area, the use of airwaves constitutes an example of ill-defined property rights. In the absence of regulation, airwaves would be a free, although not unlimited, resource. By assigning rights to airwaves, the government prevents overuse and provides for an orderly system of communication. Ill-defined property rights also justify environmental regulation since air and water are common property. However, because property boundaries are difficult to identify and divide, other forms of social regulation have substituted for the allocation of property rights.

The justifications for regulation are many; but whatever the original rationale for regulation, there is no doubt that regulation has become more rigid over time and has bred even greater regulation. This is not surprising since our increasingly complex, technologically advanced economy creates demands for regulatory controls to handle problems that did not exist in simpler times.

GROWTH OF REGULATION

For most of the nineteenth century, the U.S. economy was without much direct federal regulation. The industrial structure prior to the Civil War was mainly characterized by small business firms owned on a partnership or proprietorship basis, and the spirit of laissez-faire thrived. With the end of the Civil War, however, the economic scene began to change as the age of big business was ushered in. It was soon evident that government intervention in the economy was inevitable.

Railroads were the first industry to be regulated. The use of discriminatory pricing and other techniques associated with the possession and use of monopoly power drove the public, particularly farm communities dependent upon rail service, to seek protection. In 1887 the ICC was created to regulate railroads. The task of the ICC was to determine railroad rates, stabilize profits, and set standards for service. The ICC sought to protect railroads from themselves by stopping excessive competition and to protect the public from abuses of monopoly power.

Along with the growth of the railroad industry, other industrial firms were growing rapidly. Many had attained a large size by acquiring other firms or by forming trust organizations for the purpose of restraining trade and monopolizing markets. During the 1880s the great trusts attracted the nation's attention. The most noteworthy of the trusts were the Cotton Trust, Whiskey Trust, Lead Trust, Sugar Trust, Tobacco Trust, and—the most famous of all—the Standard Oil Trust. Again the clamor of the public against such near-monopolies resulted in the passage of federal legislation. In 1890 the Sherman Act was passed; it outlawed contracts and activities designed to create monopolies in restraint of interstate commerce. In dealing with trusts, the government's approach was to break up the trusts into smaller units and rely on market competition to protect consumers. In the case of railroads, the government allowed monopolies to continue but controlled their prices and policies.

The early advocates of government intervention primarily wanted the public sector to check the rise of monopoly power and restore market competition wherever possible. As other areas were observed where the market outcome was judged unsatisfactory, the intervention of the federal government was again requested. The scope of economic regulation expanded. In succeeding years the transportation and antitrust statutes were amended and direct economic regulation spread to other industries, including agriculture, banking, communications, and energy.

By the end of the 1930s, a total of 24 federal regulatory agencies had been legislated into existence, 10 of which were established between 1930 and 1939. The policies of these agencies were largely concerned with economics and the orderly functioning of specific industries; they constitute what economists now refer to as examples of *economic regulation*. Table 6-1 lists some of the important economic regulatory agencies, along with their functions. The list presented is not an exhaustive one, but rather includes those agencies considered the most powerful and pervasive of the federal government's economic regulators.

The number of federal regulatory agencies grew somewhat modestly until the 1960s, when 7 more were created. In the 1970s, however, a dramatic increase in the number of agencies occurred. During this period, 21 new agencies were formed and 2 others became new "off-shoots" of an existing agency.

The more recent formation of agencies represents a shift in the philosophy and approach of federal regulation. The older regulatory bodies sought to control entry, pricing, and service in defined markets; the newer agencies specialize in socioeconomic problems. Thus, the term *social regulation* has been applied to the regulatory activities of agencies that seek to correct a variety of undesirable by-products of the production and use of goods and services. Table 6-2 contains a list of major agencies engaged in social regulation, with brief descriptions of their functions.

The creation of these newer regulatory agencies occurred in response to increasing pressures to provide solutions to a number of problems,

Table 6-1

Important Economic Regulatory Agencies

Date	Agency	Functions
1887	Interstate Commerce Commission	Regulates rates and routes of railroads, most truckers, and some inland waterway carriers
1890	Antitrust Division of the Department of Justice	Regulates interstate commerce, prohibits monopolization and restraint of trade
1913	Federal Reserve Board	Regulates member banks and sets money and credit policy
1914	Federal Trade Commission Act	Regulates to prevent unfair competitive practices
1930	Federal Energy Regulatory Commission	Regulates interstate transmission of electric power, natural gas pipeline rates, and wellhead price of interstate gas
1931	Food and Drug Administration	Regulates drugs for safety and effectiveness, food for safety and purity, and labeling
1934	Federal Communications Commission	Centralizes regulation of broadcasting, telephone, and telegraph
1934	Securities and Exchange Commission	Regulates securities markets and public disclosures of company information
1935	National Labor Relations Board	Regulates labor-management practices and conducts union elections
1938	Civil Aeronautics Board[1]	Regulates domestic airlines

[1]Eliminated in 1985.

such as equal opportunity for minorities, equal treatment of the sexes, protection of the environment, protection of workers from occupational hazards, protection of consumers from shoddy merchandise, and other side effects of normal economic activity that the market often ignores. Many of these problems had been in existence for a long time but only recently did the public press for regulatory programs designed to override the market system.

Table 6-2

Important Social Regulatory Agencies

Date	Agency	Functions
1965	Equal Employment Opportunity Commission	Handles complaints of employment discrimination based on sex, race, and religion
1970	National Highway Traffic Safety Administration	Sets standards for motor vehicle safety and fuel economy
1970	Environmental Protection Agency	Regulates standards and time tables for pollution abatement
1970	Occupational Safety and Health Administration	Regulates safety and health conditions in work places
1972	Consumer Product Safety Commission	Regulates standards for product safety and hazardous substances
1973	Mine Safety and Health Administration	Regulates safety in mines
1975	Nuclear Regulatory Commission	Regulates civilian nuclear safety, including power plant licensing

The controversial nature of these newer agencies stems from the fact that the agencies affect not only the conditions under which goods and services are produced, but also the basic characteristics of the products made. As a result, the private market sector must factor additional social values into its decision-making process. Federal regulation is no longer restricted to certain industries or to exercising certain powers over the national economy. It has also become involved in the day-to-day operations of a wide range of industries. As of fiscal 1987, a total of 54 regulatory agencies were in existence.

Another way of indicating the growth of federal regulatory agencies is by examining full-time employment in these agencies. Table 6-3 presents staffing for selected years by area of regulation. The peak staffing year during the period was fiscal 1980, at which time permanent full-time positions in both economic and social regulatory agencies numbered 131,701. The rapid increase in employment during the 1970s is witnessed by the fact that the corresponding figure for fiscal 1970 was only 86,160. A reversal of the trend toward increased staffing has occurred under the Reagan Administration. By fiscal 1985, staffing for federal regulatory agencies had declined to 111,944, or 85 percent of its 1980 peak. Of this number, social regulation accounted for 87,047, or 78 percent of total full-time employment.

Table 6-3

Staffing for Federal Regulatory Activities
(Fiscal Years, Permanent Full-Time Positions)

Area of Regulation	1970	1975	1980	% Change 1970–1980	1983	1985	Estimated 1986	Estimated 1987	% Change 1980–1987
SOCIAL REGULATION									
Consumer safety and health	53,080	65,763	66,016	24%	56,350	54,395	55,065	53,755	−19%
Job safety and other working conditions	7,472	13,694	18,201	144%	15,849	14,577	14,644	14,554	−20%
Environment and energy	4,929	15,618	19,621	298%	16,391	18,075	18,858	18,599	−5%
TOTAL—Social Regulation	65,481	95,075	103,838	59%	88,590	87,047	88,567	86,908	−16%
ECONOMIC REGULATION									
Finance and banking	7,537	9,455	11,108	47%	10,730	10,896	11,757	12,016	8%
Industry-specific regulation	6,072	7,221	7,365	21%	5,921	4,969	5,035	5,015	−32%
General business	7,070	8,638	9,390	33%	8,737	9,032	9,207	9,422	0%
TOTAL—Economic Regulation	20,679	25,314	27,863	35%	25,388	24,897	25,999	26,453	−5%
GRAND TOTAL	86,160	120,389	131,701	53%	113,978	111,944	114,566	113,361	−14%

SOURCE: Center for the Study of American Business, Washington University. Derived from the Budget of the United States Government and related documents, various fiscal years.

Although staffing for social regulation continues to decline, total estimated employment levels for regulation were higher in fiscal 1986 and 1987 than for 1985. Surprisingly, the area of increased employment is that of economic regulation, in spite of the recent deregulation of many industries. Specifically, increased employment can be traced to the area of finance and banking where employment has risen 8 percent from 1980 to 1987. The difficult situation faced by many financial institutions has led to greater examination efforts by the Federal Home Loan Bank, the Federal Reserve Banks, and the Farm Credit Administration.

COSTS OF GOVERNMENT REGULATION

As a result of the increased awareness of the rapid expansion of regulatory activities in our economy, serious attention is now being directed to measuring the full costs of regulation. The task of constructing reliable cost data is by no means an easy one. Unlike the costs for providing public goods largely financed by tax revenues, the bulk of regulatory costs lie outside the administrative budgetary process. Most of the costs attributable to federal regulation are hidden because they are accounted for by higher prices paid by consumers or in lower returns to owners and shareholders. Still, these costs are as important as those included in the normal budgetary process, since both have the effect of allocating scarce economic resources that could have alternate uses in the private sector. Given the present paucity of cost information, it appears that regulation has expanded at an unprecedented rate with but minimal scrutiny as to its aggregate cost.

In response to the need for improved cost data, a number of studies have been recently undertaken by various government agencies, business firms, and university and research institutions. Cost estimates presented in this chapter are drawn from the results of several such studies.

ADMINISTRATIVE COSTS

Since costs associated with federal regulation arise in different ways, it is necessary to separate them into different categories. Expenditures for federal regulatory activities as shown in the budget of the U.S. government are used as a measure of "administrative costs" and are presented in Table 6-4. Included here are the costs of agency staffing, office supplies, and consultants' reports needed to write, manage, publish, and police regulations, as well as other operating cost items.

Table 6-4

Expenditures on Federal Regulatory Activities
(Fiscal Years, Millions of Dollars)

Area of Regulation	1970	1975	1980	% Change 1970–1980	1983	1985	Estimated 1986	Estimated 1987	% Change 1980–1987
SOCIAL REGULATION									
Consumer safety and health	$867	$1,828	$2,763	219%	$2,935	$3,195	$3,251	$3,221	17%
Job safety and other working conditions	128	364	753	488%	807	860	820	879	17%
Environment and energy	278	1,114	2,201	692%	2,261	2,967	3,539	3,591	63%
TOTAL—Social Regulation	$1,273	$3,306	$5,717	349%	$6,003	$7,022	$7,610	$7,691	35%
ECONOMIC REGULATION									
Finance and banking	$108	$188	$413	282%	$481	$685	$824	$821	99%
Industry-specific regulation	91	160	279	207%	285	289	276	287	3%
General business	115	206	354	208%	435	509	542	588	66%
TOTAL—Economic Regulation	$314	$554	$1,046	233%	$1,201	$1,483	$1,642	$1,696	62%
GRAND TOTAL	$1,587	$3,860	$6,763		$7,204	$8,505	$9,252	$9,387	
Percent change in nominal terms				326%			9%	1%	39%
Total in 1972 $	$1,646	$3,068	$3,791		$3,346	$3,680	$3,865	$3,765	
Percent change in real terms				130%			5%	−3%	−1%

SOURCE: Center for the Study of American Business, Washington University. Derived from the Budget of the United States Government and related documents, various fiscal years.

In nominal dollars, administrative costs increased from $1.5 billion in 1970 to $6.8 billion in 1980, a 326 percent increase. The estimated administrative costs of federal regulatory agencies for fiscal 1987 was $9.5 billion, a 62 percent increase from 1980. In real dollars, the increase in administrative costs is less dramatic. In 1972 dollars, administrative costs rose 130 percent from 1970–1980, but for the period 1980–1987 costs actually declined by 1 percent.

The long-term trends in federal regulatory spending can be seen in Figure 6-1. Even when expressed in 1972 dollars the increase in spending for social regulation during the 1970s is unmistakable.

Figure 6-1

Trends in Federal Regulatory Spending

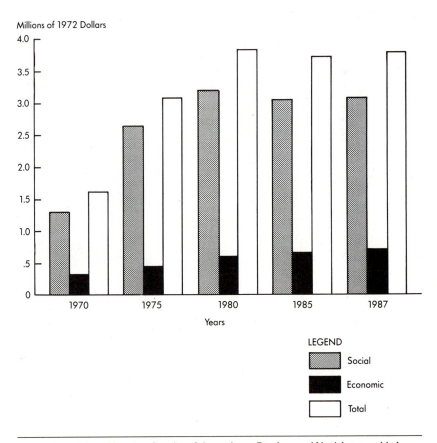

SOURCE: Center for the Study of American Business, Washington University. Derived from the Budget of the United States Government and related documents, various fiscal years.

Figure 6-2 depicts the changing distribution of expenditures within the regulatory categories. The biggest changes have been the declining budget share for consumer safety and health and the increased share for environment and energy. Economic regulation is also increasing its share of the budget because of the growth in finance and banking activities.

At a time when economists report figures for various activities in the economy in terms of hundreds of billions and even trillions of dollars, a sum of $9.4 billion for administering the regulatory arm of the federal government may seem somewhat insignificant. In fact, $9.4 billion appears to be a bargain price to pay for the undisputed benefits of a cleaner environment, safer work places, product standards, and economic fair play. However, the full cost of regulation to the economy is much larger than those costs merely attributable to administration and can be found within the areas of compliance costs and indirect costs.

Compliance Costs

The category referred to as compliance costs is comprised of expenditures incurred by business firms and state and local governments in complying with government regulations. In general, compliance costs include: (1) expenses associated with examining various possible approaches for compliance and shaping proposed standards, (2) expenses required to satisfy specific requirements established by regulations, (3) expenses attributed to collecting data and maintaining records indicating conformance with standards, and (4) expenses necessary to defend compliance efforts against legal challenges. Viewed in a more simplistic way, compliance costs are direct outlays on the part of individual economic units that would not take place in the absence of government regulation.

Attempts to measure precisely the full extent of compliance costs are obviously fraught with insurmountable difficulties. At best, any measure of compliance costs can be presented only as a crude estimate. According to data published by the Center for the Study of American Business, the economy-wide impact of complying with government regulations was approximately $120 billion in 1980. Again, it must be remembered that compliance costs are largely financed through higher prices to consumers.

Examples of Compliance Costs. Because of the imprecise nature of compliance cost estimates, some specific examples of the effects of regulatory compliance may shed some light on its overall impact. Consider as an example the ubiquitous hamburger, the cornerstone of the diet of the typical college student. The next time the reader is struggling over the mind-boggling decision of whether to dine at McDonald's, Wendy's, or perhaps Burger King, it should be comforting to know that regardless of the choice,

Figure 6-2

Changing Shares of Regulatory Spending

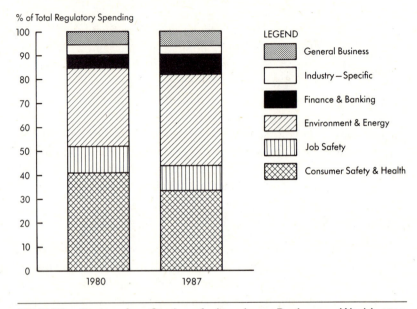

% of Total Regulatory Spending

LEGEND

General Business
Industry — Specific
Finance & Banking
Environment & Energy
Job Safety
Consumer Safety & Health

SOURCE: Center for Study of American Business, Washington University. Derived from the Budget of the United States Government and related documents, fiscal years 1981 and 1987.

the hamburger "entree" has passed numerous government regulations. In the interest of consumer protection, the government controls both the quality and the quantity of the hamburgers, as well as the numerous complementary items.

A study conducted by Colorado State University produced findings that the hamburger is subject to 200 statutes, 41,000 regulations, and 111,000 court cases associated with federal, state, and local governmental control of ground beef. Included are rules governing the grazing practices of cattle, conditions in slaughterhouses, and methods used to process meats for sale in fast-food outlets, supermarkets, and restaurants. The resulting bite out of the consumer's budget in the form of a hidden regulatory tax is estimated to be somewhere between 8 to 11 cents per pound.

If hot dogs are more to your liking, you can rest assured that the government has also mandated rules involving their manufacture and sale. Included among the many federal requirements are maximum limits for fat and moisture content, as well as preservatives and extenders, such as soy flour and cereals; there are also rules for artificial coloring and product labeling. Thus, both the hamburger and the hot dog, the staples of a quick, inexpensive meal, are subject to an extraordinary number of government rules and regulations. Yet most consumers might question whether government control has noticeably increased the quality of the final product.

Concerned by sharply rising costs due to government regulations, General Motors began compiling annual reports on the direct cost of regulation to the company. During the period 1974–1983, GM spent over $14 billion to comply with federal, state, and local government requirements. In 1982 alone, GM spent $1.8 billion for such purposes and required the equivalent of 22,000 full-time employees to comply with regulations at all levels of government. The sums mentioned are only those pertaining to research and engineering, reliability, inspections, testing, facilities, tools, and rearrangement costs and do not include the cost of any hardware required on GM products, such as safety devices or pollution control equipment. These figures do include the direct costs of government reports and administrative costs related to regulation.

GM reports that the amount of paperwork the company must file to gain government certification of its new cars for sale in a given year would make a stack 15 stories high. This illustrates another costly aspect of federal regulation—the enormous paperwork burden.

The Center for the Study of American Business recently concluded that filling out the 4,400 different federal forms, excluding tax and banking forms, requires approximately 143 million worker-hours at a direct cost of $40 billion per year.[1] The Dow Chemical Company has estimated that it costs the firm $20 million annually to complete federally mandated paperwork. In order to receive approval from the Food and Drug Administration for a muscle relaxant, one drug firm submitted 456 volumes of paperwork, each two inches thick, with a total weight of one ton. The Paperwork Reduction Act passed in 1980 was proposed by the Reagan Administration in recognition that paperwork costs were burdensome to businesses, individuals, and all levels of government. The goal of reducing paperwork costs is to be accomplished by coordinating and standardizing federal information policies.

Occupational Safety and Health Administration (OSHA). Finally, with any discussion of compliance costs, mention must be made of one of the most controversial agencies—OSHA. Passed in 1970, the Occupational Safety and Health Administration Act empowered the government to take actions necessary to ensure safe and healthful working conditions so far as possible to every working man and woman in the country. Of all the social regulatory

---

1. The Center for the Study of American Business.

agencies, OSHA has received the most sweeping criticism from the private business sector. It has been repeatedly charged with regulatory "overkill." Although the Act seeks indisputably laudable objectives, its many critics charge that the Act was passed without a systematic discussion of how much risk should be removed from the work place and without examination of how any designated optimal level can be achieved with limited economic resources.

Compliance with the agency's rules is currently enforced through direct inspection. Decisions made by inspectors are often considered arbitrary, uneconomical, and at times simply lacking in common sense. Several examples should suffice to illustrate the nature of the criticism.[2] In one instance, a miner was cited for not carrying a two-way radio. This action seems reasonable, except for the fact that the miner's job was a one-person operation and there was no one else with whom the miner could communicate. In another case, a company was fined for not having its employees wear life jackets while working on a bridge spanning a river channel, even though water normally flowing through the channel had already been diverted elsewhere during bridge construction. Finally, Dow Chemical reported that it was required by OSHA to spend approximately $60,000 per plant to lower stair railings from 42 inches to OSHA standards of 30–34 inches, despite Dow studies that indicated the higher level was in fact safer.

Although perhaps not typical, these cases serve to indicate that at least in its earlier years, OSHA engaged extensively in prescribing details rather than performance. The agency has been more responsive to criticism in recent years and has operated in a more flexible manner. However, the direct cost of complying with rules mandated by the Occupational Safety and Health Administration is thought to be in the range of $5 to $7 billion per year.

Compliance costs of $100 billion far outweigh the direct administrative costs of federal regulation. But even more costs are associated with regulatory activity, namely, costs classified as indirect costs.

Indirect Costs

Government regulation indirectly affects our economy in a variety of complex ways; but several broad generalizations can be made in terms of its impact on productivity and innovation. Since much of this discussion was

---

2. Examples presented on OSHA compliance were provided by the National Association of Manufacturers.

treated in detail in Chapter 3, only a brief treatment of indirect costs will be presented in this section. Thus, the brevity of this discussion is not meant to detract from its importance, but merely to avoid repetition.

Productivity. Productivity is decreased by government regulation when resources are allocated away from the production of measurable output, such as autos, food, or petroleum, and into nonmeasurable goods, such as cleaner air, cleaner water, or increased job safety. Economic resources directed to achieving these social goals cause a measurable loss in productivity in terms of national income statistics. However, the loss is largely the result of an inability to quantify in dollar terms the value of social benefits derived. In itself, therefore, this should not constitute a major matter of concern. But in those industries where complying with government requirements necessitates greater resource costs than resulting social benefits, national productivity is reduced, in both a measured and a nonmeasured sense. This is also true in cases in which government regulation requires the use of a larger number of costly resources than necessary to achieve stipulated social goals.

In some cases, economic regulations can bring about a direct decline in productivity. Trucking regulations mandated by the ICC, for example, traditionally required wasteful empty back hauls and circuitous routing, while in communications and finance, regulations had the effect of stifling innovations such as cable television and electronic banking.

Innovations. The flow of innovations is of critical importance in a market economy. Innovations create jobs, increase productivity, and shape the overall quality of our economy. Along with numerous other factors, government regulations have served to shift many research activities away from projects capable of producing major innovative breakthroughs. In the capital goods sector, rather than engaging in experiments involving new approaches to problems, American industry has felt compelled to retain older technology. The automobile industry, for example, is still wedded to the internal combustion engine and to controlling air pollution instead of developing more efficient and cleaner engines. In consumer industries, major innovative research is secondary to market-oriented research focusing on product differentiation. In today's climate, the introduction of another brand of frozen pizza offers greater potential payoffs at less risk than does a synthetic meat product.

Capital investment in many industries today requires lengthy delays brought about by numerous permit applications, as well as uncertainties about future regulatory requirements. Environmental impact studies and challenges delayed the completion of the Alaska pipeline for approximately five years. In another case, it required about nine years of hearings, proposals, and comments to decide whether a product labeled peanut butter should contain 90 or 92 percent peanuts.[3]

DEREGULATION OF AIRLINES

Whereas regulatory reform in the area of social regulation is likely to emphasize greater flexibility and more cost effectiveness, regulatory reform in the area of economic regulation is becoming identified with deregulation. In some industries the process of deregulation is already underway. In recent years, various degrees of deregulation have taken place in the communications, finance, and transportation industries. An examination of the domestic airline industry should prove helpful in gaining an insight into the problems associated with deregulating an industry.

CAB Regulation

Economic regulation of the airline industry dates back to 1938 with the creation of the Civil Aeronautics Board (CAB). The stated objectives of the CAB were to assure adequate, economical, and efficient air service at reasonable charges. By the end of World War II, the CAB had developed two lines of policy to achieve these objectives. It acted to control entry and exit of domestic trunk lines to prevent excessive competition, and it supervised air fares to protect the public. Although the agency was considered effective in the airline industry's formative years, by 1970 vocal critics charged that its policies had the effect of reducing efficiency, discouraging innovations in service, raising prices, and causing a severe misallocation of resources.

Under CAB regulation, airlines were prohibited from engaging in competitive pricing. Consequently, airlines were compelled to compete vigorously on the basis of customer service. Costly service competition in many cases resulted from the CAB setting prices above marginal costs. Economic theory tells us that in highly competitive, nonregulated markets, prices above marginal costs will not be sustained, since price competition among existing firms, as well as the entry of new ones, will drive prices down to the level of marginal costs. However, the CAB not only prevented price cutting but also restricted entry of new carriers. Therefore, airlines sought to increase their market shares by the only viable alternative: nonprice competition.

---

3. National Association of Manufacturers.

The most injurious form of service competition entailed flight scheduling in heavily trafficked markets. Individual airlines sought consumer identification by providing the largest number of daily flights between major cities so that customers would contact that airline first in making reservations to any destination. Many passengers carried on these flights could have been accommodated easily by fewer flights. The resulting proliferation of flights created a chronic problem of excess capacity on these routes.

To understand the economic incentive to engage in costly service competition, one can refer to the workings of a purely competitive market structure. This is not to suggest that even in the most fiercely competitive markets the industry approximates pure competition. In fact, although heavily trafficked markets are oligopolistic, elsewhere consumers are often faced with monopoly service. But pure competition provides an analytical framework for understanding the behavior of individual firms, since the airline industry possesses many of the preconditions for effective competition; namely, fairly elastic consumer demand, few economies of scale, and essentially homogeneous service, among others.

Figure 6-3 contains two panels portraying ticket prices and quantity of service for an individual airline and for the industry as a whole. Assume the market in question is the transcontinental route between New York City and Los Angeles, a route characterized by vigorous competition among carriers. Panel *a* represents the short-run market demand and market supply functions for airline service between the two cities. The equilibrium price of P_e is the market clearing price determined by supply and demand forces in an unregulated market, resulting in quantity of service Q_e. For the individual airline shown in panel *b*, the quantity of passenger service offered at the price of P_e is X_e. Since the firm in a competitive market structure is a price-taker, rather than a price-maker, the firm's demand curve, as well as its marginal revenue and average revenue curves, is the horizontal line at price P_e. Its marginal cost curve *(MC)* is the short-run supply curve and, as a profit maximizer, the firm would equate marginal cost with marginal revenue.[4] This results in an equilibrium position whereby price (and thus marginal revenue) equals marginal cost.

However, assume the CAB has set the price of air service between New York and Los Angeles at P_r, a price well above the market clearing price of P_e. Panel *a* indicates that the regulated price results in an excess supply of airline service, for the higher price has encouraged consumers to

---

4. Note that under conditions of pure competition, the short-run marginal cost curve is the firm's short-run supply curve only at points above the firm's average variable cost curve.

Figure 6-3

Pricing and Quantity of Service in the Airline Industry

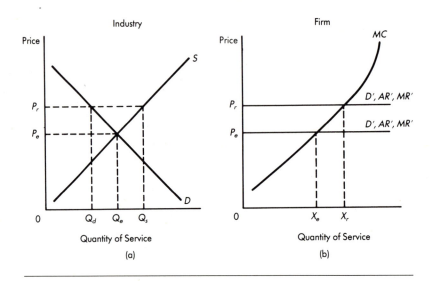

fly less while encouraging airlines to expand the amount of service offered. Why? Panel *b* shows that an individual airline faced with a price of P_r will adjust output accordingly. Acting rationally, an airline will expand service until its marginal cost curve intersects the higher regulated price. The new equilibrium quantity of service is now X_r. Since each firm behaves in the same fashion, excess capacity of $Q_s - Q_d$ results in the marketplace. The point to be stressed is that if price is prevented from falling to marginal cost in the short run, then, where healthy competition exists, airlines will tend to increase service to equate marginal cost with the higher price.

The problem of excess capacity in the airline industry tended to be chronic because the law did not allow the CAB to regulate frequency of service. The deadweight loss to society of supplying excess service at higher fares was undoubtedly substantial. To compound the problem, airlines also competed heavily in other areas of costly nonprice competition. Over the years, airlines sought to attract passengers by providing greater comfort, better first-run movies, and more exotic food and drinks than competitors. Even macadamia nuts were substituted for peanuts in the attempt to differentiate service. Such attempts to differentiate service also tended to increase the costs of providing passenger service. The feeling that the airline industry and the general public could be better served by less regulation, rather than more, became widespread during the 1970s.

The Move to Deregulation

Prior to 1976, the CAB had experimented with allowing airlines greater price flexibility, but on a relatively small scale. It had permitted the introduction of family fares, night fares, standby fares, and youth fares, among others. In 1976, the CAB took an additional step by sanctioning super-saver fares, no-frill fares, and unlimited-travel fares. Special fare reductions of this sort were given in recognition of the fact that the demand for air travel on the part of many customers was far more elastic than was historically assumed. In constructing air fares, the board had traditionally assumed an inelastic demand for all traffic. The success of special fares, although highly restrictive, complex, and discriminatory, was a major factor persuading the CAB to call for widespread deregulation of the industry.

In 1978, the Airline Deregulation Act was passed by Congress. The law called for an end to all price and route controls by 1983, as well as the termination of the CAB itself in 1985. In the interim, airlines could reduce fares by as much as 50 percent or raise them by as much as 5 percent without CAB approval.

Deregulation was not received with universal enthusiasm. Some of the major airlines vigorously opposed it. Of major concern was the possible abandonment of some markets, particularly smaller ones. Opponents of deregulation argued that increased competition in high-density markets would eliminate excess profits used to cross-subsidize service in low-density markets. Without these profits, a reduction or even elimination of service to many communities was feared inevitable. The CAB, however, presented several counterarguments. First, even under a regulated framework, airline service to many smaller communities was declining. Secondly, the CAB pointed to a large number of smaller markets that airlines could have withdrawn from but did not. Although fares for passenger service to and from these markets did not in many cases cover fully allocated costs, they did cover marginal costs and contribute to overhead expenses. From an airline's perspective, these markets may still be worth servicing. Thirdly, the CAB found very little evidence to support the cross-subsidy argument. Many smaller markets were confronted with very high air fares, while in other cases low fares went hand in hand with poor service. Finally, it was argued that where regulated carriers withdrew from markets, they would undoubtedly be replaced by commuter and regional airlines providing equivalent if not better service.

Another issue was whether deregulation would result in increased concentration in the industry. Critics contended that in a few short years the number of firms in the airline industry could easily shrink to half the existing number. The CAB responded by admitting that some increase in concentration is likely to occur in the future. But the pressure for merger activity has existed for many years and is not new. If increased concentration does occur with deregulation, it will be largely the result of inefficient management. Large airlines possess no significant cost advantages over small airlines from the standpoint of economies of scale. Both small and large airlines can acquire larger aircraft to gain lower costs per seat mile. In fact, bigness breeds inefficiency because the airline business is a personal and consumer-oriented one. Finally, as long as entry into markets remains open, the monopoly power that may result from merger activity can be checked by additional entrants.

A third point of controversy was the possibility of chaotic market conditions. This reflected the apprehension that without CAB regulation, airlines would rush into some markets while abandoning others, engage in price wars, and bring about destructive competition. A maximum of uncertainty would result, with passengers not knowing from day to day which airlines fly which routes and at what fares. The public would also experience difficulty making reservations and would be more reluctant to make long-run travel plans. In short, the adjustment process would create such trauma in the industry that the market would not function. The CAB's view was that most markets are unregulated and they function without chaos. Why not airlines?

Effects of Deregulation

As predicted, the deregulation of the airline industry unleashed previously constrained market forces that quickly impacted both airlines and consumers. With the elimination of barriers to entry, existing airlines reduced or even eliminated service to some cities while at the same time initiating service in markets from which they had been precluded. A number of new airlines were formed. Some newly formed airlines provided service to smaller communities, while others entered into direct competition with established trunk line carriers in major markets. In all, 72 new airlines were created from 1978 through 1986.

With increased mobility on the part of existing airlines and with the entry of new firms in the industry, intense price competition was inevitable. Downward pressure on airfares was exacerbated by the entry of cut-rate carriers, such as People Express, as well as by the decision on the part of several existing carriers, such as Braniff, Frontier, and Continental, to become exclusively discount-fare airlines. Introductory low-fare service was also prevalent wherever major trunk carriers, such as American, Delta, or United, entered new markets. On long-haul flights, aggressive price cutting occurred as airlines sought to deal with excess capacity resulting from CAB regulation. Competition served to keep a lid on price increases, and discount fares had become so widespread and accessible that by 1986 over 90 percent of all paying passengers were flown at a discount. Lower prices necessitated lower costs and existing carriers slashed labor costs wherever feasible.

Passenger demand proved responsive to lower fares. The total number of passengers in the United States has risen from 292 million on 14.7 million commercial flights in 1982 to an estimated 395 million on 19.2 million commercial flights in 1986. On major carrier routes, load factors increased from 56 to 64 percent, largely as a result of reduced fares. In smaller markets, passengers experienced an increase in the quality of service and in many cases an increase in the number of flights as well. Air fares, however, remained relatively high since these markets did not experience the strong competitive pressures of larger markets.

One of the most radical changes resulting from deregulation is the manner in which airlines have restructured service in order to increase market share. Whereas airlines used to fly mostly point-to-point between cities, they now operate by means of *hub and spoke* systems. Called hub and spoke because the new route systems look like spoked wheels, the system allows for increased, low-cost service between smaller cities. Figure 6-4 portrays the workings of a typical hub and spoke system. By routing some 25 flights through its connecting hub, an airline could possibly connect over 600 cities with service. From the passenger's standpoint, the system provides increased convenience, for on any given trip the need to change airlines is sharply reduced. It is now estimated that 98 percent of all passengers complete trips without changing airlines, compared to 87 percent prior to regulation.

The system is not without its problems. Passengers now experience long delays on a regular basis because of tremendous pressure on existing airport facilities. Scheduled departures and arrivals are impossible to execute at peak times. For example, Lambert Field in St. Louis is a hub for Trans World Airlines. In 1986, TWA's schedule called for 43 landings within a 50 minute period. Shortly after the last plane lands, the same 43 planes are scheduled to start taking off. When a hub experiences bad weather, the entire wheel comes apart. Annually, the loss to airlines approximates $2 billion.

The hub and spoke system also places great pressure on existing airports at a time when there is a recognized shortage of air-traffic controllers. Airports are attempting to keep pace by expanding and modernizing existing facilities. Newark, where 62 departures are scheduled between 8 a.m. and 9 a.m., is scheduled for major improvements, as are airports in Miami,

Figure 6-4

Hub and Spoke Network of Major Airlines

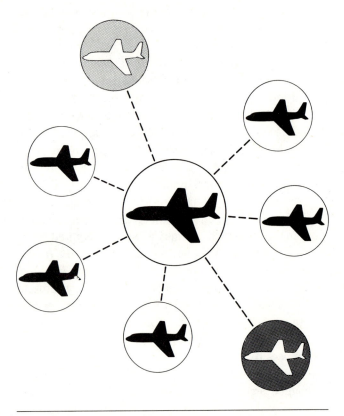

Note: Spoke flights may be those of the hub airline, a commuter airline owned by the hub airline, or an independent commuter airline with connecting flights.

Pittsburgh, Seattle, Detroit, and Washington, D.C. The Denver airport, built to handle 15 million passengers annually, now handles over 30 million annually and plans to build a completely new facility by 1995. The new airport will accommodate 100 million people annually. More airport capacity did not seem necessary until traffic exploded after deregulation.

A major concern with the hub and spoke system is its impact on competition. Strong hubs tend to reduce competition, for they not only capture connecting traffic for the airline, but they create scale economies that inhibit entry into the hub market. Pittsburgh, for example, is not an easy market to enter because it is the hub for USAir. The airline carries over 70 percent of all passengers who fly out of Pittsburgh.

The growth of market power resulting from the system is already being felt. In spite of the large number of new competitors in the airline industry since deregulation, many airlines have gone bankrupt. With discount fare wars and intense route competition, the competitive market has shown itself to be as ruthless as it was inviting on many heavily travelled corridors. At last count, 33 airlines have withdrawn from the industry since 1978. As a result of bankruptcy and mergers, the largest airlines have grown appreciatively stronger. Table 6-5 presents some of the major consolidations that have occurred in 1985 and 1986. Figure 6-5 indicates that the market share of the six largest airlines has increased from 73 percent in 1978 to 88 percent in 1987 despite the large number of new competitors that remain in the industry.

The airline industry's future appears headed for an oligopolistic market structure. The number of airlines will continue to shrink, and dominant carriers will continue to exercise greater market power. With diminished competition, the six largest airlines may claim a combined market share of over 90 percent. As the oligopoly becomes stronger, new airlines will find it increasingly more difficult to enter the market and passengers will find it much harder to purchase discount fares. A regulated oligopoly will have been replaced by an unregulated oligopoly, and the recent window of competition will be closed.

Table 6-5

Consolidations of Major Airlines, 1985–1986

Airline	Acquired Airline(s)
Texas Air[1]	Continental
	Eastern
	People Express[2]
United	Pan Am (Pacific routes only)
Northwest	Republic
Trans World Airline	Ozark
Delta	Western
American[3]	Air Cal

[1]Texas Air also owns New York Air.
[2]Purchase of People Express also included the assets of bankrupt Frontier Airlines.
[3]Pending approval.

Figure 6-5

Market Share of Six Largest Domestic Airlines 1978–1987

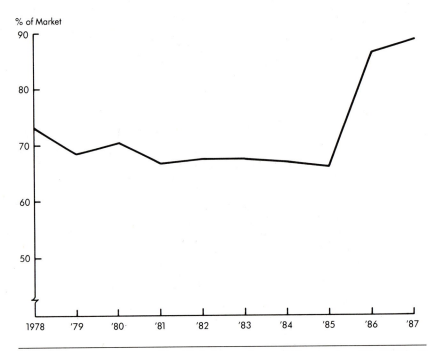

SOURCE: Adapted from *The Wall Street Journal* (September 16, 1986): 3 and *Business Week* (December 12, 1986): 52.

CONCLUSION

The general purpose of this chapter was to examine the nature and extent of the regulatory functions of the federal government. The subject of regulation continues to grow more controversial as more and more people are becoming aware of the ways in which regulatory activity affects their everyday lives. But opinions rendered as to whether there is too much or too little government regulation tend to be mixed, depending upon who is being regulated. There is hardly a person who doesn't favor some form of strong governmental regulation. In many cases, however, these individuals seek to strengthen government control over others, not themselves. As individuals, there are few among us who wish to reduce our personal and economic freedoms by submitting to government regulation. Yet we seem to have little conceptual difficulty in wanting others to be regulated. The same behavioral trait is also seen among business enterprises, except in those cases in which firms see government regulation as a means of forming a workable cartel and preventing excessive competition. Thus, it is not at all surprising to find that those who call for increased government regulation are in fact pursuing their own private interests rather than the interests of the general public. Although simplistic, this helps to explain the increased demands for regulation and the resulting increase in regulation by government.

Finally, it should be realized that the central question is not whether government regulation is bad or good, but whether social benefits exceed the social costs of specific regulations and even agencies. Despite the admittedly imperfect data, future emphasis must be placed on the development of procedures and techniques that will improve the overall efficiency of the federal regulatory process so that social benefits can be realized at minimum costs.

QUESTIONS FOR DISCUSSION

1. Develop a list containing at least five justifications for the expanding role of government regulatory activity over the last 50 years.
2. Which agencies, if any, do you believe are engaging in excessive regulatory activities? Which agencies, in your opinion, should be strengthened and given greater regulatory powers?
3. Despite recent moves to deregulate certain industries, little is heard in Washington about the likelihood of similar deregulation in the area of social regulation. What factors do you believe account for this?
4. Analyze the following statement: "Elected officials legislate new authorities and programs—and thus bigger jobs—for bureaucrats; the bureaucrats operate their agencies and programs to meet the demands of vote-producing constituency groups; the constituencies then help reelect the politicians."
5. In your view, will federal regulation continue to grow in the future? If so, at what pace and in what form? What will the relationship be in the year 2000 among consumers, business enterprises, and government?
6. Hammurabi, king of Babylonia in the eighteenth century B.C., included among his famous codes a regulation for home construction. In essence, the code stated that if a house collapsed and killed the occupant, the builder should be put to death. How does Hammurabi's building code differ from today's social regulation?

7. What changes do you foresee in the airline industry in terms of the number of firms and pricing? Under what conditions, if any, might the airline industry be regulated once again?
8. What have been the effects of deregulation in the banking, trucking, and communication industries? What industries do you believe are likely to be deregulated in the future?

SELECTED READINGS

Baumol, W. J. "Contestable Markets, Antitrust and Regulation." *Wharton Magazine* (Fall 1982).

Chilton, Kenneth, W., and Murray L. Weidenbaum. "Small Business Performance in the Regulated Economy." St. Louis: Washington University, Center for the Study of American Business, 1980.

McNeill, Charles R. "The Depository Institutions Deregulation and Monetary Control Act of 1980." *Federal Reserve Bulletin 66* (June 1980). "Reforming Regulation: Strengthening Market Incentives." *Economic Report of the President,* 1986: 159–188.

"The Worsening Air Travel Mess." *Fortune* (July 7, 1986): 50–55.

Weiss, Leonard W., and Michael W. Klass, eds. *Case Studies in Regulation: Revolution and Reform.* Boston: Little, Brown and Company, 1981.

Wilson, G. W., "Regulating and Deregulating Business." *Business Horizons* (July–August 1982).

7

SOCIAL SECURITY
WILL IT
LAST?

After many months of Congressional debate and years of political and public controversy, President Franklin D. Roosevelt, on August 14, 1935, signed into law the Social Security Act, establishing a system of social security for Americans. This event occurred decades after social security was adopted in several European countries. Bismarck had inaugurated a social security program in Germany in the 1880s. Austria and Hungary followed shortly thereafter. Great Britain established old-age pension and unemployment programs between 1908 and 1925. The hardships of the Great Depression stimulated interest in the adoption of social security in the United States. As a consequence, in June, 1934, President Roosevelt created the Committee on Economic Security to study the problems relating to economic security and to make recommendations for a program of legislation to deal with unemployment and economic security for the aged. Most of the committee's recommendations were incorporated one year later in the initial Social Security Act.

THE SOCIAL SECURITY ACT

Contributions were first paid into the Social Security System in 1937. The constitutionality of the System was upheld by the Supreme Court under the "general welfare" clause that same year, and workers began to acquire credit toward old-age insurance benefits in 1937. The first unemployment benefits under the Act were paid by the state of Wisconsin in 1936, and the first monthly payments under old-age and survivors insurance benefits were made in 1940.

Provisions of the Act of 1935

As it was originally passed, the Social Security Act contained three major provisions:

1. A federal system of old-age insurance benefits for workers
2. A federal-state system of unemployment compensation benefits
3. A program of federal financial aid to states to help them provide public assistance to the needy aged, the needy blind, and dependent children

Other programs established by the Act included maternal and child health care services, child welfare services, services for crippled children, vocational rehabilitation, and assistance to states for public health services.

At the time the Act was passed, only 3.5 million workers had any type of old-age retirement benefits, and only four states had any form of unemployment benefits. The Act provided old-age insurance for millions of additional workers. The original Act, however, covered only employees in industry and commerce. Therefore, only six out of ten members of the labor force were included in the federal old-age insurance system.

Many types of workers were not covered by the new Social Security System because of administrative or other difficulties. These included agricultural workers; domestic service workers; casual laborers; merchant ship crew members; employees of federal, state, and local governments; workers in nonprofit institutions such as schools and hospitals; and all workers over 65 years of age. Railroad workers were also excluded because they had retirement benefits under the federal Railroad Retirement Act.

With subsequent amendments, many of the initially excluded workers were brought under the Act. In 1939, for example, workers in the U.S. merchant marine service, employees of national banks, and workers in savings and loan associations and similar institutions were brought within the scope of the Social Security Act. In 1950, coverage was again broadened to include the nonfarm self-employed and the regularly employed agricultural and domestic workers. Voluntary coverage was allowed for employees of nonprofit institutions, and employees of state and local governments who were not already under a retirement system became eligible for coverage. Some civilian workers of the federal government not under an existing retirement program were included also.

Amendments in 1950 extended old-age coverage to nearly ten million additional workers, most of whom were self-employed. In 1954, coverage was extended to farmers, some professionals, and members of religious orders. In 1956, coverage was expanded to include three million or more members of the armed services currently on active duty. In 1965, self-employed physicians were included under the System. By the mid-1960s, the only large group not under Social Security coverage was employees of the federal government, who were covered by other federal pension systems. Today the Social Security System provides social insurance protection for old-age, disability, and death benefits for 90 percent of all wage and salary workers and the self-employed.

Federal Old-Age Benefits

The 1935 Act provided for monthly benefit payments, beginning in 1942, to insured workers when they reached retirement age of 65 years or more. Benefits were based upon the employee's earned wages during a qualifying period. The maximum payment was to be $85 per month and the minimum was to be $10 per month. Those reaching age 65 without qualifying for monthly benefits would receive a lump-sum payment equal to 3.5 percent of the employee's lifetime earnings. A death benefit of the same amount, less benefits previously received, was provided by the law.

In 1939 the benefit package was amended to start benefit payments in 1940 instead of 1942, benefits were increased, survivors payments were added, and a new formula for computing the benefits was adopted. In 1950, benefit payments were increased again to make it possible to include earnings up to $3,600 per year, instead of the initial $3,000 per year, when computing benefits. In addition, the formula was revised to permit the use of a 1950 starting date to calculate benefits in order to obtain a larger benefit. A major reason for this new starting date was that many of the new groups of workers being brought under the System had no earnings for Social Security credit prior to that time.

In 1955, the maximum amount of earnings taken into account in calculating benefits was raised to $4,200 per year, and by the mid-1960s it had reached $5,500 per annum. In addition to increasing the maximum earnings that could be used in calculating the benefits, the formula for determining the benefits was revised several times to provide for larger benefits.

As originally enacted, the Social Security System provided old-age benefits only for the insured workers. But amendments provided for old age payments to dependents and death benefits to survivors in the event the insuree died. Over the life of the Act, these benefits were increased. In 1956, the Act was amended to permit women to receive benefits when they reached age 62. In 1961, this amendment was extended to permit men to draw a reduced benefit at age 62. Disability benefits were added in 1956 and were subsequently liberalized. By 1987, the maximum individual primary benefit for a person retiring at age 65 had increased to $789 per month. Average old-age benefits paid had risen from $23 per month in 1940 to $488 per month.

Eligibility

Initially, according to the 1935 Act, a worker must have had earnings in covered employment over a specified length of time in order to collect benefits. In 1939, amendments provided for two types of insured status. A worker became "fully insured" only when at least $50 was earned in covered employment in each of 40 quarters (10 years). This was later modified because new groups and the self-employed were brought into the System. In 1984 workers were considered fully insured if their quarters of covered employment equalled at least the number of years between 1950 (or age 21, if later) and age 61, or to the date of death if it occurred earlier.

Under the initial Act, complete retirement was essential for the receipt of benefits. Amendments followed, however, to permit a retired worker to earn up to $1,200 per year without losing benefits. This was later boosted to $1,500, then to $3,000, and now to $8,160 per year. If a worker earns more than the stipulated amount according to the earnings test, a penalty is imposed with the loss of some benefits. At age 70 and beyond, however, there is no restriction on the earnings of a worker collecting federal old-age benefits.

Hospital and Health Care (Medicare)

In 1965, the Social Security Act was amended to provide insurance protection against the cost of hospital and related health care for covered individuals when they reached 65 years of age or older. The amendments also provided hospital insurance for persons not covered under the Act who reached age 65 before 1968. In addition, the amendments permit all persons 65 years of age and over to purchase protection against the cost of physicians' fees, with one-half of the cost paid by the federal government out of general revenue funds.

The 1965 amendments, moreover, liberalized the earnings test, the definition of disability, and the cash benefits payable for disability. The federal matching ratios for public assistance were increased, and federal grants to states for maternal and child health care and welfare services were improved. An addition was made to the Social Security payroll tax to provide financing for the new Medicare package.

Thus, after 30 years, the beginnings of hospital and health care insurance were added to the Social Security program. This concept had initially been recommended back in 1935 by the President's Committee on Economic Security for eventual inclusion under Social Security. It became the Old Age, Survivors, Disability, and Health Insurance program, OASDHI.

Supplemental Security Income

In addition to regular OASDHI benefits, the Social Security program, since 1965, has provided monthly checks to people in financial need who are 65 or older and to people of any age who are blind or disabled. People who have little or no cash income and who do not own much property may get supplemental security income (SSI), even if they have never worked or contributed to Social Security taxes. SSI is not the same as Social Security even though the program is run by the Social Security Administration. The money for SSI checks comes from general funds of the U.S. Treasury. Persons who receive Social Security benefits may also obtain SSI checks in some cases. A person, however, does not have to be eligible for Social Security benefits in order to receive supplemental security income.

THE 1972 AMENDMENTS

Prior to 1972 the Social Security law provided that a retiree's initial old-age benefit be based on earnings up to the level of the Social Security wage base existing during a person's working life. Moreover, there were no automatic increases in either the wage base, on which benefit payments were calculated, or the schedule of benefits. The only way either could be changed was by an act of Congress.

In 1972 the Social Security law was amended, however, to provide automatic increases in the wage base. This would result in a gradual rise in initial old-age benefit payments. In addition, the schedule of future benefits of current workers, along with the current benefits of current beneficiaries, was to be automatically adjusted by tying benefits to the Consumer Price Index. The primary purpose of this measure was to adjust the purchasing power of benefits to offset increases in the cost of living. As prices rose, the future benefits of current workers were adjusted upward. But higher prices were also in large part responsible for higher earnings, which would yield them greater benefits upon retirement. This double adjustment is commonly referred to as "coupling." This coupling helped cause the "replacement rate," the ratio of a worker's retirement benefit to previous average monthly earnings, to rise substantially, putting an added burden on the Social Security System.

With the highly inflationary conditions of the 1970s, the coupling process resulted in substantial increases in old-age and survivors benefits payments. Between 1972 and 1976, for example, the average monthly benefit for a retired worker increased by 39 percent in current dollar value.

Table 7-1 shows changes in federal old-age benefits vis-à-vis changes in the Consumer Price Index. It can be noted that since payments first started in 1940, the average monthly benefit for a retired worker increased from $22.60 to $481.20 in July, 1986, an increase of 2,029 percent. In terms of constant or real dollars, the average monthly benefit increased 171 percent, as shown in the final column of Table 7-1. For a more recent period it can be observed that between 1970 and 1987 the average monthly benefit rose from $118 to $488, an increase of 314 percent in current dollars, as shown in Figure 7-1. During that same period the average benefit in constant dollars rose 45 percent. After the 1977 amendments, however, constant dollar benefits remained more stable.

Table 7-1

Social Security Average Monthly Benefit for a Retired Worker
in Current and Constant Dollars, 1940–1986

	Average Monthly Benefit	Cumulative % Increase	CPI (1967 = 100)	CPI (1983 = 100)	Average Monthly Benefit (in 1983 $)	Cumulative % Increase
1940	$22.60	—	42.0	14	$161.42	—
1945	24.50	8	53.9	18	136.11	− 15
1950	43.86	94	72.1	24	182.75	15
1955	61.90	174	80.2	27	229.26	42
1960	74.04	228	88.7	30	246.73	53
1965	83.92	271	94.5	32	262.25	62
1970	118.00	422	116.3	39	302.56	87
1975	207.00	816	161.2	54	383.33	137
1980	341.41	1,441	246.8	83	411.34	155
1983	440.77	1,850	298.4	100	440.77	173
1985	478.62	2,018	322.2	108	443.17	175
1986 (July)	481.20	2,029	328.0	110	437.45	171

SOURCE: *Statistical Abstract of the United States: 1986* and previous issues, and *Social Security Bulletin* (October 1986).

Figure 7-1

Average Monthly Social Security Benefit to Retired Workers, 1970–1987

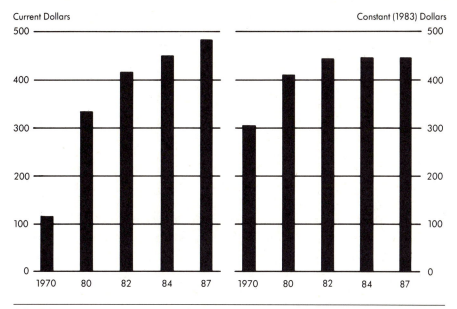

SOURCE: Chart prepared by U.S. Bureau of the Census.

FINANCING SOCIAL SECURITY

According to the 1935 Act, old-age and survivors insurance was to be financed solely through a payroll tax levied on both employers and employees in covered employment. The initial tax rate was set at 1 percent each on the employer and employee on the first $3,000 of yearly earnings. An increase of 0.5 percent was to be made every three years until 1949, when the total contribution by both employer and employee was to become 3 percent. Subsequent amendments by Congress, however, postponed or eliminated these increases until 1950, when the tax rate was increased to 1.5 percent and the tax base was raised to $3,600 per year. The tax rate was retained at that level until 1954 when it was raised to 2 percent. It was held there through 1955 and 1956, but the tax base was increased to $4,200 per year. The next year the rate was increased by 0.25 percentage point to cover the cost of disability insurance. Increases subsequently raised the rates to 4.2 percent on $6,600 of earnings by 1966. At that time scheduled increases were to raise the rates of 4.85 percent by 1973. Moreover, additional taxes were to be added to finance hospital insurance, beginning in 1966. That rate, initiated at 0.35 percent on employer and employee in 1966, was to rise to 0.80 percent each in 1987.

According to the Act, receipts from payroll taxes were to be deposited in a special reserve account or trust fund with the U.S. Treasury. Future benefits were to be paid from the trust fund without the need to use general revenue funds from the federal government. The law required that the funds in the account be invested in securities of the federal government, and interest from these securities was to be used also for the payment of benefits.

Even though amendments that increased the number of covered workers and raised the size of benefits increased the cost of the Social Security program, the trust fund grew more rapidly than anticipated. This was due to the increase in employment and the payment of higher wages, especially because of World War II. As a result, scheduled increases in the tax rates were eliminated or postponed.

Tax contributions plus earned interest on government securities continued to exceed disbursements for old-age and survivors insurance until 1957. In that year disbursements for benefits exceeded net income for the first time in the short history of the Social Security program. Net assets of the trust fund declined thereafter.

Between 1970 and 1975 total assets (reserves) of the OASI trust fund fell from 111 percent of annual expenditures to 62 percent of annual expenditures, as shown in Table 7-2. Moreover, by 1977, reserves equaled less than 50 percent of annual expenditures. In addition, in the period 1970–1976, the number of retired workers increased 28 percent. It was also projected that in future decades a greater percentage of the population would be in the retirement-age bracket.

With an increasing number of workers retiring with higher initial old-age benefits, the Social Security Administration foresaw the eventual depletion of the OASI and DI trust funds within eight years. One projection for the status of the trust funds is shown in Table 7-3. This depletion of the trust funds and the pending deficit was to occur despite the scheduled increases in the payroll tax wage base and the tax rates.

A number of factors can be cited to explain the substantial decline in the levels of the OASI and DI trust funds. First, benefits for persons already retired had been indexed in recent years to the Consumer Price Index. This resulted in a sharp increase in benefit payments with the higher rates of inflation in the 1970s. Secondly, the high levels of unemployment during the 1970s, especially during the 1974–1975 recession, resulted in a slowdown in the growth of payroll tax receipts. Thirdly, the number of beneficiaries receiving disability insurance payments had been higher than anticipated. In fact, the drain on the disability trust fund was so rapid that

Table 7-2

Social Security (OASDHI) Trust Funds, 1970 to 1985
(In millions of dollars, except percent)

	1970	1975	1980	1985
Old-age and survivors insurance (OASI):				
Net contribution income	30,256	56,816	103,456	180,165
Transfers from general revenue	449	425	540	2,203
Interest received	1,515	2,364	1,845	1,870
Benefit payments	28,796	58,509	105,074	167,248
Administrative expenses	471	896	1,154	1,591
Assets, end of year	32,454	36,987	22,824	35,842
Reserve ratio[a] (percent)	111	62	21	21
Disability insurance (DI):				
Net contribution income	4,481	7,444	13,255	17,413
Transfers from general revenue	16	90	130	1,017
Interest received	277	502	487	807
Benefit payments	3,067	8,414	15,437	18,827
Administrative expenses	164	256	368	608
Assets, end of year	5,614	7,354	3,629	6,321
Reserve ratio[a] (percent)	174	84	23	34
Hospital insurance (HI):				
Net contribution income	4,881	11,509	23,866	47,576
Transfers from general revenue	874	669	871	47[b]
Interest received	161	671	1,116	3,362
Benefit payments	5,124	11,318	25,067	47,580
Administrative expenses	157	263	476	834
Assets, end of year	3,202	10,517	13,749	20,499
Reserve ratio[a] (percent)	61	91	54	42

[a]Assets at the end of the year as a percentage of benefit payments and administrative expenses during the year.
[b]Does not include $371 million transferred from the railroad retirement account.

SOURCE: *Social Security Bulletin* (October 1986).

it was estimated in 1977 that the fund would be depleted by 1979. Lastly, the unintentional process of coupling benefits to a higher wage base and to the CPI caused initial benefits of retirees to rise more rapidly than wages.

Wage rates were expected to rise because of inflation. Therefore, the wage base for Social Security payroll taxes was increased. Since retirees' benefits were based on earnings in their most recent work years (years that, incidentally, had the higher Social Security wage bases compared to earlier years), their initial retirement benefits were higher than they

Table 7-3

Projected Status of Trust Funds as of 1976

Year	Trust Fund Reserves (Billions)
1977	$41.1
1978	35.5
1979	28.6
1980	20.7
1981	11.6
1982	0.1
1983	− 14.8

would have been under the older law. Benefits were also tied to the CPI to offset the adverse effects of inflation on their purchasing power. This meant that benefits would increase for two reasons: first, because the wage base increased, and second, due to the rise in the CPI. But since both the wage base and the CPI increased because of inflation, this was tantamount to giving a double adjustment in benefits for inflation.

Looking at the long run, the OASI and DI trust funds were in even more serious trouble. Using the wage base, payroll tax, and age and retirement assumptions existing in 1977, the Social Security trustees estimated that average OASI and DI expenditures over the next 75 years would exceed payroll taxes by an amount equivalent to 8 percent of taxable earnings. Avoiding this situation would require a tripling of Social Security tax rates by the year 2050 just to finance the schedule of future benefits provided for in the existing law.

It was calculated that half the projected long-range deficit was due to the technical flaw caused by the double adjustment for inflation resulting from the coupling process. In fact, it was calculated that double indexing would eventually cause benefits to exceed preretirement wages for some workers. The other half of the projected deficit was due to a continued increase in disability payments and a sharp rise in the number of retired persons relative to the working population after the year 2000. In 1977, for example, there were 19 persons age 65 or over for every 100 persons in the 20–64 age bracket. It is estimated that by the year 2030 there will be 34 persons 65 or over for every 100 persons in the 20–64 category. In still other terms, today the ratio of current workers to retirees, their survivors, and their dependents is 3 to 1. In another generation or two the ratio could drop to 2 to 1, putting further stress on the Social Security System.

Based upon the findings of an Advisory Council on Social Security appointed in 1975 and Congressional hearings on the subject, Congress in 1977 struggled hard to find a solution to the Social Security problem. One major point of controversy revolved around the issue of whether OASDHI benefits should be financed solely through payroll taxes or be financed in part from general revenue funds of the federal government. This led to a broader question regarding the nature of the Social Security System. Should it be an actuarially sound system with benefits paid at retirement based strictly on an individual worker's tax contributions (and those of the employer) made during the worker's life? It was evident that the System long ago had moved away from this concept and to a "pay-as-you-go" system in which tax contributions from current workers were used to finance benefit payments to current retirees. Consequently, some analysts suggested the use of general revenue funds to help finance Social Security benefits. Others thought that such a practice would lead to widespread liberalization of benefits and abuses in the System.

It was also pointed out during the hearings that, for a number of reasons, the Social Security System was in large part a huge "transfer payment" mechanism instead of an insurance system. For example, benefits paid to higher-income wage earners were not in proportion to the greater amount of taxes they paid during their working years. On the other hand, many low-income wage earners were receiving benefits greater than were merited by their tax contributions. This was especially true of those receiving the minimum benefit. In fact, some of the tax funds were used to provide benefits to persons who contributed nothing to the Social Security funds.

A further issue arose regarding the dual-wage-earner family. The retirement benefit to a worker and nonworking spouse was as good as that going to a family in which both the husband and wife worked prior to retirement. In short, many working wives thought they were being cheated since they would receive the same amount of benefits by not working. In effect, the wife's contribution counted for nothing extra in benefit payments for the family.

The question of whether the employer-employee tax rates should be identical also was debated. Some advocated that the employer ought to pay a larger portion of the total tax. The ratio of the tax rate for the self-employed worker was also given consideration.

Much was discussed, too, regarding the best way to increase contributions to avoid future deficits. Some strongly suggested that tax rates be increased substantially. Others advocated increasing the tax base. Those objecting to an increase in the tax rates did so in large part because the higher rates would hit hardest the lower-wage-earning groups. Since taxes are paid only on earnings within the tax base, it would mean that some of the earnings by higher wage earners would not be subject to the tax. On the other hand, practically all of the lower-income and many of the middle-income wage earners would have their entire earnings subject to higher Social Security taxes.

Those advocating an increase in the tax base did so with the notion that the additional funds needed could be raised in this manner. This would tax those wages in excess of the current base earned by some middle-income people and particularly by high-wage-earning groups. At the same time it would leave the taxes paid by the lower-income group unchanged.

The problems caused by the coupling process came in for much discussion and debate. It was pointed out that coupling had increased the replacement rate, which is the ratio of a worker's Social Security benefit to that worker's final working pay. This replacement rate was 43 percent in 1977, compared to an average of 31 percent between 1950 and 1969. The Advisory Council on Social Security recommended that it be kept between 40 and 45 percent.

Other measures discussed and debated included the feasibility of taxing Social Security benefits; raising the retirement age; incorporating civilian employees of the federal government, along with their existing pension fund, into the Social Security System; and making any future national health insurance program a part of the Social Security System.

THE 1977 AMENDMENTS

After much discussion, and faced with a pending deficit in both the short run and the long run, Congress amended the Social Security Act in 1977 to alleviate the System's financial plight. Since Congress has generally operated on the presumption that Social Security benefits should be financed primarily via payroll tax receipts, it was essential to raise Social Security tax rates and/or raise the wage base subject to Social Security taxes if deficits were to be avoided. The major changes brought about by the 1977 amendments include the following:

1. A decoupling of the double adjustment in benefits (wage base increases plus COLA) for inflation. This change was expected to cut the pending long-run deficit by one-half.
2. Beginning in 1979, Social Security tax rates for employers and employees were increased.
3. The schedule of taxes for the self-employed was likewise increased.
4. To prevent the reserves of the disability insurance fund from being exhausted, the 1977 law provided for a reallocation of current tax rates among the trust funds in 1978.

5. The taxable wage base for employers, employees, and the self-employed was increased. The base was raised from a maximum of $25,800 in 1983 to $43,800 per annum for 1987.

The 1977 law also made a number of changes in the benefit structure. Regarding the retirement earnings test, for example, under the prior law a retiree between the ages of 65 and 72 could earn up to $3,000 annually without suffering a benefit penalty. For every $2 in earnings above the $3,000 level, however, the retiree's benefit was reduced by $1. There was no restriction for retirees aged 72 or more. Under the 1977 law the ceiling on earnings for a retired worker between the ages of 65 and 69 rose to $4,000 per year in 1978, and then by $500 increments up to a ceiling of $6,000 by 1982. When earnings in any year exceeded the ceiling, Social Security benefits were reduced by $1 for every $2 earned in excess of the ceiling.[1] According to the 1977 law, no earnings ceiling was imposed on beneficiaries after they reached 69 years of age. Now that age is 70.

On the other hand, a bonus was provided for those who postponed retirement beyond age 65. Benefits received rose from 1 to 3 percent for each year an eligible retiree did not elect to receive benefits between the ages of 65 and 72.

THE 1983 AMENDMENTS

Despite the 1977 amendments, by the end of the 1970s it was evident that our economic growth was significantly less favorable than had been anticipated and that the Social Security program would experience significant difficulties in the 1980s. In October, 1980, Congress made specific provision for the reallocation of tax revenues from the DI part of the program to the OASI part of the program for the years 1980 and 1981. This was to give Congress additional time to deal with OASI financially in the early 1980s.

A subsequent provision, enacted in December, 1981, authorized borrowing among the OASI, DI, and HI trust funds through December, 1981 to meet benefit expenditures for no more than the first six months of 1983. Also in December, 1981, President Reagan established the National Commission on Social Security Reform (NCSSR) with the charge to review the

---

1. The ceiling was prorated and imposed on a monthly basis during the first year of retirement. Thereafter, it was on an annual basis.

current and long-range financial condition of the Social Security trust funds and report its finding to the President and Congress by December 31, 1982.

In the meantime, the 1982 Annual Reports of the Boards of Trustees of the OASI, DI, and HI trust funds showed that both the OASDI and the HI programs faced serious financing difficulties and that without remedial legislation the OASI trust fund would be unable to make timely benefit payments after June, 1983.

After nearly two years of Congressional debate and consideration of several different proposals, Congress passed, and on April 20, 1983 President Reagan signed, Public Law 98-21, the Social Security Amendments of 1983. The 1983 Amendments are described in the text that follows.[2]

Universal Coverage Provisions

1. All federal employees hired after January 1, 1984, including executive, legislative, and judicial branch employees, are covered. This includes all members of Congress, the President, the Vice-President, federal judges, and most executive-level political appointees of the federal government.
2. Current and future employees of private tax-exempt nonprofit organizations are covered effective January 1, 1984 on a mandatory basis.
3. States are prohibited from terminating coverage of state and local government employees if the termination has not gone into effect by April 20, 1983.

Benefit Computation

1. The July, 1983 COLA was delayed until January, 1984. Future automatic COLAS are effective on a calendar year basis, with the increase payable in January rather than July of each year.
2. Beginning with the December, 1984 Social Security benefit increase payable in January, 1985, future automatic increases will be limited to the lesser of the increase in wages or prices if the ratio of combined OASI and DI trust fund assets to estimated outgo falls below a given percentage. The "triggering" trust fund percentage is 15 percent through December, 1988 and 15 percent thereafter.
3. For workers who are first eligible after 1985 for both a pension based on noncovered employment and Social Security retirement, a different method of computing the Social Security benefit will apply.

---

2. *Social Security Bulletin* (July 1983), p. 25–40.

4. Beginning in 1990, the earnings test benefit reduction will decrease from $1 for each $2 of earnings over the annual exempt amount to $1 for each $3 of excess earnings for individuals who attain full-benefit retirement age. (In 1987, pensioners below age 65 could earn $6,000 with no cut in benefits; those 65 to 69 could earn $8,160; and there was no limit for those over 70.)
5. The delayed retirement credit (DRC), payable to workers who postpone retirement past the full-benefit retirement age (currently age 65) and up to age 70, will be gradually increased.

Income Tax Treatments of Benefits

Beginning in 1984, up to one-half of Social Security benefits received by taxpayers whose incomes exceed certain base amounts will be included in taxable income. The base amounts are $25,000 for a single taxpayer, $32,000 for married taxpayers filing jointly, and zero for married taxpayers filing separately.

Revenue Measures

1. Changes in Social Security tax rates and allocation of tax income: Table 7-4 shows the Social Security tax rates for employees and employers under Public Law 98-21 and under prior law. The previous scheduled tax increase for 1985 was shifted to 1984, and a part of the scheduled increase for 1990 is to take effect in 1988.
2. Social Security tax credits and deductions: For 1984 only, Public Law 98-21 provided a credit for employees against their Social Security tax liability of 0.3 percent of their wages.

Financial Effect

Under the economic assumptions used for the 1983 Amendments, Public Law 98-21 is expected to provide a total of $166.2 billion in additional revenues or reduced cost during the period 1983-1989 and provide a surplus in each year through 1992.

Mechanisms to Ensure Continued Benefit Payments — "Fail Safe"

1. The new law establishes accounting procedures for crediting the OASI, DI, and HI trust funds at the beginning of each month with estimated revenues for the entire month.
2. Interfund borrowing: Authority for interfund borrowing among the OASI, DI, and HI trust funds is reinstated and extended for calendar years 1983–87 with the provision for repayment of the

Table 7-4

Social Security Tax Rates under Public Law 98-21 and under Prior Law

Year	Employer and Employee Rates				Self-Employed Rates			
	OASI	DI	HI	OASDHI	OASI	DI	HI	OASDHI
Public Law 98-21:								
1983	4.775	0.625	1.3	6.7	7.1125	0.9375	1.3	9.35
1984	5.2	.5	1.3	7.0	10.4	1.0	2.6	14.0
1985	5.2	.5	1.35	7.05	10.4	1.0	2.7	14.3
1986–87	5.2	.5	1.45	7.15	10.4	1.0	2.9	14.3
1988–89	5.53	.53	1.45	7.51	11.06	1.06	2.9	15.02
1990–99	5.60	.6	1.45	7.65	11.20	1.2	2.9	15.3
2000 and later	5.49	.71	1.45	7.65	10.98	1.42	2.9	15.3
Prior law:								
1983	4.575	.825	1.3	6.7	6.8125	1.2375	1.3	9.35
1984	4.575	.825	1.3	6.7	6.8125	1.2375	1.3	9.35
1985	4.75	.95	1.35	7.05	7.1250	1.425	1.35	9.9
1986–89	4.75	.95	1.45	7.15	7.1250	1.425	1.45	10.0
1990 and later	5.1	1.1	1.45	7.65	7.6500	1.65	1.45	10.75

SOURCE: *Social Security Bulletin* (July 1983).

principal, with interest, at the earliest possible time but no later than the end of calendar year 1989.

3. If the Board of Trustees determines at any time that the OASI, DI, or HI trust fund reserve ratio may become less than 20 percent for any calendar year, the board must promptly submit to Congress a report recommending statutory adjustments affecting the receipts of, and disbursements from, the trust fund(s) necessary to achieve a 20 percent ratio.

CURRENT STATUS OF THE TRUST FUNDS

In July, 1986, there were over 33 million persons receiving Social Security benefits of one kind or another. The average monthly benefit for a retired worker was $481.20. For a retired worker and spouse, the average monthly benefit amounted to $729.05.

In 1985 net receipts to the OASI trust fund from tax contributions, transfers from general government revenues, and interest income amounted to $184.3 billion. During the same year, expenditures from the trust fund for benefit payments and administrative expenses amounted to $169.5 billion. There was $62.6 billion in the combined OASI, DI, and HI trust funds. The $133 billion in combined benefit payments was about 14 percent of the total federal government budget outlays. The total benefit payments were equivalent to 3.1 percent of the GNP. In 1987, the Social Security program was financed by a payroll tax of 14.3 percent—7.15 percent paid by the employer, 7.15 percent paid by the employee—on earnings up to $43,800 per year. The tax for both will increase to 7.51 percent on earnings up to $45,600 in 1988. About 90 percent of all wage and salary earners as well as the self-employed are covered by the Social Security program and are subject to its compulsory contributions (tax).

DO WE PAY TWICE FOR SOCIAL SECURITY?

Some critics of Social Security claim that participants are paying twice for their benefits, once when the payroll tax is paid and again when they receive their benefits. This happens, they say, because of the way in which Social Security contributions are funneled through the trust funds.

Monies once collected and held in the trust funds by law have to be invested in federal government securities. Thus, the initial payroll tax funds are invested in U.S. government securities and held as assets in the respective trust funds. In short, the Social Security Administration lends its reserves to the U.S. Treasury. Interest collected on these securities adds to the value of the trust funds. When the Social Security Administration needs funds for distribution, however, it must redeem the securities to have money to meet Social Security benefit payments. When the securities are redeemed, the Treasury must raise taxes, or borrow funds elsewhere, in order to pay for these securities. Thus, say the critics, we pay once again for Social Security when the government taxes us to obtain the money to pay off the securities redeemed by the Social Security Administration for the purpose of paying Social Security benefits.

Even if we were taxed directly to raise money to pay off the government securities, however, it would not be a double tax for Social Security. What actually happens is that as a result of the Social Security Administration purchasing government securities, the Treasury can levy fewer taxes. The money borrowed from the Social Security trust funds is used by the Treasury in lieu of tax funds to purchase certain goods and services. Some taxes that should have been collected are thus not collected, or the collection is postponed. Money borrowed from the Social Security trust funds may have been used, for example, to build a dam or other such project. As a result, no taxes are collected for that purpose. When the borrowed funds are paid back at the time the Treasury redeems the securities held by the trust funds, the tax we may pay to give the Treasury the money it needs to pay off the securities is a tax to pay for the dam. In short, the tax we should have paid earlier for the dam was postponed until the time when the borrowed funds had to be repaid. The true purpose of the tax at that time is to pay for the dam, not to pay again for Social Security.

WHO ARE THE DOUBLE-DIPPERS?

Much concern has been expressed in the Social Security controversy regarding the so-called double-dippers. This term refers to a person who is collecting a dual pension from the federal government, one of which is from the Social Security program.

Most persons work and pay Social Security taxes on the first dollar earned and on every dollar thereafter, within the Social Security wage base, until retirement. If the person retires or quits early, he or she can collect no pension benefits until age 62. Upon official retirement, the participant collects only one federal pension—Social Security.

On the other hand, employees of the federal government hired before January 1, 1984, excluding military personnel, did not contribute to the Social Security System because most federal employees are covered by their own pension system. The federal employee, however, may get into the Social Security System by holding a second job part time (moonlighting) that is covered by Social Security. Another way a federal employee may attain coverage is by working again after retirement. Many of them do this since they frequently retire at an earlier age than 65. They can collect their federal pension and at the same time be working on a Social Security–covered job. However, the regular worker covered by Social Security is restricted from working and collecting Social Security benefits simultaneously by the earnings test. Moreover, when the federal employee retires again at age 62, 65, or later, Social Security benefits can be collected in addition to the federal pension—thus, the concept of the double-dipper.

Double-dipping would not be a serious problem except that Social Security benefits are weighted in favor of those at the low end of the income scale. This is done on the theory that the lower-income workers are poor people who had limited earnings during their work lives and who need a little more help upon retirement. Social Security benefits are proportionately higher for low-income workers. The double-dipper who worked for only a few years, or at part-time work, shows up on the Social Security records as having had limited wage earnings. Consequently, a windfall is received as a result of having minimal Social Security coverage. A post-retirement federal employee who has ten years of coverage under Social Security and has a generous federal pension will show up on the Social Security rolls as poor, when in fact that is not true.

It has been estimated that if double-dipping were terminated, it would save the Social Security System more than $1 billion per year. There are several ways to eliminate this problem. One way is to bring all federal employees under the Social Security System. The 1983 Amendments provide for all newly hired federal employees to become part of the System and pay the tax. A second method would be to exempt retired federal employees from Social Security coverage whenever they moonlight or begin a second career after retirement. Although not currently done, this could be accomplished by exempting them from Social Security taxes on secondary jobs. A third way (currently being used) is to establish a method of computing Social Security benefits that results in lower benefits for persons who also receive pensions based on earnings from noncovered employment.

CONCLUSION

When the Social Security program was first inaugurated in 1935, it was intended that the program would be self-financing and operated on an actuarially sound basis. Contributions from payroll taxes were expected to exceed benefit payments in the early years and result in an accumulation of assets in the various trust funds.

A retiree's benefits were to be closely related to prior work-life earnings, except for special treatment given to those receiving a minimum benefit. Moreover, it was stated clearly that the purpose of the Social Security System was to provide partial aid for beneficiaries. Social Security was not originally designed nor intended to provide complete or full living costs for retirees. Many individuals in the early years thought, too, that their payroll taxes plus their employers' contributions were going into individually numbered accounts, out of which they would receive benefits upon retirement. This was not the case, however, since the payroll taxes of all persons and employers go into general trust funds.

Amendments to the Social Security Act as early as 1939 began to change the characteristics of the Social Security program by providing that some individuals retiring early in the life of the program would receive benefits greater than the actual value of the individuals' combined employer-employee contributions and that dependents of retired workers would receive benefits without any additional payroll taxes required.

In 1950, amendments moved the Social Security System farther away from a fully funded, actuarially sound system toward the type of "pay-as-you-go" system that it is today. Today, those who are currently working pay for the benefits of those who are retired.

Over the years, additional workers have been brought under the Social Security System and benefits have been increased periodically by Congress. In 1972, amendments provided for adjusting benefit levels with changes in the Consumer Price Index so that benefits would rise automatically with price changes. In addition, the maximum taxable earnings base was indexed roughly to consumer prices. Therefore, it also increased over time. This provided a "coupling effect" or a double adjustment for inflation.

All these changes put a strain in recent years on the trust funds, which have accumulated over the past 30–35 years. Depletion of the trust funds and eventual deficits were projected for the early 1980s by the Social Security trustees. This led Congress in 1977 to make changes designed to strengthen the trust funds and prevent the projected deficits from occurring. The major changes enacted by Congress regarding financing were the increases in the Social Security tax rates and the increase in the base of covered earnings.

The changes made by Congress in 1977 removed the immediate threat that the cash benefit programs would run out of funds, corrected the obvious flaw in benefit computation, and reduced the projected average deficit over the next 75 years from 8 percent of estimated payroll to 1.5 percent.

Nevertheless the situation worsened. In 1983, because the trust funds were running out of money, additional amendments were enacted. These amendments included some expansion of mandatory coverage (especially to new federal employees), a delay and lessening of the COLA adjustment, a decrease in the earnings test benefit withholding rate, an increase in the delayed retirement credit, the imposition of income taxes on certain Social Security benefits, the authorization of interfund borrowing, and an increase in the Social Security tax rates.

Some of the long-run issues, however, are still unresolved. The differences in treatment between single- and dual-earner households; the problems between the cash benefit programs and private retirement programs; the question of whether all government employees and their own pension reserve fund ought to be brought under the Social Security System; and the major issue of whether a larger part of all of the funds for Social Security benefits should come from federal general tax revenues remain as issues to be resolved. Moreover, there is concern about the impact from the rise in Social Security payroll taxes on the status of the general economy.

QUESTIONS FOR DISCUSSION

1. Should Social Security benefits be financed wholly by general revenue funds?
2. Should all civilian employees of the federal government be forced to join the Social Security System?
3. Should the Social Security System revert to being fully funded and actuarially sound?
4. If a system of national health insurance is adopted in the United States, should it be incorporated into the Social Security System?
5. Should future Social Security contributions, if needed, be generated through increasing the tax base or the tax rate?
6. Should working spouses in dual-earner households receive their own Social Security benefits independent of the other spouse's earnings and benefits?
7. Do you think that Social Security ought to be used as a transfer payment mechanism?
8. Congress, in early 1978, raised the compulsory retirement age from 65 to 70 years. Do you think that Social Security benefits should thus start at age 70 instead of the present 65 or 62 years of age?
9. Do you favor taxing Social Security benefits as regular income?
10. Should double-dipping be eliminated? If so, how?

SELECTED READINGS

Ackley, G. "Social Security: Raise the Retirement Age and Reduce the Benefits." *Dun's Review* (February 1980).

Bandow, Doug. "Faulty Foundations of Social Security." *Wall Street Journal* (April 25, 1983).

Bladen, A. "Shocking Shape of Things to Come." *Forbes* (May 26, 1980).

Chapman, Steven H., M. P. LaPlante, and Gail Wilensky. "Life Expectancy and Health Status of the Aged." *Social Security Bulletin* (October 1986).

Ferrara, Peter J. *Social Security: Prospects for Real Reform.* Washington, D.C.: Cato Institution, 1985.

"How to Save Social Security." *Business Week* (November 29, 1982).

Juster, F. Thomas. "Social Security Entitlements: The Economics and the Politics." *Economic Outlook USA.* University of Michigan (Autumn 1982).

Kopits, G., and P. Gotur. "Influence of Social Security on Household Savings." *IMF Staff Paper* (March 1980).

Levy, M. D. "Case for Extending Social Security Coverage to Government Employees." *Journal of Risk and Insurance* (March 1980).

Martin, Linda Gray. "The Social Security System: Should You Withdraw?" *The New England Economic Review.* Federal Reserve Bank of Boston (September–October 1981).

1987 Guide to Social Security. Louisville: Meidinger, Inc. (December 1986).

Svahn, John A., and Mary Ross. "Social Security Amendments: Legislative History and Summary of Provisions." *Social Security Bulletin* (July 1983).

Weaver, Carolyn. "Understanding the Sources and Dimensions of Crisis in Social Security." *Fiscal Policy Council* (June 1981).

"What to Do about Indexing Social Security." *World Report.* First National Bank of Chicago (March–April 1981).

"Reconsidering Social Security." *Weekly Letter.* Federal Reserve Bank of San Francisco (February 28, 1986).

8

THE NATIONAL DEBT
WILL IT EVER
BE REPAID?

Large federal deficits in the past several years and a national debt in excess of $2.3 trillion have generated renewed interest in the problems of our large national debt. The debt ceiling has been raised substantially in recent years to accommodate large planned federal deficits in connection with major income tax reductions and accelerated government spending. In a six-year period, 1981–1986, the federal debt doubled from $1 trillion to more than $2 trillion, and today it is not uncommon to have an annual federal deficit of $100 to $200 billion. How did the debt arise? To whom do we owe it? How long can the debt continue to rise? Can it reach a point where it will bankrupt the nation? Is it fair or possible to pass the debt on to future generations? Why have we not paid it off? Will we ever pay it off? How much interest are we paying on the national debt? What happens when the debt comes due and the government does not have the money to pay? These are but a few of the many questions in the minds of citizens that we will try to answer here.

HISTORY OF THE DEBT

Since our nation first began, the national debt has been a source of heated debate. A brief review of the past will shed some light on the present and reveal that many of the current issues regarding the debt are not new.

The Early Years

There was much argument and opposition in 1789 to Secretary of the Treasury Alexander Hamilton's proposal for the new Union to assume the debts of the various states, as well as the debts of the Confederation. Most legislators agreed that the foreign debt should be taken care of as soon as possible, especially since the United States would in all probability become a debtor nation in its early years and would want to borrow from foreign nations. There was no such ease of agreement on what to do about the domestic debt, however.

Hamilton suggested that the current debts be disposed of by having the new government issue its own bonds in exchange for existing forms of indebtedness. He recommended that the federal government give full service to the national debt so that the government's ability to fulfill its domestic and foreign obligations would never again be questioned. He suggested that interest payments always be met fully at the designated periods and that the bonds of the federal government be sustained at a premium and not sold at a discount.

The entire plan and its funding method, according to Hamilton, would serve three purposes: (1) it would serve as a common bond to unify the people of the nation and draw the states closer together, (2) it would establish a sound basis for further credit expansions both here and abroad, and (3) the federal instruments of debt would serve as a basis of circulating capital to help alleviate the shortage of currency in the new nation. After Hamilton won his point to pay the old debts off at their original price, to pay the interest in arrears, to maintain the 6 percent interest, and to assume the state debts, the new nation began its fiscal career with a total outstanding debt of more than $77 million.

Although in the early years of our nation it was thought that the new federal debts would be gradually retired out of federal surpluses, little effort was made in this direction as government expenditures proved to be larger and revenues smaller than anticipated. In its first decade of operation, the Treasury had difficulty balancing the budget; and the national debt, instead of being reduced, had risen to nearly $80 million by the time the Federalists left office in 1801 and Albert Gallatin became the new Secretary of the Treasury under the Jefferson Administration. During the 12 years of the

Washington and Adams Administrations, $7.5 million of the original debt had been liquidated, but $10.6 million in new debt had been incurred.

In the next 12 years, the national debt was reduced by payments of more than $46 million, including repayment of practically all the foreign debt, leaving the unretired amount of debt at $34 million. But the addition of $11 million of new indebtedness for the Louisiana Purchase left a total federal debt of $45 million. Nevertheless, Gallatin was able to show a remarkable 44 percent net reduction in the national debt during the years that he managed our federal finances.

This debt reduction did not continue since the Treasury found its outlays increasing rapidly in the next few years, especially as a result of the War of 1812. The Madison Administration was forced to issue bonds at a substantial discount to raise money for the unpopular war and also to issue short-term Treasury notes. Even so, expenditures exceeded revenues and the national debt skyrocketed to a new high of $127 million by the end of the Madison Administration in 1816.

The next 21 years, 1816–1836, saw a steady and substantial reduction of the national debt. Surpluses experienced in each of these years, some of them quite large, were applied to the retirement of the national debt. Increased trade after the War of 1812 brought about an increase in government revenue from customs duties. Increased tariff rates plus large-scale sales of public land further added to federal revenues.

Reduction of the debt was so successful that during the early 1830s there was widespread discussion about what should be done with budget surpluses when the national debt would be extinguished within a few years. Secretary of the Treasury Roger Brooke Taney in his report of December, 1834 declared that on January 1, 1835 the remaining balance of the federal debt would be paid or provided for, and he stated that "the United States will present that happy, and probably in modern times unprecedented, spectacle of a people substantially free from the smallest portion of the debt."

Not only was the debt completely paid off by January, 1835, but a large surplus had accumulated. The Treasury surplus by mid-1836 approached $40 million. After much debate and controversy, a bill was finally signed by President Jackson calling for the Treasury to retain a surplus balance of $5 million and deposit any excess surplus as of January 1, 1837 with the states on a pro rata basis according to their representation in Congress. Since the surplus on that date amounted to $42 million, $37 million was available for deposit with the states. Although the distribution of the funds to the states was in the form of non-interest-bearing loans, it was never intended that the federal government would recall these loans. Consequently, they became outright gifts to the states. The distribution among the states was to be made in four installments. After making the first three installments, in

1837 the Treasury found itself in financial difficulty. The fourth installment of $9 million was postponed until 1839 but was never actually paid, and the surplus balance in the Treasury became a thing of the past never to be attained again.

Civil War Debt and Retirement

For most years until 1851, the federal government ran into deficits. One big item of expense during this period was the Mexican War. It cost approximately $64 million, of which $49 million was financed through the sale of bonds and notes. The maximum federal debt during the period between 1835 and the Civil War occurred in 1851, when it reached a level of $68 million. Thereafter, surplus years and debt repayment brought the national debt down to $28 million by 1857, but it rose to nearly $65 million by the outbreak of the Civil War.

Naturally the financing of the war, which cost an estimated $3.5 billion, necessitated an unprecedented amount of borrowing. Since $2.5 billion, or 70 percent of the war cost, came from loans, the national debt rose to an astronomical $2.77 billion by 1866, the year after the war. This made the per capita burden of the national debt about $80. The war certainly cost more than the $400 million that the President had asked for at the beginning of the conflict so that the war might be "a short and decisive one." Most of the borrowing was financed at interest rates in excess of 6 percent, except for the $450 million in non-interest-bearing U.S. notes, otherwise known as "greenbacks," issued during the period. Unfortunately, the large debt and the depreciation of the greenback, which at times had a market value as little as 40 or 50 cents, caused prices to rise tremendously.

Expansion of business activity after the war brought about increased tax revenues. In fact, many taxes imposed during the war did not begin to become effective revenue sources until after the war had terminated. In practically every year from 1866 to 1893, government receipts exceeded disbursements and the government was able to consistently reduce its debt. By the end of this period, the national debt was less than $1 billion. Although the debt rose above the $1 billion mark the following year, it remained fairly constant until World War I, fluctuating between $1 billion and $1.4 billion. In April, 1917, our net national debt stood at $1,207,827,886, or $11.59 per capita. This was a reduction of $1.5 billion from the maximum of $2.7 billion of debt existing at the end of the Civil War.

World War I Debt and Depression Spending

Financing of the First World War resulted in a twentyfold increase in the national debt, which rose to $25.4 billion by 1919. Although Secretary of the Treasury McAdoo originally hoped to finance 50 percent of the war cost through taxation, he was soon convinced that his figure was too high and lowered his estimate to 33.3 percent. With the greater-than-expected increases in government disbursements and the lag between new tax measures and the collection of revenues therefrom, the need for money was staggering; the only way of obtaining it quickly was through borrowing. An outstanding feature of the war financing was the sale of liberty bonds to the general public in small denominations, with some as little as $50. The success of the first liberty loan campaign led to others. In total, the four liberty loan campaigns plus the postwar victory loan of 1919 netted the federal government $21.5 billion.

After the war, expenditures, which had reached $2 billion per month, dwindled impressively by 1920. The prosperous Twenties, which brought sizeable surpluses, resulted in a reduction in the debt each year until 1930, when the total national debt was down to $16.2 billion, a reduction of 36 percent during the decade.

In the 1930s we witnessed the Great Depression, with the gross national product falling from $103 billion in 1929 to $56 billion in 1933, while unemployment increased from 1.6 million to 12.8 million. With one-fourth of the labor force unemployed and many more partially employed, with homes being lost through mortgage default and businesses collapsing through bankruptcy, the federal government began to use the federal budget as a stabilization device. Deficit spending on public works and relief became the order of the day. New Deal deficits ran as high as $4.4 billion in 1936 and totaled more than $26 billion during 1932–1940. As a result, the national debt reached a new peak of $43 billion in 1940.

World War II and Subsequent Debt

Our entry into World War II in December, 1941 occasioned increases in our national debt. The cost of the war at that time was considered stupendous. In fact, toward the last few weeks of the war it was estimated that we were spending $1 billion a day on the war effort. The total cost of the war is estimated at $288 billion. Although some of this cost was defrayed through extended and higher taxes, approximately three-quarters was financed through borrowed funds. Deficit budgets ran in excess of $45 billion a year in the last three years of the war. The national debt surpassed the $100 billion mark in the second year of the war; reached $200 billion during 1944; and by February, 1946, reached a staggering $280 billion figure. This was a burden of $2,000 per capita. The war financing program was known for its sizeable increase in direct taxes, the widespread use of withholding taxes, the levying of excess profit taxes, the advent of war bond campaigns, and a large program of lend-lease to foreign nations.

Although the debt was subsequently reduced to $252.2 billion by 1948, no substantial effort has been made to pay off or even to reduce the national debt since the end of the war. In fact, in the more than 40 years that have elapsed since that time, we have had budget surpluses in only 8 years. Deficits incurred during the period of the Korean conflict, plus the $12.9 billion deficit incurred in connection with the recession of 1958, moved the debt toward the $300 billion mark. The sizeable deficits of the two decades after World War II, and the even greater deficits of the past few years, as shown in Table 8-1, pushed the total debt up to the current figure of over $2 trillion; it is expected to reach $2.5 trillion by the end of fiscal year 1988.

The last budget surplus—a modest $3.2 billion—occurred in fiscal 1969. With the use of deficit spending to stimulate the economy during the 1974–1975 recession, the deficit rose by over $50 billion. Deficits of $100–$200 billion in the years 1982–1987 stirred much controversy as they pushed the national debt beyond $2 trillion.

DEBT CEILING

The statutory limit or ceiling on the national debt was first established in 1917 when Congress passed the Second Liberty Bond Act. This Act authorized the Treasury to issue bonds not to exceed $7,538,945,460 and to issue certificates of indebtedness up to $4 billion. As borrowing to finance World War I continued beyond expectations, Congress merely amended the Second Liberty Bond Act to accommodate new debt authority as needed. After the war the same procedure continued, as it has carried on to the present day. Thus, the rise in the national debt ceiling occasioned by Depression spending of the 1930s, World War II, the Korean conflict, the war in Vietnam, and intermittent peacetime deficits was permitted through extensions of the Second Liberty Bond Act. During World War II, however, there was some change in structure in the debt ceiling. Prior to that time Congress set individual ceilings on the various types of government indebtedness. But in 1941 it abolished the individual debt ceilings and created one ceiling on the total debt outstanding.

This is the type of debt ceiling we have today. The debt ceiling has been raised dozens of times in the past 33 years, increasing from $275 billion in 1954 to a projected $2,500 billion in fiscal 1988. On a number of occasions in the past decade, proposals have been made to eliminate the statutory ceiling on the national debt. Arguments for and against the debt ceiling abound. In 1969 the Nixon Administration tried to get around the ceiling problem by proposing that certain debts, especially those held by government agencies, be removed from the statutory limitation. This would have left the Treasury with ample flexibility to contract more debt without having to request Congress to raise the debt ceiling. This proposal, however, was not approved by Congress. The Treasury asked that the ceiling be eliminated, and in June, 1969, repeal of the ceiling was recommended by a group of 67 leading academic economists, but to no avail.

Arguments Against the Debt Ceiling

Many arguments can be marshaled for and against the debt ceiling. Opponents of the ceiling maintain that it may at times limit needed expenditures on important government programs, such as defense or depression spending, whenever tax revenues are not up to expectations or the government has failed to increase taxes sufficiently to meet its spending obligations.

It is claimed also that a debt ceiling results in fiscal subterfuge by the Treasury. The statutory limit is on a defined portion of the total federal debt that is usually associated with the annual federal budget. The federal government, however, has many nonbudgetary financial obligations. Many federal agencies, such as the Commodity Credit Corporation, the Federal National Mortgage Association, the Postal Fund, and the Rural Electrification Fund, which normally borrowed funds from the Treasury, are now empowered to borrow from private financial institutions and investors if they desire. Frequently when the Treasury is pinched for funds and is approaching the debt limit, it will request a particular agency to borrow in the financial markets rather than borrow from the Treasury. This, of course, relieves the Treasury of the task of borrowing the funds and avoids raising the national debt to service the agency.

Most of this off-budget financing is done through the Federal Financing Bank, which raises money for other agencies. For fiscal year 1986 these agencies borrowed $16.7 billion. Their borrowing is estimated to grow to $51.7 billion in 1989.

Critics of the debt ceiling contend further that it restricts the freedom of the Treasury to manage the debt efficiently, especially when the debt is close to the ceiling. In such circumstances the Treasury may have to wait until old securities mature before issuing new ones for fear of going over the debt ceiling. Critics of the ceiling argue that it would be better for the Treasury to experiment with new issues sometime before the expiration of the old to test the interest rate and have time to make any necessary adjustments to obtain the best price. Otherwise the Treasury will be at the mercy of the market if it must wait until the day that old issues expire before issuing new securities to replace them. Frequently it will have to pay a higher interest rate.

Table 8-1
Federal Budget Receipts and Outlays for Fiscal Years 1929–1988
(Billions of dollars)

| Fiscal Year | Budget | | Surplus or Deficit (−) | Off-Budget Outlays | Gross Federal Debt |
	Receipts	Outlays			
1929	3.9	3.1	0.7	—	16.9
1933	2.0	4.6	−2.6	—	22.5
1939	5.0	8.8	−3.9	—	48.2
1940	6.5	9.5	−3.1	—	50.7
1941	8.6	13.6	−5.0	—	57.5
1942	14.4	35.1	−20.8	—	79.2
1943	23.6	78.5	−54.9	—	142.6
1944	44.3	91.3	−47.0	—	204.1
1945	45.2	92.7	−47.5	—	260.1
1946	39.3	55.2	−15.9	—	271.0
1947	38.4	34.5	3.9	—	257.1
1948	41.8	29.8	12.0	—	252.0
1949	39.4	38.8	0.6	—	252.6
1950	39.5	42.6	−3.1	—	256.9
1951	51.6	45.5	6.1	—	255.3
1952	66.2	67.7	−1.5	—	259.1
1953	69.6	76.1	−6.5	—	266.0
1954	69.7	70.9	−1.2	—	270.8
1955	65.5	68.5	−3.0	—	274.4
1956	74.5	70.5	4.1	—	272.8
1957	80.0	76.7	3.2	—	272.4
1958	79.6	82.6	−2.9	—	297.7
1959	79.2	92.1	−12.9	—	287.8
1960	92.5	92.2	0.3	—	290.9
1961	94.4	97.8	−3.4	—	292.9
1962	99.7	106.8	−7.1	—	303.3

Year					
1963	106.6	111.3	− 4.8	—	310.9
1964	112.7	118.6	− 5.9	—	316.8
1965	116.8	118.4	− 1.6	—	323.2
1966	130.9	134.7	− 3.8	—	329.5
1967	148.9	157.6	− 8.7	—	341.3
1968	153.0	178.1	− 25.2	—	369.8
1969	186.9	183.6	3.2	—	367.1
1970	192.8	195.7	− 2.8	—	382.6
1971	187.1	210.2	− 23.0	—	409.5
1972	207.3	230.7	− 23.4	—	437.3
1973	230.8	245.6	− 14.8	0.1	468.4
1974	263.2	267.9	− 4.7	1.4	486.2
1975	279.1	324.2	− 45.2	8.1	555.1
1976	298.1	364.5	− 66.4	7.3	631.9
1977	355.6	400.6	− 44.9	8.7	709.1
1978	399.6	448.4	− 48.8	10.4	780.4
1979	463.3	491.0	− 27.7	12.5	833.8
1980	517.1	590.9	− 73.8	14.2	914.3
1981	599.3	678.2	− 78.9	21.0	1,003.9
1982	617.8	745.7	− 112.9	17.3	1,147.0
1983	600.6	808.3	− 207.8	12.4	1,381.9
1984	666.5	851.8	− 185.3	16.2	1,576.7
1985	734.1	946.3	− 212.3	14.8	1,827.5
1986	769.1	989.8	− 220.7	16.7	2,112.0
1987[a]	842.4	1,015.6	− 173.2	19.5	2,320.6
1988[a]	916.6	1,024.3	− 107.2	39.7	2,509.0
1989[a]	976.2	1,069.0	− 98.2	51.7	2,684.3

[a]Estimate

SOURCE: *Economic Report of the President, 1987* and *The United States Budget In Brief, 1988.*

Arguments for the Debt Ceiling

Proponents of the statutory limit stress the fact that the debt ceiling is needed to restrain government spending and that it prevents the national debt from becoming dangerously high. Although the debt ceiling has been raised liberally by Congress in the past several years, the presence of the debt ceiling does tend to make Congress look a bit closer at the budget and decide whether it really wants to vote for appropriations that will necessitate borrowing and raising the debt ceiling.

It is frequently pointed out that since the interest on the national debt exceeds $135 billion annually the debt is costly to the American taxpayers. It might also be argued that insofar as the ceiling limits deficits in the annual budget, it makes the taxpayers more conscious of the total cost of government services. Many taxpayers may not balk at government expenditures of $1,050 billion when taxes are scheduled to be $950 billion; but if they were taxed $1,050 billion to cover the total cost of government spending, they might very well decide to do without some government services. In short, deficits and a rising national debt can mislead the taxpayers about the true cost of government services.

INTEREST RATE CEILING ON NATIONAL DEBT

In addition to a statutory limit on the national debt, there is also a ceiling on the interest rate that may be paid on government securities. This interest rate ceiling also was originally established by the Second Liberty Bond Act of 1917, which authorized the Treasury to issue certain amounts and kinds of bonds at interest rates up to 4 percent. Later the Third Liberty Bond Act raised the ceiling to 4.25 percent to ensure a successful sale of that issue. From the early 1920s until the 1950s, the cost of federal borrowing was far below the legal interest rate ceiling. With the boom and inflationary conditions of the mid-1950s, however, the interest rate ceiling became a handicap to the Treasury. With interest rates of 15–20 percent in the late 1970s and early 1980, the problem became more pressing. Congress did remove the interest rate ceiling on U.S. savings bonds, which are often purchased by individuals as a method of personal saving.

Whether the interest rate ceiling is good or bad is a matter of judgment. Proponents of the ceiling maintain that it keeps down the cost of federal borrowing. Opponents of the interest ceiling claim that it prevents the Treasury from being competitive in the market for funds. Furthermore, since there is no ceiling on short-term government obligations, only on long-term bonds, critics of the ceiling maintain that in a competitive market in which the going rate of interest in the long-term market is higher than the interest rate ceiling, it can force the Treasury into the short-term market where it will have to pay an even higher interest rate than it would in the long-term market.

The elimination of the interest rate ceiling has been an issue for years and again became a debatable issue in Congress in the 1960s with the general rise in interest rates throughout the economy. In 1969, for example, the federal Treasury was paying interest rates of 8 percent or more for money it borrowed in the short-term market. With the 4.25 percent ceiling, it was almost impossible to compete for funds in the long-term market against high-grade corporate securities paying 7–8 percent interest. At that time the Nixon Administration requested that Congress review the 4.25 percent ceiling on long-term government bonds. The problem eased somewhat in 1971 when Congress authorized the Treasury to issue as much as $10 billion in bonds at rates of interest exceeding 4.25 percent. But problems with debt financing arose again in the late 1970s when interest rates were rising, ultimately reaching a range of 15–20 percent. The question of the interest rate ceiling has again been highlighted in the 1980s since the Treasury has the problem of financing new deficits in excess of $100 billion annually as well as refunding another $100 billion or more of maturing debt.

STRUCTURE OF THE NATIONAL DEBT

The national debt is composed of various types of government obligations and securities, both long term and short term. These various types of obligations, and their dollar amounts, are shown in Table 8-2.

Treasury Bills

About 19 percent of the debt is made up of Treasury bills, which are short-term securities with 3-month to 12-month maturity dates. Bills issued as frequently as each week are sold to obtain funds to retire portions of the debt that may become due at a given time. These bills, such as the 91-day variety, sell at current interest rates and are sought by financial institutions as a short-term outlet for their reserve cash.

Although Treasury bills have no stated interest rate or par price, they are sold by the Treasury through sealed-bid auctions in which investors are able to submit written bids at prices of their own choosing. Individual bidders may submit multiple prices; that is, they may offer to buy different amounts of an issue for different prices. Once the bids are received, they

Table 8-2

Total Debt by Type of Security, March, 1986

	Amount (Billions)		Percentage of National Debt
Total gross debt		$2,059.3	100.0%
Marketable total		1,498.2	72.7
Bills	$396.6		19.1%
Notes	869.3		42.3
Bonds	232.3		11.3
Nonmarketable total		558.5	27.2
State and local gov't series	98.6		4.8
Savings bonds and notes	82.3		4.0
Gov't account series	372.3		18.1
Foreign issues	5.3		0.3
Non-interest-bearing debt		2.6	0.1

SOURCE: *Federal Reserve Bulletin* (January 1987).

are listed in order of decreasing price and filed accordingly. Beginning with the highest price offered or bid, the Treasury sells to the successively lower bidders until the issues are all sold. Since individual bidders may pay different prices, the yield for each of them will be different, with the lowest bidders, of course, obtaining the best yield. Some of the bills, however, are reserved to be sold on a noncompetitive basis to small buyers. These bills are sold at the average price of those sold on the competitive-bidding basis. Since the price remains rather stable and they can be disposed of readily, they are generally considered to be near-money by holders.

Treasury Notes

Investors who desire a higher return but are willing to wait longer for maturity dates can purchase a second form of government obligation, Treasury notes, which may run as long as five years and make up about 42 percent of the national debt. At times both of these forms of securities, bills and notes, are sold as tax-anticipatory issues. The Treasury will sell them to obtain funds to tide it over until it receives anticipated tax revenues in the immediate future. Businesses frequently will buy such issues with reserves they may be accumulating for tax payments in the immediate future. This gives the taxpayer the opportunity to earn some interest on the money being held in anticipation of tax payments. Furthermore, when the tax payment deadline arrives, the taxpayer's tax obligation may be discharged by turning in the securities to the government.

Treasury Bonds

About 11 percent of the national debt is in the form of long-term obligations known as Treasury bonds. These are marketable bonds payable to the bearer at a particular maturity date. This means that they can be bought and sold by persons other than the original holders. These bonds have maturity periods of 5, 20, and even 40 years. One issue of these bonds paying 4.25 percent interest matures in 1993; still another issue paying 3.5 percent does not mature until 1998.

These bonds are sold at a fixed price and a stated interest rate. One disadvantage of this method, compared to the auction technique of selling short-term government securities, is the fact that the Treasury has to set an interest rate or yield on the bond. If for some reason the yield is set too low, the Treasury runs the risk of a financial failure if investors do not purchase the bonds. On the other hand, if the Treasury sets the interest rate too high, it may pay an unnecessarily generous rate of interest to the investors. Holders clip off the coupons attached to the bonds and periodically turn them in to the Treasury, or a designated agent, in exchange for interest payments. Although they are issued and sold at a fixed price and a stated interest rate, the market price of the bonds fluctuates as they are bought and resold in the bond market.

Usually when market interest rates are higher than the coupon rate, these bonds will sell in the market at a discount. This in effect raises the real rate of interest or the yield on the bonds. When market interest rates are lower, the bonds will sell at a premium, which in turn lowers the yield on the bonds. The holder at maturity date, however, is always paid the full face value of the bond. Consequently, the resale market price of the bond will bear some relationship to the proximity of the maturity date of the bond. As a result of the relatively low interest rate ceiling on these bonds, in recent years their sale has been declining. They now comprise about 11 percent of the federal debt, compared to one-third of the debt in 1965.

Savings Bonds

Another instrument of national debt is the savings bond, issued in large amounts during and since World War II. They are often purchased by individuals using payroll-deduction plans. Savings bonds make up about 4 percent of the total debt. These are nonmarketable securities since they can be cashed only by the person to whom they are issued. Series EE bonds are purchased at a price less than maturity value, and interest accumulates in the form of the higher value on the bond as it approaches maturity. Series EE bonds have 10-year maturities, but they can be redeemed for cash at any time after 60 days, either at the Treasury or through banks. Series HH bonds are purchased at face value and pay interest semiannually. The interest is sent by check directly to the bondholder. They have a 10-year maturity.

Government Account Series

Another important part of the debt is the government account series, issues of which are reserved for government agencies and trust funds. These are issued to federal government investment accounts and comprise about 18 percent of the federal debt. The Federal Old-Age and Survivors Insurance trust fund, for example, has certain revenues from the collection of Social Security taxes that it does not need immediately. These are deposited with the Treasury in exchange for special-issue government obligations. When the Social Security Administration needs its funds, the government obligations are redeemed by the Treasury and the cash is transmitted to the trust fund. This arrangement provides a source of funds for the Treasury and, at the same time, a source of investment for the trust fund. Interest receipts on the special issues, of course, help to pay some of the cost of Social Security.

Foreign Series

Two types of special nonmarketable securities have been issued by the Treasury to foreign governments and monetary authorities. One is a dollar-denominated security designated "Foreign Series." The other is the "Foreign Currency Series," which is denominated in the currency of the country or institution making the purchase. Currently about $5 billion of the foreign series (dollar-denominated) is still outstanding.

Maturity Dates of the Debt

Another interesting aspect of the national debt, especially in regard to repayment or refunding, is the maturity dates of the existing debt. As of April, 1987, for example, more than 44 percent of the total debt was due within 1 year. Another 33 percent was due in 1–5 years. The remainder was stretched out over a 20-year period. In addition to the marketable securities, there is $82 billion in savings bonds that can be redeemed at the will of holders.

Most of the debt is refunded when due, rather than paid off, either by offering existing debt holders new securities in exchange for the old or by selling securities to new debt holders to pay off the old. To this extent, it means that debt payments and reduction are merely postponed. Many of the new issues offered to replace the old are obligations of one year or less. Consequently, the Treasury is faced with a perennial problem of refunding a large portion of the national debt.

FEDERAL BUDGET AS A STABILIZER

If used properly, debt policy can help to stabilize the economy and modify business cycles. A surplus budget can help prevent inflation during a prosperity period, and a deficit budget can help boost the level of economic activity during a depressionary period. Use of a budgetary policy is shown in Figure 8-1.

Balancing the Budget over the Cycle

When the budget is used as a tool for economic stabilization, it is desirable to balance the budget over the period of the cycle. To accomplish this, it would be necessary for the surplus of prosperity to equal the deficit of the recession. However, this would be difficult to accomplish. A question might arise as to whether we should start such a practice by building up a surplus during prosperity and then spending it during the next recession, or whether we should incur the deficit during the recession and then repay the debt with the surplus obtained during the subsequent prosperity.

Assuming that the surplus-first method was utilized, a second problem would arise. How much surplus would need to be accumulated during

Figure 8-1

Budget Used as an Anticyclical Device

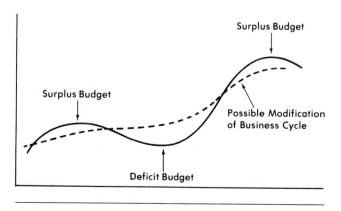

The National Debt—Will It Ever Be Repaid?   **213**

the prosperity period? This would depend not only upon the inflationary pressures of prosperity but also upon the estimated deficit that would occur during the subsequent recession in the economy. It is practically impossible, however, to determine what the duration and the intensity of the prosperity will be, let alone the duration and the intensity of the subsequent recession. Therefore, it is usually suggested that it is more feasible to run the deficit first. This also has its weaknesses. How can we be assured that the subsequent prosperity will be long enough or strong enough to permit the accumulation of sufficient surplus to pay off the deficit incurred in the previous recessionary period?

Another weakness of the deficit-first method is that most administrators and legislators are willing to use deficit spending during a recession to help alleviate unemployment, but many of them are reluctant to support measures to accumulate the offsetting surplus during prosperity years. From a political point of view, the emphasis is often on tax reduction rather than higher taxes during a prosperity period, especially if there is a surplus in the federal budget. In short, deficit spending when used to bolster the level of economic activity can be very popular with the public during the recession, but increased taxes and budget surpluses to combat inflation during a prosperity period are seldom popular.

It should be remembered also that a surplus acquired during a prosperity period should be held in reserve for best results. It can be used to pay off the debt subsequently, but not until the level of economic activity begins to decline. If the surplus obtained during prosperity is used immediately to pay off the debt incurred by the deficit spending of the recession, it will merely result in putting back into the economy an amount of money equivalent to the surplus. Thus, the reduction of spendable incomes through taxation will be offset by government expenditures plus debt repayment.

This means that the total spending of the economy will remain the same, provided the recipients of debt repayment spend or invest the money received from the government. In such a case, the surplus budget will have a neutral effect instead of being anti-inflationary. The better practice would be to hold the surplus funds until economic activity begins to decline. Repayment of the debt at such a time could give a boost to the economy when recipients of debt repayment spend or invest these funds.

The Full-Employment Budget

In recent years there has been a tendency to look at the *full-employment budget* instead of the actual budget for the purpose of analyzing the fiscal effects of the budget. Regardless of the state of the actual budget, the full-employment budget is a measure of the potential government revenue and expenditure that would result from full employment. It is said by some that the actual budget may be misleading. Let us say, for example, that the existing budget shows a deficit (or fiscal stimulus) of $30 billion and that the rate of unemployment is 6 percent. Projection may indicate that if the economy were at full employment (5 percent unemployment or less), the budget would show a surplus of $8 billion. Thus, if the economy expands toward the full-employment level, it will encounter a fiscal drag, which will impede the attainment of a full-employment objective.

Proponents of "functional finance," who look at the budget as a tool of stabilization and growth rather than as something to be balanced annually or even periodically, would contend that the inherent drag of the full-employment surplus should be eliminated if the economy is to attain its goal of full employment. This, of course, could be accomplished by increasing the size of the current fiscal stimulus (deficit) by reducing taxes or increasing government spending.

Carrying the analysis one step further, proponents of this theory claim that once full employment and a balanced budget have been reached, care must be taken to prevent the development of a subsequent surplus (fiscal drag on the economy). It is pointed out that with a given tax rate, total revenues will increase by $20 billion or more annually as a result of the normal forces of growth in our economy. To prevent this from occurring, it is suggested that a *fiscal dividend* be declared in the form of either a tax reduction or an increase in federal spending. Advocates of the full-employment balanced budget approach place little emphasis on the notion that surpluses should be accumulated during prosperity to offset the deficits of previous periods.

An example of this type of budgeting was used by President Carter in presenting his budget for 1980. It showed deficits (fiscal stimuli) for 1980 and 1981 with a surplus of $4.8 billion for 1982, resulting from the two-year fiscal stimulus.

Although President Carter's budget missed the target, President Reagan used a similar type of budgeting. Starting with an assumed deficit of $54.5 billion for 1981, budget deficits were to decrease to $45 billion and $23 billion respectively in the next two years. Finally, a surplus of $0.5 billion was to result in 1984 followed by larger surpluses in 1985 and 1986. Unfortunately the deficit mushroomed to $208 billion in 1983. The budget for 1984 showed a deficit of $185 billion but again sustained deficits over $200 billion both in 1985 and 1986. This type of budgeting, of course, assumes that Congress will go along with the budget and that everything will develop as planned with no substantial changes in the economy other than those forecast. Thus far, however, it appears that the concept of full-employment budgeting has not been successful.

PROBLEMS OF THE DEBT

Our experience with budgetary policy as a means of stabilizing business activity is rather limited. It is difficult, therefore, to determine whether we can time deficits and surpluses accurately and have them of proper size to act as stabilizers of the economy. Furthermore, we have not had sufficient experience in the past 40 years to determine whether, in the absence of emergencies, the deficits and the surpluses can offset each other sufficiently to prevent a growing debt. We incurred a sizable debt during the Depression of the 1930s as a result of our deficit-spending program. Without having had a chance to diminish this debt, we entered World War II, which increased the debt to about $280 billion.

Our opportunity to reduce the debt was further hampered by the outbreak of the Korean conflict in 1950. Since then we have made very little headway in reducing the debt. In fact, the national debt, now in excess of $2 trillion, has grown to such proportions that it presents several problems. Bankruptcy, redistribution of income, debt burden, size of the debt, refunding, and the burden of interest payments are a few of these problems.

Bankruptcy

The average person often thinks that the debt may become so large that it will bankrupt the nation. It is commonly believed that the government may get into a situation where it will be unable to pay the debt. This misunderstanding arises from the failure to distinguish clearly the true nature of government financing compared to the normal method of business financing. When the government borrows and repays funds, it is more like the financial transactions taking place within a family than the type of financing practiced by private enterprises.

Comparison to Business Debt. From an accounting point of view, whenever a business has total financial liabilities in excess of total assets, it is insolvent. In short, it does not have sufficient cash to pay its current debts in the immediate future. When creditors press for payment, the firm may voluntarily and legally have itself declared bankrupt or the creditors may force the company into bankruptcy. In either case, the court will decide whether the business should continue under receivership, that is, under a court-appointed manager, or whether the assets of the company should be liquidated to pay off the creditors. Whenever the firm pays off its debt, it decreases the total assets of the company. Money paid out actually leaves the firm, thereby reducing the assets.

Comparison to Family Debt. The national debt is more like an internal family debt than a business debt. Consider the family as a spending unit, and suppose that a daughter borrows $600 from her dad over the period of the school year and that she intends to repay it from the money she earns from summer employment. When the daughter borrows, she does so within the family unit. Likewise, when she repays the $600 in the summer, the money remains within the family.

When the daughter pays her debt, her individual assets are decreased by $600, but her father's assets are increased by $600. Therefore, the net assets of the family remain the same. Unlike the debt repayment of the firm, in the family situation no money leaves the family as a result of the debt repayment. There is merely a transfer of cash (assets) from one member of the family to another. There is no net reduction of assets, nor is there any money leaving the family.

The Federal Debt. When the government borrows money, it borrows primarily from individuals, businesses, and banks within the economy. When it makes repayment on the debt, the money stays within the economy. There is no reduction in the total assets of the economy when the government makes repayment on the debt. Furthermore, the government's ability to repay is governed by the total assets of the economy or, more immediately, by the total income of the economy and the government's ability to tax. For example, the national debt in 1986 was $2,112 billion. Considering that the gross national product (GNP) for 1986 was about $4,200 billion and that the total national income was about $3,500 billion, it is easy to see that the total income of the nation was sufficient to take care of debt repayment if the government decided to raise taxes sufficiently to obtain the funds required to pay it off. Theoretically, but unrealistically, the government could tax a sufficient amount to pay the debt off in the course of one year. If the government were to do this, it would not in any way reduce the total income or assets of the nation as a whole. The taxation and repayment of the debt would merely cause a redistribution of income, or cash assets, inside the economy. The income given up by individuals and firms in the form of taxes would be offset by payment to those holding the debt. Thus, the total income or assets of the economy would be the same after payment of the debt as before. The major difference is that income and cash assets held by various individuals and firms would be changed. It must be pointed out, however, that foreigners now hold 11 percent of the national debt compared to less than 5 percent as late as 1970.

Although a tax rate sufficient to pay off the debt in one year would be prohibitive, certainly over a relatively long period the government could operate at a surplus sufficient to pay off the debt. Surpluses obtained during periods of prosperity could be used to pay the debt during periods of contraction in the economy. As far as ability to pay is concerned, in a ten-year period, for example, a national income of over $45 trillion (in 1986 dollars) would be available to repay a $2.1 trillion debt.

Effect on Redistribution of Income

The question naturally arises: Why does the government not take more positive steps to pay off the debt? Reluctance to reduce the debt by sizeable amounts stems not only from the fact that the larger tax burden necessary to do so would be politically unpopular, but also from the fact that it would cause disruptive economic repercussions. One important problem involved would be the redistribution of income brought about by repayment of the debt.

If the debt were to be paid off on a large-scale basis, heavy taxes would reduce total effective demand, especially among the lower-income groups. Whether such reductions in effective demand would be offset when the government used tax money to pay off the debt would depend on what the recipients of debt repayments would do with the money they received. Since it is quite probable that the total propensity to consume or to invest of the debt holders who receive repayment would be less than that of the taxpayers in total, the net effective demand of the economy could easily be reduced by repayment of the debt. The possibility of this occurring becomes evident when we look at the ownership of the debt. It is generally agreed that the lower-income groups do not hold much of the federal debt. It is held primarily by banks, businesses, government agencies, and individuals in the higher-income groups. This is shown in Table 8-3.

Table 8-3

Ownership of the U.S. National Debt, 1986 (Percentage)

U.S. government agencies and trust funds		18.1%
Federal reserve banks		8.9
Private investors		73.0
Commercial banks	9.6	
Money market funds	1.1	
Insurance companies	4.8	
Other companies	2.9	
State and local governments	11.5	
Individuals	7.6	
Foreign and international	11.6	
Miscellaneous[1]	23.9	
TOTAL		100.0%

[1]Includes savings and loan associations, nonprofit institutions, credit unions, corporate pension funds, and certain U.S. government deposit accounts.

SOURCE: *Federal Reserve Bulletin* (January 1987).

Of course, if the debt holders would spend the income they received at the time the debts were repaid, it would not have an adverse effect on the economy. This would be the case if the debts were repaid during a full-employment period. It would be best, however, to pay off the debt during the periods of less than full employment with the money obtained through taxation during a prosperous or inflationary period. In this way, the debt could be used as a tool for economic stabilization.

Burden of the Debt

It is often thought that when the debt is not paid during the period in which it is incurred, the burden of paying the debt is passed on to future generations. The extent to which this may be true depends upon whether we are considering the effect on the total economy or on individuals and firms; it also depends upon the source of the debt.

Effect on the Total Economy. If we are considering the total economy, it is impossible to pass the real cost of the debt on to future generations. The real cost of the debt to the total economy can be measured by the cost of goods and services that individuals and firms must forego when they give up their purchasing power to buy government bonds. When consumers and investors purchase such bonds, they not only buy fewer goods and services, but they also give the government revenue to make its purchases. For example, during World War II citizens and firms gave up the purchase of automobiles, homes, food, clothing, machinery, raw materials, and the like when they purchased bonds. In the meantime the government, with its borrowed purchasing power, bought tanks, planes, ships, ammunition, and other necessary war materials. The decrease in consumer production was, in effect, the real cost of the debt. The people in the economy at the time the debt was incurred shouldered the real burden of the debt through the loss of goods and services.

For the economy as a whole the debt repayment, whether repaid immediately or postponed until future generations, does not cost anything in terms of goods and services. As a result of the redistribution of income that takes place at the time the debt is repaid, some individuals and firms may suffer a loss of purchasing power; but this will be offset by gains to others, and no net decrease in purchasing power in the economy will take place. For example, if the debt were to be paid even in a period of one year, the total tax necessary to pay off the 1987 debt would be well over $2 trillion. This would decrease the purchasing power of taxpayers. It would reduce effective demand and result in decreased production. When the government paid out the $2 trillion to debt holders, however, it would tend to offset the adverse effect of the tax. Total purchasing power in the economy would remain the same. The effective demand, and therefore production, would remain the same, provided the propensity of the debt holders to consume and to invest was the same as that of the general taxpayers. There would be no loss of total goods and services at the time the debt was repaid. Thus, in the sense that there is no cost or loss for the economy as a whole when the debt is repaid, it is impossible to pass the burden of the debt on to future generations.

Effect on Individuals. Although the burden of the debt cannot be passed on to future generations from the viewpoint of the total economy, the burden for individuals and firms can be passed on to future generations. If the government were to pay off the debt in a relatively short period, say within the generation in which the debt occurred, the particular individuals taxed to pay the debt would have to give up purchasing power. Thus, each would be burdened with the cost of the debt to an individual extent. If payment on the debt is postponed for a generation or two, however, the tax will fall to a large extent on the descendants of those individuals and businesses in the economy at the time the debt was incurred. Thus, even though the net cost or burden of the debt cannot be passed on to future generations, the individual burden can be passed on.

For example, during the mid-1980s, the United States incurred a sizeable debt. If the debt were to be paid off within the generation in which it was incurred, Frank Delgado, a taxpayer of the period, might have to pay $2,000 in taxes to provide the government with the money to repay Heather Morgan, who, we will assume, is a holder of bonds and therefore an owner of the debt. This payment would decrease the purchasing power of Mr. Delgado and would increase the purchasing power of Ms. Morgan.

If debt repayment were postponed until 2000, however, Mr. Delgado's niece or someone else in the economy would have to pay the taxes. The individual burden of the debt would have been passed on from Mr. Delgado to someone in a subsequent generation. Ms. Morgan, who made a personal sacrifice to lend the government money in return for the bonds, would be deprived of repayment until a later date. In fact, if she passed away, her descendants, not Ms. Morgan, would receive the individual gain at the time repayment was made. In actual practice, however, Ms. Morgan could at any time eliminate this situation by transferring her ownership of the debt to someone else by selling the bonds.

Furthermore, even if the debt were passed on to future generations, it can always be argued that, since future generations receive some benefits as a result of the government incurrence of previous debt, they should help repay it. Future generations, for example, reap certain benefits from the use of hospitals, schools, dams, highways, and medical and other research financed through deficit spending.

Effect on the Money Supply. Another problem involved in the repayment of the debt, which tends to strengthen our reluctance to pay it off, is the effect of the repayment on the money supply. We know that when an individual or a business loans the government money, there is no increase in the money supply. For example, if Mr. Pulaski buys a bond for $1,000, he generally will pay cash for it. Therefore, there is merely a transfer of cash from the individual to the government with no change in the total money supply. If a bank lends the government money, however, it can pay for the bonds in cash or through the creation of a demand deposit against which the government writes checks. Demand deposits brought about by the creation of credit increase the money supply. Therefore, if a bank were to buy $250,000 worth of bonds and pay for them by using a demand deposit, it would increase the money supply accordingly. This process is referred to as *monetizing the debt*.

Changes in the money supply can affect the level of economic activity and/or the price level. When the government goes into debt by borrowing from the banks, it increases the money supply and thus increases the level of economic activity and/or the price level.

Since 1970 the money supply has increased from $215 billion to $535 billion, a large portion of which came into existence as a result of the sale of government bonds to the Federal Reserve and commercial banks. Therefore, the national debt today is supporting a sizeable part of the total money supply.

We know that a decrease in the money supply will tend to decrease the level of economic activity and/or decrease the price level unless it is offset by some other force such as an increase in the velocity of money. Just as the debt was monetized when the government borrowed from the banks, the money supply will be decreased when the debt is paid off. This is known as *demonetizing the debt*. For example, if the government redeemed the $100 million in bonds held by the banks, demand deposits would be reduced by that amount and the money supply reduced accordingly. Thus, if the government were to reduce the national debt by sizeable amounts over a relatively short period of time, it could reduce the money supply to such an extent as to have an adverse effect on the level of economic activity. Payment of the debt supported by the bank credit would be beneficial during a period of full employment since it would reduce inflationary pressures. During periods of less than full employment, however, such debt repayment could be harmful to the economy as a whole.

Size of the Debt

The mammoth size of the debt is in itself sufficient to discourage many people regarding its repayment. It might be pointed out, however, that although we have not reduced the debt absolutely, increased productivity and higher income have reduced the size of the debt relative to our annual income. For example, in 1946 the national debt was $271 billion. Our total income (GNP) for that year was $210 billion. Since the debt was considerably larger than our total income, it could not have been paid out of current income within a period of one year even if we chose to do so. As a matter of fact, to pay off at the rate of $20 billion per year would have exerted quite a hardship.

Although the debt was substantially larger in 1986, the total production of the nation had increased to $4,200 billion. Since the annual income of the nation exceeded the national debt, it was possible to pay off the debt within a period of one year or less. Although possible, of course, it would not have been feasible to do so. With a national income of $3,500 billion, however, if the decision were made to pay the debt off at the rate of $20–$25 billion annually, less of a hardship would be created on the economy than would have been in 1946 when the gross national product was substantially lower.

In effect, through our increased productivity and higher price level, the monetary income of the nation increased in the period 1946–1986. Since the absolute amount of the debt increased about 510 percent during this period, the burden of the debt relative to gross national product was reduced by well over two-thirds. Actually, in 1946 the GNP was only 75 percent of the outstanding national debt. Today, however, it is approximately 200 percent of the debt. For this reason, those who were worried about the size of the debt 25–30 years ago have less cause to worry about it today. Another way to state this is to say that the ratio of national debt to the GNP was 135 percent in 1946 compared to 50 percent in 1988. It should be remembered, however, that decreasing income resulting from either a falling price level or a drop in production or employment would increase the size of the debt relative to income.

The suggestion has occasionally been made that we should postpone payment on the debt since it becomes less burdensome as the years go on. To the extent that we increase income as a result of increased productivity, this suggestion has some merit. But if the higher GNP and therefore the greater income are brought about primarily by higher prices, the suggestion is a poor one since greater problems than that of debt retirement will result from rising prices. Furthermore, if due to continuous inflation the purchasing power of a $100 Savings Bond at maturity is of less value than the $50 price of the bond, the purchase of bonds by individuals and firms could be discouraged when the government needs money in the future.

Refunding of the Debt

Since government debt obligations may reach maturity at a time when the United States Treasury does not have the money to pay them, the problem of refunding the debt arises. At such a time, the federal government generally issues and sells new securities to raise money to pay off the matured obligations. This, however, may not be accomplished easily, especially when billions of dollars worth of bonds and other securities may be maturing within a short period of time. Furthermore, the government may be forced to pay a higher interest rate when it borrows funds for this purpose.

For example, billions of dollars worth of long-term bonds paying 4.25 percent interest come due annually. Today, however, it is possible for an investor to buy at discounts many nonmatured long-term government bonds that effectively have a higher yield than their coupon rate. In addition, many high-grade corporate bonds are yielding more than 10 percent.

Under these circumstances it is difficult for the United States Treasury to sell new bonds at 4.25 percent when existing bonds yielding more than 10 percent can be purchased in the open market. It is for this reason that the Treasury Department occasionally requests that the 4.25 percent interest rate ceiling on long-term government bonds be removed to permit the selling of long-term government bonds at a higher interest rate. Because Congress has generally refused to grant this request, the Treasury is forced to sell short-term government obligations on which there is no interest rate ceiling.

When the Treasury redeems the matured obligations in this manner, the total cost of the debt increases because the interest rate on the new obligations is higher than that on the refunded portion of the debt. Furthermore, it puts itself in a position where it must pay off or refund again this portion of the debt in another relatively short period. If the Treasury Department were able to issue long-term securities at a competitive interest rate, it would prolong the payment or refunding date for 15, 20, or even 30 years; on the other hand, there could be times when the debt might be refunded at a lower cost. The situation on refunding worsened in the 1970s, however. In 1979, for example, the Treasury was at times paying more than 14 percent to refund billions of dollars of debt coming due at that time. Fortunately, the rate paid on many government securities dropped below 10 percent by late 1982, and was below 7 percent by 1986.

Burden of Interest Payments

Included each year in our federal budget is $135 billion or more for payment of interest on the national debt. Although taxation for the payment of this interest does not impose a net burden or cost on the economy as a whole, it does cause an annual redistribution of income and, therefore, a specific burden to individuals and firms in the economy. If the government had originally increased taxes instead of going into debt or if the government had paid off the debt shortly after it had been incurred, it would have imposed a smaller total burden on the individuals than it does when the debt repayment is postponed. With postponement of the current debt, the total redistribution of income necessary to retire the debt is not only in excess of $2 trillion, the principal amount, but also $135 billion or more annually for interest on the debt. This interest comes only as a result of postponement of the payment of the debt.

In the last five years there has been an interest burden of more than $550 billion because the debt has been outstanding. It is a matter of judgment whether individuals and firms would prefer the hardship of paying off the debt in a relatively short period, as opposed to giving up more total income but spreading the hardship or inconvenience in smaller doses over a longer period of time. There has also been the suggestion that some type of interest-free financing for federal borrowing would ease the debt burden.

The Balanced Budget Amendment

In the hope of halting deficits and stemming the growth of the national debt, a Constitutional Amendment requiring a balanced budget was introduced in the House of Representatives in October, 1982. Although the vote was 236 to 187 in favor of the amendment, it did not obtain the two-thirds majority needed for passage. At the same time 31 of the 34 states needed to call a Constitutional Convention indicated a desire to do so for the purpose of drafting a balanced-budget amendment.

Subsequently, in order to limit debt accumulation Congress enacted the Balanced Budget and Emergency Deficit Control Act of 1985 (Gramm-Rudman-Hollings Act). This act called for the federal government to reduce the federal deficit each year and finally produce a balanced budget by 1991. Accordingly, the deficit is to be no more than $144 billion in fiscal 1987, $94 billion in 1988, and is to reach zero by 1991. Nevertheless, President Reagan and others as late as 1987 were still seeking more support for a balanced-budget amendment.

CONCLUSION

It is unlikely that the national debt will be reduced by any substantial amount in the near future. In 1982 Congress removed the interest rate ceiling on Savings Bonds. In 1985 it passed the Gramm-Rudman-Hollings Act requiring that the federal deficit be narrowed to zero by 1991. In 1986 it increased to $250 billion the amount of long-term bonds the Treasury could sell without regard to the statutory 4.25 interest rate ceiling. In August, 1986 the U.S. Treasury asked Congress to raise the ceiling on the national debt to $2,323 billion. The increase was needed to finance the ongoing deficit of 1986 and prevent the Treasury from dipping into the Social Security Trust Fund to meet current obligations, as it did in 1985.

Obviously the problems of the national debt will continue. The issues may become more or less pressing, depending on economic measures adopted in the future to alleviate the problems of the debt. Since it does not appear that the debt is going to be repaid any time soon, the most we can hope for is that future increments in the debt will be less than the additions to our national income, which would reduce the relative burden of the debt.

QUESTIONS FOR DISCUSSION

1. Will the national debt bankrupt the nation?
2. Should the national debt be paid off either in part or in full?
3. Should the statutory ceiling on the national debt be removed or tightened?
4. Should our national debt be financed through interest-free financing, that is, by sale of non-interest-bearing bonds to the banks?
5. Should the interest rate ceiling on the national debt be removed?
6. Do you recommend that the government avoid going into debt by printing more currency to cover its expenses?
7. Should the federal budget be used as a device for stabilizing the economy?
8. What do you think of the "full-employment budget" concept?
9. Would it be feasible for the government to issue shares of stock in the U.S. economy and declare dividends each year, instead of selling bonds and paying interest on them?

SELECTED READINGS

Ackley, Gardner. "You Can't Balance the Budget by Amendment." *Challenge*. (November–December 1982).

Bennett, James T., and Thomas J. DiLorenzo. "Off-Budget: Federal Spending Booms." *Economic Review*. Federal Reserve Bank of Atlanta (April 1983).

The Budget in Brief, 19–. Washington D.C.: U.S. Government Printing Office, annually.

The Budget of the United States Government, 19–. Washington D.C.: U.S. Government Printing Office, annually.

Cunningham, Thomas J. "The Long-Run Outcome of the Permanent Deficit." *Economic Review.* Federal Reserve Bank of Atlanta (May 1980).

———— and Rosemary Thomas Cunningham. "Projecting Federal Deficits and the Impact of the Gramm-Rudman-Hollings Budget Cuts." *Economic Review.* Federal Reserve Bank of Atlanta (May 1986).

"The Deficit Puzzle." *Economic Review.* Federal Reserve Bank of Atlanta (August 1982).

Handbook of Securities of the United States Government and Federal Agencies. Boston: First Boston Corp., 1986.

Pechman, Joseph A. *Setting National Priorities: The 1988 Budget.* Washington D.C.: The Brookings Institution, 1986.

Special Analysis: Budget of the United States Government, 1988. Washington D.C.: U.S. Government Printing Office, 1987.

9

HEALTH CARE
WILL IT
BANKRUPT
US?

One of the most difficult problems facing the American people in the years ahead is that of meeting their medical expenses. The increased demand for improved health care and the lack of sufficient funds to pay for sharply rising health costs have made health care a critical economic issue for health care consumers and suppliers, employers, private health insurers, and public agencies.

In the 1960s, when President Johnson initiated his Great Society, Americans were told that health care was a right and not merely a privilege for those who could afford it. The passage of the Medicare and Medicaid programs in 1965 was seen by many as but the first step in arriving at a national health insurance program guaranteeing "cradle to grave" protection against catastrophic health care costs.[1]

Today the nature of the health care debate is decidedly different. National health insurance is, if not a dead issue, certainly in a comatose

---

1. For summaries of major national health insurance proposals offered in the 1970s, see the fifth edition of this text.

state. Rather than promising comprehensive health care, policymakers in Washington are now warning of the possibility of "rationing" health care. Federal and state budgets strain to meet the demands that existing programs are exerting on them. This chapter examines economic aspects of the health care industry along with the approaches to abate rising costs.

SCOPE OF THE PROBLEM

In the broadest view, four out of five people have some form of illness each year. Of those who become ill, one out of seven is incapacitated for more than 90 days. In addition, there are 350,000 citizens who suffer permanent disability each year. Approximately one person out of every three families is hospitalized annually, and an estimated 60 percent of these people require surgery. The cost of this care is astounding. Over 500,000 families will have medical bills in excess of six-month earnings, while nearly seven million families will have out-of-pocket medical expenses exceeding 15 percent of their annual income.

In 1983, over 35 million Americans were without medical insurance of any kind, either public or private. During the same year, it is estimated that another 15 million people had inadequate insurance coverage. Keeping these figures in mind, it should not be surprising to discover that close to 50 percent of all personal bankruptcies each year are the result of medical debts.

The essential problem of medical care today is how to pay the bill when costs are soaring. The United States is the only industrialized nation without a universal health insurance program, yet spending is greater for health care on a per capita basis than in any other country. Table 9-1 presents data on national health expenditures for selected years. In 1986, Americans paid approximately $454 billion for medical care. This sum represents a 53 percent increase over what was paid in 1981 and a 506 percent increase over the level of health expenditures in 1970. Table 9-1 reveals that hospital care accounted for $181 billion of the projected $454 billion spent in 1986. The second highest expenditure was for the services of physicians, which totaled nearly $91 billion. Expenditures for hospital care and physicians' services are the components of medical care that are experiencing the greatest cost pressures.

On a per capita basis, America's medical bill has climbed relentlessly during the past three decades. In 1950, per capita national health care expenditures stood at $82. By 1986, per capita expenditures were estimated to be $1,876.

Table 9-1

National Health Expenditures for Selected Years
(Billions of dollars)

Type of Expenditure	1950	1960	1970	1980	1986[1]
Health services and supplies	11.7	25.1	69.6	236.1	437.2
Personal health care	10.9	23.7	65.4	219.1	402.9
Hospital care	3.9	9.0	27.8	101.3	180.8
Physicians' services	2.7	5.7	14.3	46.8	90.5
Dentists' services	1.0	2.0	4.7	15.4	30.7
Other professional services	0.4	0.9	1.6	5.6	10.9
Drugs and medical sundries	1.7	3.7	8.2	18.5	31.0
Eyeglasses and appliances	0.5	0.8	2.0	5.1	8.8
Nursing home care	0.2	0.5	4.7	20.4	38.9
Other health services	0.5	1.1	2.1	5.9	11.4
Program administration and net cost of insurance	0.5	1.0	2.8	9.2	21.9
Government public health activities	0.4	0.4	1.4	7.7	12.3
Research and construction of medical facilities	0.9	1.7	5.3	11.9	17.0
Research	0.1	0.7	1.9	5.4	7.8
Construction	0.8	1.0	3.4	6.5	9.2
TOTAL	12.6	26.8	74.9	248.0	454.2

[1]Estimated.

SOURCE: *Health Care Financing Review* (Spring 1986) and *Statistical Abstract of the United States,* various years.

Figure 9-1 depicts national health expenditures as a percentage of gross national product for selected years. The percentage of GNP attributed to health care has risen from 4.4 percent in 1950 to an estimated 10.8 percent in 1986. The mere size of the health care industry is indicative of the scope of the economic problem.

OUR PRESENT HEALTH CARE SYSTEM

Americans pay for their medical care in a combination of ways. In addition to direct or out-of-pocket payments, many individuals and families purchase

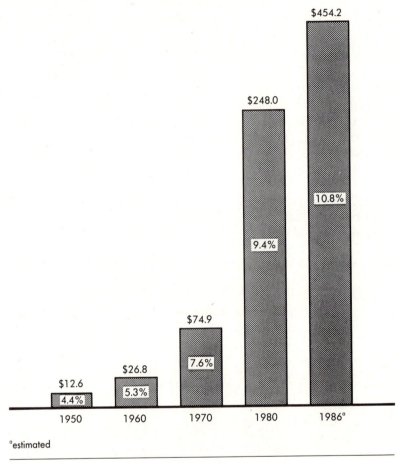

Figure 9-1

National Health Expenditures as a Percentage of Gross National Product
(Dollars in billions)

$454.2

$248.0

$74.9

10.8%

9.4%

$26.8

7.6%

$12.6

5.3%

4.4%

1950   1960   1970   1980   1986ª

ªestimated

SOURCE: *Health Care Financing Review,* (Spring 1986).

medical insurance in the form of individual or group plans. Because of
the high cost of medical care, a fairly high level of income is required to
make direct payments or to purchase insurance. Unless totally or partially
employer-paid group insurance is provided, insurance is unaffordable for
many. To extend the availability of proper medical attention, in 1965
Congress passed two major health programs—Medicare and Medicaid.
Medicare is for those over 65 years of age or eligible for Social Security
disability benefits. Medicaid is provided for those who qualify as medically
poor.

Medicare

The Medicare plan (Title 18 of the Social Security Act), which began paying benefits on July 1, 1966, covers almost everyone over 65 years of age. The plan is made up of two kinds of health insurance: hospital insurance and medical insurance. Participation in the hospital insurance is obligatory, and the medical insurance is voluntary. The hospital insurance is financed by a compulsory payroll tax on all employees covered by Social Security.

Benefits of the hospital insurance (Part A) include partial payment for up to 90 days of inpatient hospital care, 100 days in a skilled nursing facility, and an unlimited number of home health service visits by medical personnel other than doctors. These benefits are the maximums for each benefit period, which is defined as the first day of hospital treatment to 60 days after release from the hospital or skilled nursing facility. In addition, each person is eligible for a "lifetime reserve" of 60 days of hospital care to apply to situations in which the number of hospital days during a benefit period exceeds 90.

In 1987, for each day up to 60, hospital insurance covered all inpatient hospital costs after the patient paid the first $520. The patient cost for days 61 through 90 was $130 per day. The patient had to pay $260 a day for each of the lifetime reserve days. Costs covered by insurance include bed and board in a semiprivate room, ordinary nursing services, drugs, supplies, and diagnostic services. The total costs for the first 20 days in a skilled nursing facility were covered by Medicare. The patient cost for days 21 through 100 was $65 per day.

The other part of Medicare is voluntary medical insurance (Part B). This plan is financed by monthly premiums from those who wish to be covered. Premiums are deducted from the Social Security benefit checks of those who participate. The federal government matches the payments from general tax revenues. Monthly premiums rose from $3 in 1966 to $17.90 in 1987. The medical insurance program pays 80 percent of Medicare-recognized charges for covered services, except for the first $75 in a calendar year.

Services covered include physicians' and surgeons' services, home health visits, diagnostic tests, and other health services, regardless of where rendered. In 1985, about 30 million persons were enrolled in the Medicare hospital insurance program, and of those, nearly all were enrolled in the medical insurance program. As shown in Table 9-2, over $18 billion of general tax revenues were used to finance the Medicare program in 1985. On an annual basis, Medicare has grown from a $3.3 billion program in 1967 to an estimated $70 billion program in 1985.

Table 9-2

Payments into Medical Trust Funds, 1985
(Dollars in billions)

	Hospital Insurance (Part A)	Medical Insurance (Part B)	Total	Percent
Payroll taxes and voluntary premiums	$47.7	$ 5.6	$53.3	70.0%
General revenues	0	18.2	18.2	24.0
Interest	3.4	1.2	4.6	6.0
TOTAL	51.1	25.0	76.1	100.0

SOURCE: *Statistical Abstract of the United States: 1987.*

Medicaid

Medicaid (Title 19 of the Social Security Act) is a medical assistance program rather than an insurance program. It is a federal-state program administered by the states individually. It is financed out of the general revenue at the federal level along with state and local tax money. The federal share of Medicaid programs in a given state is derived from a formula based on the state's per capita income. Each state determines its eligibility requirements and operating rules under broad federal guidelines. Thus, the number of people eligible or types of benefits received vary considerably from state to state. One state, Arizona, has no Medicaid program.

The medical assistance program (Medicaid) was designed to help unify the nation's health care by bringing the medically poor (those with enough money for daily needs but not enough to pay for health care) up to par in terms of benefits with welfare recipients. After some states defined income levels for the medically poor at unrealistically high levels, Congress, in 1969, set income levels for Medicaid at 133 percent of the income level needed to qualify for welfare payments under Aid to Families with Dependent Children.

The manner in which Medicaid is implemented has resulted in wide variations in the amount, quality, and types of services provided among the various states. According to federal law, states having Medicaid programs are required to provide care to those who are enrolled in the two major programs of federally aided cash welfare assistance—Aid to Families with Dependent Children (AFDC) and Supplementary Security Income (SSI) for the aged, blind, and disabled. Other individuals may or may not be included, depending on the type of state program. The major groups that tend to be excluded from Medicaid programs are nondisabled poor adults under age 65 and children in intact families with an employed father in the home.

Not only does coverage vary among states, but the quality of service varies as well. All states must provide inpatient hospital services, outpatient hospital services, lab and X-ray services, skilled nursing services, and physicians' services. The law does not require the provision of dental services, prescription drugs, home health care services, private duty nurses, or long-term institutional care. However, states may provide the aforementioned services if they so choose. Also, the law does not specify the amount of coverage to be provided to various groups. For example, some states use approximately 50 percent of their Medicaid funds for nursing home care, while other states spend little for this purpose. Because of these many factors, there has been a growing recognition that the Medicaid program has not been totally successful in providing quality care to all the poor and medically indigent.

Compared to Medicare, the Medicaid program finances more long-term, nonacute, institutional care using nursing facilities, mental hospitals, and home health agencies. Long-term care expenditures amount to nearly 50 percent of all Medicaid program expenses. In 1985, Medicaid paid medical bills for an estimated 22 million persons eligible under public assistance standards. On a cost basis, total expenditures in 1985 were approximately $38 billion.

Private Health Insurance

The size of the private health insurance industry has been growing, reflecting the increased demand for protection from health care costs. In 1984, 79 percent of the population had some private health insurance for hospital care and surgical expenses. Approximately 50 percent of private spending for health care that was not covered by public programs was reimbursed by private insurance. Typically, coverage is some sort of group health insurance associated with employment.

Extension of coverage beyond surgical procedures in recent years has led to a higher share of physicans' services being reimbursed by private insurance. In 1985, private insurance reimbursements covered 44 percent of total expenditures for physicians' services, while for hospital care expenses the corresponding figure was 36 percent. Insurance coverage has been more limited for other health care services. Dental care is one area in which coverage is growing. Private insurance reimbursed about 34 percent of all dental expenditures in 1985.

Another rapidly growing form of health insurance is major medical care. Currently about one-half of the population is covered. Such coverage is typically tied in with other insurance since it pays when basic medical insurance stops paying or pays for certain illnesses not covered by basic plans. After a stipulated deductible is met, major medical pays 80 percent of the charges up to a designated maximum, the exact amount depending upon the type of plan.

The most common criticism of private health insurance is that it does not cover a high enough portion of the total bill. A similar charge can be made against Medicare. However, when one considers the high costs involved, it is economically inefficient for any plan to cover 100 percent of personal health expenses.

Figure 9-2 shows the sources and uses of a single dollar of health care spending on a national basis. Private health insurers and other private third parties contributed 32 cents of each dollar spent on health care in 1984.

CURRENT PROBLEMS IN THE HEALTH CARE INDUSTRY

Despite increased public and private insurance participation in the health care field, major problems continue to challenge the health care industry. Two such problems relate to the supply of physicians and the increasing costs of medical services.

Supply of Doctors

One of the major controversies concerning the health industry is whether there is an adequate number of physicians to provide for the medical needs of the population. Patients who have waited in doctors' offices for long hours beyond the time of their appointments are likely to support those who decry the acute shortage of medical personnel. However, many health care experts claim that there is a sufficient number of medical doctors, but what is needed is a more rational medical care system. They point out that the supply of doctors is increasing with the demand for more health services. Each year the number of doctors graduating from our medical schools is growing at a faster rate than the population as a whole. This is happening despite monopolistic factors on the supply side, such as laws on licensed practice and certain restrictive practices of professional associations of doctors.

The underlying basis for such opposing views on the need for additional doctors can be traced to problems associated with defining the term *need*. According to economic analysis, need is incorporated into consumer demand since it is a measure of the amount of medical care consumers will purchase at various prices. For the most part, however, need is defined in the health industry in a social sense and is related to the standards of the

Figure 9-2

The Nation's Health Dollar in 1985

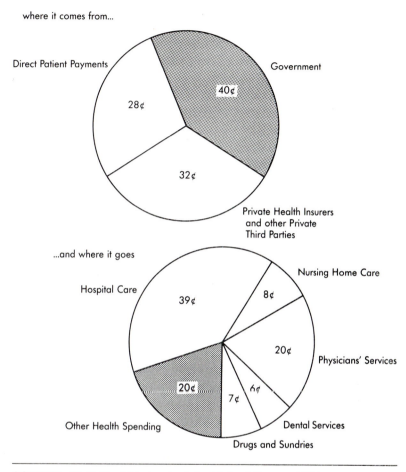

where it comes from...

Direct Patient Payments

Government

40¢

28¢

32¢

Private Health Insurers
and other Private
Third Parties

...and where it goes

Nursing Home Care

Hospital Care

39¢

8¢

20¢

Physicians' Services

20¢

7¢

6¢

Other Health Spending

Dental Services

Drugs and Sundries

SOURCE: *Statistical Abstract of the United States: 1987.*

health profession. This social approach fosters controversy because of the lack of easy quantification.

Although disagreement exists over the need to increase the supply of doctors nationally, there is a general consensus that there are distribution problems in the delivery of medical services. The maldistribution of physicians on a geographical basis has long been recognized. Although the ratio of physicians to population is not a perfect measure of the adequacy of medical service, it is a widely used criterion for designating areas that are underserved. As with most professions, physicians tend to gravitate toward urban areas within more heavily populated states and away from states that are primarily rural. For example, Massachusetts and Maryland (295 physicians per 100,000 population) had the highest physician-to-population ratio of any of the 50 states in 1983. On the other hand, the lowest ratios of physicians to population existed in Idaho (112 per 100,000) and Mississippi (115 per 100,000).[2]

The higher incomes and the amenities derived from urban residence are major factors accounting for the lack of geographical balance in the supply of physicians. But there are other important considerations as well. Urban areas offer the benefits of professional interaction among physicians. This affords a greater opportunity to keep abreast of the constantly changing state of medical science. In addition, urban medical centers provide better research accommodations and access to the most advanced facilities and equipment. Finally, having spent more than a few years studying medicine, many graduating doctors no longer feel strong ties to their rural hometowns.

The maldistribution among medical specialties has also hampered the provision of effective health care. Of particular concern is the growing disparity in the number of primary-care physicians compared to general surgeons. The relative decline in primary-care physicians can be seen by the sharp decrease in the number of general practitioners as a percentage of the total number of physicians. In 1970, 17 percent of physicians providing health care were general practitioners. By 1983, this figure had declined to approximately 12 percent. This trend toward medical specialization is the result of several factors. Perhaps foremost is that, on the average, specialists earn higher incomes than do primary-care physicians. In addition, hospital staffs encourage specialization. An intern seeking a staff position in a good hospital quickly realizes that the chances for acceptance are enhanced if one is a specialist.

Consequently, medical care delivery has become not only more city oriented, but more specialized as well. These shifts have altered the structure of both the supply of and the demand for the services of physicians and have resulted in increased medical costs.

Rising Costs

In 1984, the average citizen was paying almost 11 times the dollar amount that was paid in 1960 for medical services. Many factors accounted for such a sharp increase in medical expenditures. Three major causal factors, however, stand out as most worthy of our attention: price increases as measured by the Consumer Price Index, changes in the health care system, and changes in the size and age distribution of the population. It should be

---

2. *Statistical Abstract of the United States: 1987,* 92.

noted that the latter two factors do not represent price increases per se, but rather increasing expenditures for health care.

Price Increases. Price increases for various categories of medical service are compared to changes in the cost of living in Figure 9-3. If we use 1977 as the base year, the cost of living as measured by the CPI increased by 178 percent between 1977 and 1985. Hospital charges, however, increased at much faster rates than the overall price index. Figure 9-3 indicates that hospital

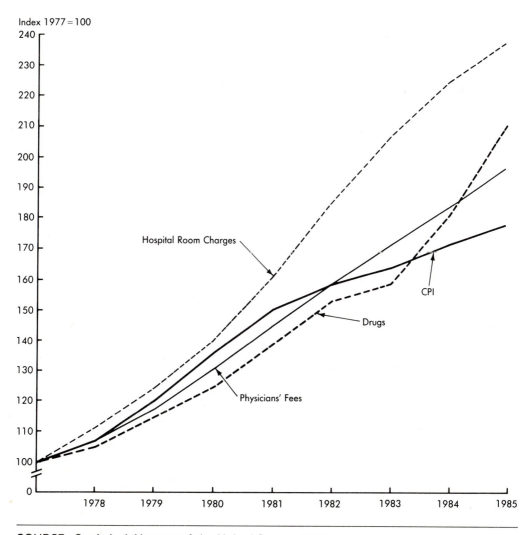

Figure 9-3

Medical Cost Increases (1967=100)

SOURCE: *Statistical Abstract of the United States: 1987.*

charges have risen by 237 percent since 1977, whereas the physicians' fees index and the CPI have shown a closer relationship. Until very recently the costs of drugs and prescriptions increased at a relatively slow pace, but sharp increases in 1984 and 1985 have resulted in drug price increases in excess of the CPI.

Changes in the Health Care System. The effects of changes in the composition of medical services and goods as well as changes in utilization rates are referred to as changes in the health care system. This category contains a variety of interlocking factors whose effects on health costs are difficult to measure independently of each other. These factors include changes in technology and treatment that alter the mix and frequency of services used. Also included are changes in access to medical care, either by removing financial barriers or by increasing the supply of services, which affect utilization rates.

Undoubtedly the quality of health care service in the United States has improved appreciably with increased use of previously known methods and techniques, coupled with the introduction of new ones. Chemotherapy treatments, kidney dialysis, and open-heart surgery are very expensive and add significantly to national health expenditures. However, these examples represent major progress in medical care. In addition, the quantity of health care services used has increased with growing demand, increasing total spending for health care. One relatively recent phenomenon serving to increase the per capita use of health services has been the sharp increase in malpractice suits. The fear of malpractice suits has led many physicians to perform as many diagnostic tests as are available to prevent legal charges of medical incompetence. These extra tests are costly and may necessitate longer stays in the hospital—factors that add still more to the total cost of health care.

Size and Age of Population. The third factor contributing to increased health expenditures is the changing size and age distribution of the population. As our nation's population becomes older, it is logical to expect greater expenditures on health care. Per capita expenditures for physicians' services, for example, are nearly three times greater for the age 65 and over population than for those under age 65. Both the number of physician visits per capita and complexity of services per visit are relatively higher for the aged. Counterbalancing an older population is the fact that increases in the total population have declined during the past ten years. This has reduced demand pressures on certain segments of the health industry, particularly those segments servicing infants and youths.

Supply and Demand Analysis. Supply and demand analysis serves as a useful framework for analyzing price inflation in the health care industry. Critics of the health care industry have long argued that consumers demand more medical care than they would otherwise because most do not bear the full cost of treatment. In most cases third-party payments, either private insurance or public taxes, cover expenses. Insurance can be treated as a shift in the demand curve for health care to the right as seen in Figure 9-4a. The increase in demand is also strengthened by a fiscal policy that excludes employer-provided health insurance as taxable income.

Figure 9-4

Demand and Supply in the Health Care Industry

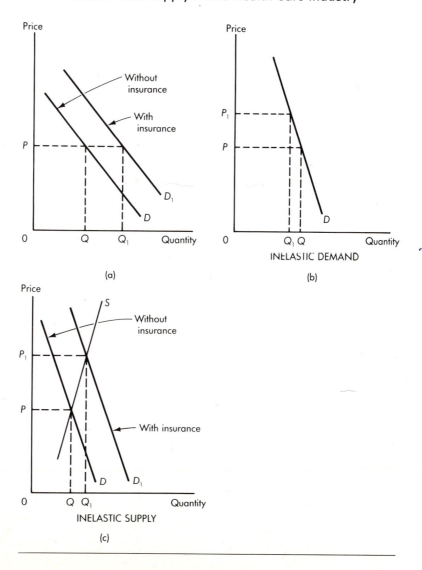

(a)

(b)

INELASTIC DEMAND

(c)

INELASTIC SUPPLY

Health insurance also reduces the price elasticity of demand for medical services by making consumers less sensitive to price increases. Since insurance covers most if not all of the medical bill, consumers are not prone to decrease the quantity demanded of health care services commensurate with the increase in price. Figure 9-4*b* indicates the result in quantity demanded from a price increase when the demand curve becomes highly inelastic.

It must be recognized that an increase in demand would not automatically increase the cost of medical service. The impact that an increase in demand and inelastic demand curve would have on price also depends on the elasticity of supply. If supply were price elastic, then an increase in demand might result in only a small increase in price and a much larger increase in quantity. However, the supply curve for the health care industry tends to be highly inelastic, and an increase in demand quickly pushes the industry to capacity and forces higher prices as seen on Figure 9-4*c*.

The supply of medical care output does not respond to higher prices largely because of substantial barriers to entry. Monopolistic restrictions on the part of professional associations have limited increases in supply. The power of the American Medical Association to control the licensing of physicians, to certify medical schools, and to regulate hospital internship and residency requirements has restricted the quantity of health care services supplied. Consequently, higher prices for medical care result in higher income for suppliers more than they ration demand or increase output. When combined with the rise in household incomes, the wide availability of private health insurance, and the huge outlays on public health service programs such as Medicare and Medicaid, the steady increase in health care prices is not surprising.

Figure 9-5 shows the effects of several major factors on national health care expenditures from 1974 to 1984. General inflation accounted for 55.6 percent of the increase, while health care prices, over and above the general inflation rate, accounted for 13.6 percent. Population growth accounted for a 7.9 percent increase, and changes in real per capita expenses reflecting the quantity and/or composition of goods and services accounted for the remaining 22.9 percent increase.

Rising costs constitute a critical problem in the health care industry, and little in the way of relief appears in sight. As costs continue their upward climb, alternatives to the present system are gaining momentum.

NEWER APPROACHES TO HEALTH CARE

Most Americans still receive medical treatment in the traditional way, seen by a private practitioner on an acute, episodic basis and admitted

Figure 9-5

Factors Accounting for Growth in Total National
Health Care Expenditures, 1974–1984

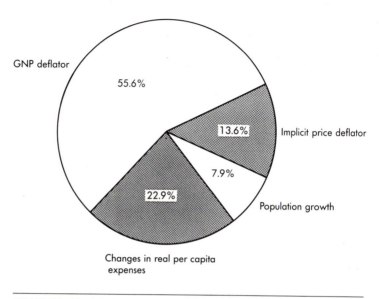

SOURCE: *Health Care Financing Review* (Spring 1986).

to a hospital if necessary. Charges for such medical attention usually are paid wholly or in part by third-party insurers. In recent years, however, numerous variations to the traditional system have proved to be successful.

Health Maintenance Organizations

One of the most widely used alternatives to traditional medical care is the Health Maintenance Organization (HMO), consisting of groups of people who band together to receive a wide range of health services. Members prepay for medical service. The goal of the HMO is to provide satisfactory medical service under one plan at the lowest possible cost.

As members of an HMO, individuals and families have access to an entire team of physicians. In addition to primary-care physicians, such as general practitioners and pediatricians, an HMO typically includes specialists on its staff such as radiologists, surgeons, obstetricians, gynecologists, and neurologists. In large part, the number of specialists included on a given HMO physician team is a function of the size of the HMO. The HMO's team of doctors, nurses, laboratory technicians, and administrators usually works together in a single building and provides 24-hour medical care, 365 days a year. Many HMOs also have their own pharmacies. If

Health Care—Will It Bankrupt Us?   **241**

hospitalization is deemed necessary, the patient is admitted and treated at an affiliated hospital at no additional out-of-pocket expense to the individual. The annual cost of membership depends on the extent of service provided and the area of the country. Although not inexpensive, HMO membership does provide financial protection against the burdens of catastrophic illness. Consequently, most, if not all, of the economic uncertainty resulting from unexpected large medical expenses is precluded. This protection is usually not included in standard insurance programs and is only acquired at additional expense.

The Health Maintenance Organization is based on the concept of the efficiency of preventive medicine. By stressing preventive and outpatient care, HMOs hope to reduce the economic waste associated with the overuse of inpatient hospital services. Since members have prepaid for medical service, they are more prone to seek medical care in the early stages of illness. This keeps HMO members healthier at reduced costs. Also, HMO doctors are paid on a salary basis, which removes the incentive to perform unnecessary services.

As a result of a 1973 federal law, most employers of 25 workers or more with health insurance plans must offer an HMO plan as a voluntary option. In 1970, there were only 25 HMOs operating in the United States. By 1985, the number of HMOs had increased to 393 with nearly 19 million persons enrolled.

Despite rapid growth in membership, the HMO system of health care delivery has experienced increased criticism in recent years. Prior to 1985, the most-often-mentioned concerns included the limited choice of physicians and hospitals, the impersonality of the system (particularly the doctor-patient relationship), and the possible lack of coverage outside the service area. Now, however, concern centers on costs and effectiveness. Some HMOs, with their broader scope of benefits, now carry higher premiums than traditional insurance plans. In part, this has been brought about by the success of HMOs, for they have forced regular insurance plans to become more competitive. In addition, many employers are now requiring certification of hospital stays and employee co-payment of premiums, while higher deductibles have made some employees in traditional plans more cost conscious. HMOs also face heavy competition from one another, as large corporations and small operators continue to enter the industry.

The future of HMOs is viewed with guarded optimism by health care analysts. Where costs are rising and regular insurance plans are heavily used, HMOs have more room to cut costs and underprice standard plans. Resistance is being met in markets where competition has already reduced medical costs. Consequently, as long as HMOs can show they can be more cost-effective than traditional private insurance programs, membership growth should continue.

Medical Centers

Because of lower costs and greater convenience, privately owned and operated emergency medical centers are entering the health care scene throughout the United States. Often operated as a chain of centers, they are usually open from early morning to late evening seven days a week and provide walk-in services for most routine health care and medical emergencies.

Most centers are equipped to stabilize serious and dangerous conditions resulting from accidents or illness. However, they deal mostly with problems that make up the majority of cases handled in hospital emergency rooms, such as broken bones, earaches, and the like. If hospitalization is required, centers will transport patients to nearby hospitals.

From but a handful several years ago, there are now approximately 1,000 centers in operation. Most are independent, but some are associated with hospitals. Naturally, emergency medical centers are not without their critics, some of whom refer to the centers as "doc in a box." A major criticism of the centers is that a patient who underestimates the seriousness of his or her condition may waste valuable time by not going directly to a hospital. The impersonality of the operation is also criticized, but the key complaint appears to stem from the fact that centers are taking nonurgent cases away from hospital emergency rooms. Emergency rooms are sources of much-needed revenue for hospitals. Medical center visits can cost up to 50 percent less than equivalent treatment in hospitals. As a result they are beginning to attract industry support. Businesses are now sending more employees to centers for routine examinations and minor complaints, taking away insurance-paid cases from hospitals.

As a form of emergency room competition, centers will probably continue their rapid growth. The lifestyle of American families has changed. Convenience in health care appears to be as much in demand as convenience in stores; health care at a cheaper price is certainly attractive.

Home Health Care

The marked growth in the home health care industry in recent years can be traced to a combination of factors working together: the need to constrain health care costs, an aging population, advanced technology, and an increased desire by patients for treatment outside of hospitals. Billions of dollars are being spent each year on home health care for such sophisticated services as intravenous feeding, cancer chemotherapy, and kidney dialysis.

In large part, the increase in demand for home health care is a function of costs. The average home care visit costs about $60 compared with an average cost of $411 for a day's hospital stay. Health insurers have noticed the cost differential and encourage home care, whenever feasible, as a viable alternative to hospital confinement.

Home health care reflects the national trend for out-of-hospital treatment, as witnessed by the increasing number of home births, birthing centers, and hospice centers. The majority of home health care patients are over the age of 65.

The industry was assisted by a provision of the 1980 Omnibus Reconciliation Act that expanded Medicare reimbursement for home health treatment. Medicare now covers part-time skilled nursing care, health aids, occupational therapy, medical social services, and supplies and equipment.

Quality of home health care continues to be of prime concern. To ensure quality, personnel and patients must be thoroughly trained. Reimbursement can be another problem, since Medicare does not include all home therapies and coverage by private insurance varies widely. The general opinion, however, is that home care is equal to or better than hospital care for the nonacute stages of many diseases.

Hospice Centers

Terminally ill patients also have increased the demand for out-of-hospital care through use of hospice centers. These centers are designed for individuals whose life expectancy is usually six months or less. The major aim of hospice centers is to provide greater personal comfort and dignity to those who need medical attention, but for whom a cure does not exist. Hospice care has also been shown to be more cost effective than traditional hospital confinement.

Beginning in late 1983, Medicare hospital insurance began providing financial assistance to individuals who elect to receive hospice care instead of other Medicare benefits, except services of the attending doctor. Covered services include hospice care therapies, medical social service, nursing care, short-term inpatient care, and outpatient drugs for pain relief. Special benefit periods, daily co-insurance amounts, and coverage requirements apply to hospice care benefits in the same manner as other Medicare benefits.

GOVERNMENT'S INCENTIVE APPROACH TO CUTTING HEALTH CARE COSTS

With the demise of the numerous multifaceted and extraordinarily expensive national health insurance bills of the 1970s, the federal government has taken a different approach to public health programs. Now the emphasis is on cost control.

Encouraged by cost reductions resulting from increased competition in health care, a large number of legislators support proposals that seek to change the incentives in an insurance system that has hidden the real cost of health care from physicians and patients alike. In the long run, they hope to alter consumer attitudes about how much health care they need.

In 1983, Congress enacted a law that many predict is just the beginning of a system that will include strong competitive forces and new incentives for holding down health care costs.

Reducing Expenditures

In 1983, Congress passed the "prospective payments" system for hospitals. Since Medicare's inception, hospitals had been paid for all costs incurred in treating the elderly. This approach gave hospitals the incentive to retain patients longer and run more tests than sound medical practice might dictate. With the 1983 law, Medicare's fixed payments for specific hospital treatments are determined in advance. Hospitals that can perform a procedure for less than the stipulated rate are allowed to keep the difference. Hospitals with costs above the fixed rate must absorb the loss. A major advantage of the new rate-setting system is that the government can better estimate what its costs will be in any given year. A weakness in the prospective payment plan is that only hospital costs are predetermined. Physicians' fees and other medical costs under Part B are to continue to be reimbursed as incurred.

CONCLUSION

Over the past two decades, Americans have come to expect the best medical care money can buy. By 1983 it was apparent that unless changes were made, Medicare was headed for bankruptcy. The prospective payment system reflects the collective concern over the political fallout that would occur if Medicare became insolvent. With the enactment of this system, health care analysts now believe that Medicare will be financially solvent at least to the year 2000.

As a result of recent federal approaches to containing medical costs, the delivery of health care in the United States has undergone a significant change. Hospitals have cut costs, slashed prices, turned to creative marketing, and adopted cost-effective technology in diagnosis and treatment. Although Americans are still spending more than a billion dollars a day on health care, they are visiting doctors and hospitals less often. Hospital stays have also become noticeably shorter.

The transformation in health care delivery has produced heated controversy. Evidence exists that hospitals are releasing patients before they are ready to be discharged; readmitting patients with the same diagnosis; improperly classifying patients; and unbundling services to shift the source of payment from Medicare's Part A to Medicare's unregulated Part B. Another controversial issue concerns the charge that some people are being denied treatment at private hospitals because they cannot afford to pay and are being sent to overburdened public hospitals. It is contended that without changes, the present system will make low-income patients and patients with complex medical problems undesirable to treat. The new system will continue to be a barrier for these patients and will drive public and inner-city hospitals further into deficits.

In the private sector, difficult choices will continue to confront businesses and their employees. Private businesses spend billions of dollars annually on health insurance premiums and have experienced sharp increases in premium costs. Company-paid insurance programs are now a major bargaining item in many industries as employers seek to prevent overuse of health care services. Cost sharing by employees is now seen as the most efficient approach to overuse. Businesses are presenting a wide variety of options, including HMOs and PPOs (preferred provider organizations), that involve choice and flexibility in selecting medical care.

The health care industry is the third largest industry in the economy. In both the public and private sectors, the undeniable trend is toward greater competition and free-market incentives to guide decision making.

QUESTIONS FOR DISCUSSION

1. What role should the federal government assume in guaranteeing adequate health care?
2. How might competition be increased in the health care field?
3. Are hospitals already too competitive in that the same expensive technology is duplicated by hospitals in the same marketplace?
4. To what extent should employees contribute to employment-related health insurance premiums?
5. Given longer life expectancy, improved medical technology, and rapidly increasing health costs, how does economic analysis enter into a decision to prolong the life of a seriously ill aged patient?
6. In your opinion, is there truly a "doctor shortage"?
7. With the implementation of Medicare's prospective payment system

and the increased popularity of HMOs, do you believe the quality of medical care has been affected? If yes, how?

8. What steps do you think could be taken to reduce the costs of Medicare and Medicaid?

SELECTED READINGS

"Deregulation: What the Doctors Ordered." *Economist* (December 11, 1982).

Duncan, S.J. "What's Next on Health Cost Control." *Nation's Business Review* (January–February 1984).

Fuchs, Victor R. *The Health Economy*. Cambridge: Harvard University Press, 1986.

Russell, Louise H. *Is Prevention Better Than Cure?* Washington D.C.: Brookings Institution, Inc., 1986.

Sorkin, Alan L. *Health Care and the Changing Economy*. Lexington, MA: Lexington Books, 1986.

———. *Health Economics: An Introduction*. 2nd ed. Lexington, MA: Lexington Books, 1984.

10

CLEAN AIR
ARE WE
WILLING TO
PAY THE
PRICE?

The problem of pollution attracted national attention in the 1970s and became one of our most hotly debated issues. The passage of comprehensive environmental legislation gave evidence of our national commitment to cleaner and healthier surroundings. Today, the debate no longer centers on the question of whether the environment should be protected, but rather on how environmental protection can be attained in the most cost-effective way.

In spite of all that has been accomplished, more can be achieved. These continuing efforts will prove increasingly costly as more difficult choices must be made. The need for an integrated approach to pollution has long been recognized. The role of economists has become increasingly important as federal policy has shifted toward free-market incentives to preserve the environment.

While recognizing the importance of water, solid waste, noise, and radioactive pollution problems, this chapter focuses on the problem of air pollution since it is presently the most widespread and well known of all pollution problems.

AN ECONOMIC PROBLEM

Air is a resource provided by nature without charge and is available in abundance to anyone wanting to use it. Since air is so copious, it is rather difficult to place a price tag on it. For this reason, air has been classified as a "free" good by the economist, and free goods are outside the realm of economic analysis. Public acceptance of this classification actually stimulates pollution. Because air is a free good, the detrimental social consequences caused by misuse of air are not directly allocated to the costs of producing goods and services.

Air, then, is underpriced; and because it is used without regard to its purification, clean air is becoming increasingly scarce in many urban areas. The contamination of our atmosphere forces a change in the concept of air as a free good. On the contrary, clean air must be classified as an "economic" good. The scarcity of clean air is evident by the fact that under certain conditions a price must be paid to acquire it.

The mere presence of air pollution represents a marked departure from the workings of a perfect market economy. If perfect competition were operative, all economic resources would be directly allocated and the welfare of society would be maximized. Prices would reflect the true costs of production and provide automatic, socially valid guidelines for investment and production.

The existence of air pollution, however, shows a misallocation of resources. Pollution implies that some portion of the total production costs are being externalized and are being borne by other economic units in our society. Externalities arise because the private marginal cost function that dictates the behavior of a profit-seeking firm is less than the marginal cost to society. Unless all costs of production are internalized, a firm's cost structure will not accurately determine the firm's product price and its optimum output level. The firm will underprice its product, overproduce, and reap greater profits as a result. Other firms in the area, however, are forced to account for costs stemming from pollution because of damaged crops and deteriorated materials, structures, and machines. Consumers of products produced by the pollution-causing firm also absorb these costs, for in purchasing these products they forego the opportunity of acquiring other products that could have been produced with these additional resources. The pricing mechanism is not reflecting the alternative uses to which these resources might be put; thus, there is a misallocation of resources.

However, if all costs of production, including the external cost of pollution, are fully internalized, the firm's product will sell at a higher price and fewer units will be sold. Now the additional resources can be channeled into producing alternative products. The value of these alternative products, which could not have been produced if the cost of pollution had not been internalized, represents the external cost of pollution to the consumer.

In addition to actual cost outlays by firms and foreclosure of consumer opportunities, air pollution harms our health, affronts our senses, and lessens our enjoyment of life. In analyzing the imperfections of the real world, however, the economist does not advocate the universal elimination of externalities, because the net benefits may be negative. Also, in many cases the costs of preventing air pollution may exceed the costs of depollution. Rather, the economist is mainly concerned with a method whereby all costs and benefits of an economic activity are included in the firm's economic decision-making process. Thus, the economist is indeed very much concerned with the problem of air pollution.

POLLUTION-CAUSING FACTORS

Air is a gaseous combination of oxygen, nitrogen, and argon. Also included to a much lesser extent are helium, hydrogen, krypton, neon, xenon, and carbon dioxide. These gases constitute the earth's atmosphere, the habitable portion of which is but a relatively thin layer tightly encompassing the earth. The earth's total air supply seems vast; it is estimated by scientists to be in the neighborhood of six quadrillion tons. But still people have, in their own inimitable way, managed to crowd, dirty, and deplete this resource to the extent that the air we breathe is injurious to human, plant, and animal life. The crux of the problem is that the total volume of air is much less a concern than is the availability of sufficient quantities of fresh air in a given place on the earth's surface at a given time.

Air is never really pure. Even excluding people-related pollution activities, some contamination of the air from natural processes occurs all the time. Natural occurrences, such as volcanic eruptions, forest fires, dust storms, and vegetation decay, discharge a variety of gases that pollute the air. But the pollution resulting from such natural phenomena rarely looms large in the atmospheric pollution problem. What is a major concern is the pollution of the air stemming from the activities of people, whether they be farming, manufacturing, or simply moving about. By emitting a vast number of gases, fumes, and particles into the sky, humanity has caused a significant pollution overload. When combined with certain environmental factors, this overload might produce major ecological consequences.

Industrialization

Air pollution is directly related to a nation's level of output and use of economic goods and services. Thus, it is a by-product of affluence. The most affluent nation in the world, the United States, is therefore confronted with an air pollution problem of great magnitude. Extended economic growth requires greater participation in traditional pollution-causing activities. More electric energy results in greater coal consumption; more automobiles result in greater use of gasoline; more garbage results in a greater number of public dumps; and more materials processing results in greater industrial waste emissions.

Technological advances are constantly being made that not only increase the variety and number of goods at our disposal but also load the air with thousands of new pollutants. In areas with a concentration of such industrial activity, there is likely to be an acute air pollution problem.

Urbanization

Air pollution also tends to be directly related to population density. More than ever before, Americans are crowding themselves together in less and less land. At present, over 76 percent of the population is concentrated on just over 16 percent of the land. Urban areas, which have tripled their populations since 1940, now account for over 73 percent of the nation's population.

This increased urbanization process has inevitably resulted in increased atmospheric pollution levels in our cities. As more and more of our people congregate in urban areas, they cause larger amounts of pollution and, in turn, are exposed to larger amounts of pollution without a corresponding increase in the available air supply. As a result, no major American city is without an air pollution problem. The essence of the problem is that it is becoming increasingly more difficult for us to continue to crowd together and, at the same time, survive amid our own wastes.

Weather

Air movement is essential to pollutant dispersion. As a rule, the earth's atmosphere is capable of cleansing itself of the various pollutants discharged from heavily populated, industrialized areas as long as either horizontal or vertical air currents prevail. Should both of these air currents be absent, however, an air pollution disaster may be in the making.

Without horizontal wind movement, the sole means of dispersing pollutants in the atmosphere is vertical currents. Ordinarily, atmospheric temperature is inversely related to height. The temperature of air falls by 5.4 degrees Fahrenheit for every 1,000 feet above the earth's surface. This temperature gradient allows the atmosphere to rid itself of pollutants because the warm polluted air, being lighter, can rise into the cooler air and disperse. However, if the temperature decrease is less than 5.4 degrees Fahrenheit per 1,000 feet, the warm air, unable to rise because of the existence of even warmer air above it, then hovers over the sources of pollution. Pollutants concentrated in the lower stratum thus become trapped. This situation is known as a *thermal inversion*.

There are two basic ways in which a thermal inversion can occur. One type of inversion occurs at night when the earth's surface loses the heat radiated by the sun during the day. As the earth cools, so does the air in the lower stratum. The upper air, however, remains warm and acts as a ceiling to prevent the cooler air from rising. The lower stratum of air, then, remains polluted. The inversion will normally persist until the sun's rays of the following day warm the earth's surface, making the lower air once again warmer than the air above it. Unfortunately, this "radiation" inversion traps pollutants emitted during the peak hours of pollution-causing activities. It begins in the evening when auto transportation is at a peak and continues long enough in the morning to overlap the morning rush hour.

The second type of thermal inversion is of a greater latitude and longer duration. It stems from windless high-pressure systems, which can blanket an entire section of the country. As such a mass of cold air approaches, it moves beneath a layer of warmer air, creating an inversion that can last for weeks. The danger will continue until a moving storm or weather front arrives to break up the inversion. In the eastern part of the United States, inversions of this nature often occur in late summer or early autumn, creating the hazy "Indian summer" weather. U.S. Weather Bureau studies indicate that these inversions are occurring about 25 percent of the time throughout the United States. The role of inversions in air pollution crises is a major one. They have been present in every air pollution disaster involving death and serious illness.

POLLUTANTS AND SOURCES

Transportation, industrial processes, solid waste disposal, and fuel combustion from stationary sources are the major sources of air pollution. Table 10-1 compares air pollutant emissions by pollutant and source for the years 1970 and 1983. It is obvious from the data presented in the table that a great deal has been accomplished in reducing the amount of harmful pollutants since the Clean Air Act of 1970 was enacted. How much more can be achieved will depend upon the economic benefits and costs associated with alternative solutions.

Historically, public concern over pollution centered primarily on discharges of smoke and visible particles. In recent years, however, greater attention has been paid to gaseous emissions. Transportation, dominated by the automobile's internal combustion engine, is clearly the largest single source of carbon monoxide and lead emissions. Stationary-source fuel

Table 10-1

Air Pollutant Emissions, by Pollutant and Source: 1970 and 1984

(In millions of metric tons, except lead in thousands of metric tons)

| | Total Emissions | Transportation | | Controllable Emissions Stationary Fuel Combustion | | Industrial Processes | Solid Waste Disposal | Misc. Uncontrollable |
		Total	Vehicles	Total	Electric Utilities			
1970:								
Carbon Monoxide	98.3	71.8	62.7	3.9	0.2	9.0	6.4	7.2
Sulfur oxides	28.2	0.6	0.3	21.3	15.8	6.2	z	0.1
Volatile organic compounds	27.0	12.3	11.1	0.9	z	8.7	1.8	3.3
Particles	18.0	1.2	0.9	4.5	2.3	10.1	1.1	1.1
Nitrogen oxides	18.1	7.6	6.0	9.1	4.5	0.7	0.4	0.3
Lead	203.8	163.6	156.0	9.6	0.3	23.9	6.7	z
1984:								
Carbon Monoxide	69.9	48.5	41.4	8.3	0.3	4.9	1.9	6.3
Sulfur oxides	21.4	0.9	0.5	17.4	14.5	3.1	z	z
Volatile organic compounds	21.5	7.2	6.0	2.6	z	8.4	0.6	2.7
Particles	7.0	1.3	1.1	2.0	0.5	2.5	0.3	0.9
Nitrogen oxides	19.7	8.7	6.8	10.1	6.6	0.6	0.1	0.2
Lead	40.1	34.7	32.6	0.5	0.1	2.3	2.3	0.3

z = Less than 50,000 metric tons.

SOURCE: U.S. Environmental Protection Agency, *National Air Pollutant Emission Estimates, 1940–1984.*

combustion constitutes the second largest source. Electric generating stations that use coal and oil as fuel account for most of the sulfur oxide emissions produced by stationary sources. Industries of various sorts and sizes contribute a major share of both gaseous and particulate matters found in the atmosphere. Iron and steel mills, petroleum refineries, chemical plants, smelters, sawmills, and rubber manufacturers have traditionally been among our worst offenders. Solid waste is the fourth major source of air pollution.

However, data contained in Table 10-1 are presented in terms of weight emitted, and conclusions drawn on this basis can be misleading. In evaluating environmental quality, emphasis should be directed to the effects of various emissions and not merely the amount emitted. It can be shown that a ton of sulfur oxides is likely to have a greater environmental effect than a ton of carbon monoxide. Also, the presence of some pollutants in the atmosphere will influence the effects of others. Sulfur oxide effects, for example, are worsened by the existence of particulate matter in the air. In addition, some pollutants may react with others to form new substances. The effect of these reactions can be influenced by factors such as temperature, relative humidity, sunlight, and pollution concentration. Finally, national figures can be misleading because they do not indicate the regional differences in pollution levels. An uneven distribution of population and industrial activity may result in pollution problems that differ in severity.

EFFECTS OF AIR POLLUTION

For years, air pollution was generally thought to be a small price to pay for continued economic progress. Relatively few scientists questioned its impact on property and health. Major air pollution episodes were considered isolated events caused in large part by unfavorable climatic conditions. With an increase in the frequency and severity of such episodes in metropolitan areas, however, it became evident that air pollution was more widespread and costly than ever imagined. The conditions of "smog," "fog," and "haze" were viewed with suspicion and found to be in many cases euphemisms for dangerously contaminated air.

Major Episodes

The first recorded air pollution catastrophe in modern times occurred in the Meuse River Valley in Belgium from December 1 to December 5, 1930. Heavy industry characterized the economic structure of the valley, and substantial amounts of sulfur dioxide and particulate matter were being regularly discharged into the air. During the first week of December, a static air mass hung over the valley. This stationary air mass was accompanied by heavy fog. The result was a thermal inversion. Trapped by the ceiling of warmer air, industrial wastes became concentrated in the motionless air, causing a serious pollution overload. By the time the inversion lifted—four days later—63 persons had died and approximately 5,000 had become seriously ill.

A similar episode occurred in Donora, Pennsylvania, during October, 1948. Donora is a small industrialized town situated on the banks of the Monongahela River, some 30 miles south of Pittsburgh. The daily radiation inversion, which normally cleared around noon, did not lift on October 26. A windless high-pressure system had blanketed the entire eastern section of the country, and in Donora a recent rainfall added fog to the inversion. Particulate matter and large amounts of sulfur dioxide, discharged by industrial plants, riverboats, and trains, saturated the atmosphere. This low-hanging air mass continued for four days. During that time, 20 persons died and nearly 6,000 of the 16,000 residents became ill. Like the incident in the Meuse River Valley, this disaster was caused by the combination of pollutants and thermal inversion accompanied by fog.

On December 5, 1952, a killer smog settled over London, England. Pea-soup fog and coal smoke combined in the inverted atmosphere. Since it was December, the situation was immeasurably aggravated by the widespread use of soft, smoky coal in household furnaces. As the black smoke belched from chimneys throughout London, the city's air became inundated with pollutants. This disaster lasted a full four days, killing approximately 4,000 people and resulting in numerous respiratory illnesses. But unlike the Meuse Valley and Donora episodes, household coal burning, rather than industrial wastes, was the most probable source of the pollution.

Some areas in the United States experience pollution overloads all too frequently. In November, 1971, Birmingham, Alabama, a steel-producing center, was severely affected by a thermal inversion lasting three days. It was not the first air pollution crisis for the city. In fact, there had been a similar crisis only eight months before. The November crisis, however, was more severe. On the second day of the inversion, the pollution count had climbed to 771 micrograms of particulate matter per cubic meter of air. This is three times the level at which negative health effects are produced. Particularly affected were the elderly, the young, and those people suffering from cardiovascular or respiratory ailments. It was the sixty-sixth time in 1971 that the danger level in Birmingham had been passed.

Emissions declined in the following days as industrial plants curtailed operations and a westerly wind broke up the inversion. During the inversion, 5,000 workers were laid off with a loss of $400,000 in wages. The costs to human health have not been ascertained, but they were undoubtedly large.

These major air pollution episodes are unique. But, as deadly as they are, persistent air pollution may be more harmful in the long run. Persistent pollution involves the daily low-grade contamination of our atmosphere. Its effects are insidious, and its costs are extremely high.

Costly Effects

It is difficult to accurately assess exactly what low, persistent levels of air pollution do to human beings. But air pollution has been linked to a number of respiratory ailments, including lung cancer, emphysema, asthma, and chronic bronchitis. Particulates, sulfur oxides, and nitrogen oxides have all been associated with acute respiratory disease. Other pollutants, such as asbestos fibers and lead particles, are known to be extremely hazardous, even in small amounts. The total number of dangerous pollutants and their effect on human health is unknown, and the introduction of new chemicals and new uses for existing chemicals make measurement even more difficult.

Property is also subject to immense damage from air pollution over time. Air pollution is responsible for abrasion, corrosion, tarnish, soil, cracks, and weakening of materials, structures, and machines. Sulfur compounds in the atmosphere are particularly damaging. They are known to attack and destroy even the most durable of materials. Sulfur dioxide attacks iron and steel; rots leather; destroys cotton, wool, and nylon fabrics; and harms upholstery. Sulfuric acid in the air causes sulfates to form on the surface of stone. These sulfates dissolve in water, wearing away buildings and statues. Limestone, marble, roofing slate, and mortar are especially vulnerable to attack from sulfuric acid. Hydrogen sulfate, another sulfur compound, reacts with lead compounds and blackens homes painted with lead-based paints. It also tarnishes both copper and silver.

Ozone, a product of photochemical smog, cracks rubber rapidly. Ozone produces heavy costs to car owners and the telephone and electrical industries. Particulate matter soils cars, homes, clothing, and buildings. This necessitates the steam cleaning of buildings, additional cleaning in the home, and more frequent cleaning of cars. In addition, it has a negative effect on real estate values.

Every metropolitan area in the country now experiences some damage to vegetation from air pollution. Livestock and vegetation losses stemming from currently known pollutants probably amount to hundreds of millions of dollars each year. Trees, shrubs, flowers, vegetables, fruits, and grain are all being damaged by air pollution. Sulfur dioxide, ozone, and fluorides are known to be major destroyers of plant life.

Sulfur dioxide unites with water contained in leaf cells to form sulfate, which in turn kills off plant cells. Cotton, wheat, barley, and oat crops are especially susceptible to sulfur dioxide poisoning. Ozone enters the underside pores of a leaf and begins destroying cells under the leaf's surface. Grapes, tobacco, and spinach are examples of the dozens of crops injured by ozone. Fluorides destroy plant life by accumulating on the tips of leaves, causing them to wither. With increased accumulations of fluoride, the entire plant may die. Corn, peaches, and flowering plants of many varieties are severely damaged by fluorides.

Air pollution also represents a definite safety hazard to land and air transportation because it reduces visibility. When combined with fog, pollutants from industry and smoldering refuse can present extremely dangerous driving conditions. A motorist entering a cloud bank of smoke and dust is inclined to swiftly apply the brakes for lack of visibility. As other automobiles enter the cloud bank, a chain collision results. Several major accidents of this sort have occurred along the New Jersey Turnpike as well as on other major thoroughfares throughout the country.

Air pollution can also be hazardous and costly to air transportation. Air pollution mixed with normal fog conditions can result in costly delays to both travelers and airlines. Worse yet, this combination may also be responsible for tragic airplane crashes.

COMBATING AIR POLLUTION

Increased knowledge of the dangers of air pollution, along with changing public opinion, caused Congress to enact federal legislation aimed at regulating air quality. Undoubtedly, the Donora episode served as a catalyst in making air quality a major national priority. In hindsight, however, the federal-state partnership to preserve the environment would have been more effective if greater attention had been given to research into the problem. Federal legislation was passed while the technology needed to control pollution was still unknown or unproven and the costs of cleanup uncertain.

Federal Programs

The federal government's fight against air pollution was launched in 1955 with the passage of the Air Pollution Act. In retrospect, the act was but a modest beginning toward cleansing our nation's air. The act primarily dealt with research into the nature and extent of the nation's air pollution problem. The Public Health Service was authorized to prepare or recommend research activities, conduct studies, disseminate information, and provide limited funding to private and public agencies for surveys, research training, and demonstration projects.

Although not of great importance in itself, the Air Pollution Act of 1955 provided the basis for a series of landmark amendments to the Act in subsequent years. The Clean Air Act of 1963 authorized the Public Health

Service to take corrective action in areas in which air pollution was an interstate problem. It could also grant money to local agencies to initiate or expand their control programs. Local areas initiating or expanding control programs were eligible to receive a two-thirds subsidy from the federal government toward the cost of a program. In 1965, amendments to the Clean Air Act gave the federal program authority to curb motor vehicle emissions. Federal standards were first applied to 1968 motor vehicles.

Current federal government activity in air pollution abatement and research stems from the Air Quality Act of 1967 and the Clean Air Act of 1970. The Air Quality Act represented a systematic effort to deal with air pollution problems on a regional basis. It called for states to set air pollution standards on a regional basis and for regional standards to be enforced, locally if possible. It also substantially strengthened the powers of local, state, and federal authorities in matters of pollution. The work accomplished under the 1967 legislation paved the way for enactment of the Clean Air Act of 1970.

The Clean Air Act is undoubtedly the most controversial and comprehensive federal pollution control program. The 1970 act was the first law to call for national, uniform air quality standards based on geographic regions. The Environmental Protection Agency has the authority to enforce two sets of standards, primary and secondary. Primary air quality standards concern the minimum level of air quality that is necessary to keep people from becoming ill. These levels are based on the proven harmful effects of individual pollutants. Secondary standards are aimed at the promotion of public welfare and the prevention of damage to animals, plant life, and property in general. Within each geographic region, state governments may determine how national air pollution objectives are to be reached.

Automobile emissions received particular attention in the 1970 act. New cars must meet EPA emission standards, which are applicable to vehicles and engines for their useful life, five years or 50,000 miles, whichever comes first. The effect of these amendments was to require a virtually emission-free automobile by 1976. Since leaded gasoline has been shown to impede the effectiveness of pollution control devices and is a danger to human health, the EPA required that nonleaded gasoline be made available for all 1975 automobiles.

The Act required the EPA to set standards of performance for new and modified stationary sources of pollution. This has resulted in direct emission limitations on all major pollutants from specified types of sources. For all existing unmodified sources in specific categories the states are required to set state performance standards.

The Clean Air Act was amended again in 1977. These amendments included an extension of the auto emission abatement schedule and required the use of the best available control technologies for new manufacturing plants and electric utilities in order to limit the emission of harmful pollutants.

Faced with the possibility that no new 1978 model cars could be produced because of the inability to comply with pollution control laws, Congress substantially relaxed the schedule for abatement of auto emissions. Also, the 1977 amendments increased the amount of nitrogen oxides that would be legally permissible in 1981 and thereafter. The law allowed nitrogen oxide emissions to remain at 2½ times the levels permitted under the original standards of the 1970 Act.

In addition, the 1977 amendments provided that stationary sources of pollution would be given an extension until 1979 to meet clean air requirements, after which they would become liable for penalties calculated to remove the economic incentive for noncompliance. In any area where air quality standards have not been fully attained, no new industrial plant can be built unless the state has adopted an acceptable air pollution control plan that will ensure compliance. Any new source of pollution is required to install the best available control technology as defined by the federal government.

Under the Act, citizens are specially authorized to take civil court action against the private or governmental officials who fail to carry out the provisions of the law. Public hearings are required at various steps in the standards setting, enforcement, and regulatory procedures so that all interested persons can make their feelings known.

State and Local Programs

To be effective, national programs to prevent and abate air pollution must function at all levels of government. Many states have tightened pollution control standards or expanded their coverage to new pollutants or activities. State governments possess regulatory authority to combat air pollution and often set a precedent for federal action. California's automobile emission laws, stemming from air pollution problems in the Los Angeles basin, are an example. Air pollution laws in California set the stage for national legislation in this area. The federal government has traditionally looked to the states for effective control over pollution in order to encourage comprehensive regional programs.

With the federal Clean Air Act, the costs of state control programs have risen even more steeply than the costs of local programs, since the act places the primary control responsibilities on the states. However, it should be noted that dollars expended by states do not fully depict the adequacy of state efforts. In determining the extent of such efforts, factors such as population, pollution sources, past accomplishments, and organizational efficiency must be weighed heavily.

At the local level, early efforts to combat air pollution centered on only one aspect of air pollution—smoke emissions from fossil fuels, primarily coal. Chicago and Cincinnati led the way with smoke control laws in 1881. By 1912, most of our largest cities had similar laws. Although a few states involved themselves directly in control programs, regulation for the most part remained a local concern until the mid-1950s. But even on the local level, air pollution control continued to be primarily a matter of controlling smoke through local ordinances. Even with increased knowledge of gaseous pollutants, coupled with the realization that the problem should no longer be thought of as essentially local in character, much of the authority for setting air quality standards and for translating them into emission limitations and compliance schedules is still largely delegated to the local level.

The success of both state and local efforts is mixed, but there is a marked trend toward improvement. Although state and local governments spend billions of dollars annually for air quality control, only with continued federal funding is it likely that states and localities can sustain and increase their pollution control activities.

OPTIMIZING AND CONTROLLING POLLUTION

Two very difficult problems faced by environmental policymakers concern the amount of pollution that is acceptable to society and the methods by which predetermined pollution limits may be achieved.

Optimal Pollution Levels

Recognizing that a pollution-free environment is not a rational objective, government must set standards that allow for the existence of tolerable amounts of pollution. Not unexpectedly, the lack of an "all-or-nothing" standard has generated heated controversy over the levels of pollution that should be permitted. As pointed out in this chapter, polluted air and pollution control are both costly choices for society. The objective, therefore, is to minimize the sum of both costs and in so doing determine the optimum amount of pollution.

Figure 10-1 graphically presents the economic approach used to arrive at the optimum level of pollution. The curve labeled *PC* represents the increased cost to society of additional amounts of pollution and is read from left to right. The *CPC* curve refers to the increased cost to society of controlling additional pollution emissions and is read from right to left on the graph. By adding the cost of pollution to the cost of pollution control, a total cost curve can be constructed. The minimum point of the total cost

Figure 10-1

Optimizing Pollution Levels

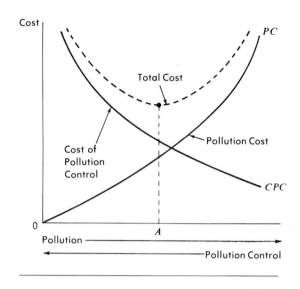

curve indicates that the optimum level of pollution occurs at point *A*. To reduce the pollution level below point *A* would be noneconomic, since the extra cost of pollution control would exceed the extra cost to society of the additional pollution. On the other hand, at pollution levels greater than point *A*, the social cost of the additional pollution is greater than the additional cost of preventing pollution. In both cases, the total cost function would be higher than the minimum cost at point *A*.

The major drawback in using this analysis to make public policy lies in the measurement of costs. Although it is relatively easy to measure the control costs of preventing pollution, it is far more difficult to calculate the costs resulting from additional levels of pollution.

Benefits and Costs

Keeping such measurement problems in mind, studies conducted for the Council on Environmental Quality presented "reasonable estimates" of both pollution control costs and reduced-pollution benefits to society for 1978. The best estimates indicate that compliance costs associated with federal air pollution programs totaled $16.6 billion, with $7.6 billion expended on air pollution stemming from mobile sources (cars, trucks, buses, etc.) and $5 billion for the control of air pollution from industrial plant sources. Utilities spent approximately $2.8 billion for pollution control in the same year. Of the total $16.6 billion incremental pollution control costs, $7.5 billion was for capital expenditures and $9.1 billion was for operating and maintaining pollution equipment.[1]

On the other hand, the benefits realized by society from reduced levels of pollution in 1978 were valued at $21.4 billion.[2] These benefits of reduced pollution damage reflect measured improvements in air quality since 1970. Of the $21.4 billion total, $17 billion represented reductions in mortality and morbidity, $2 billion in reduced soiling and cleaning costs, $0.7 billion in increased agricultural output, $0.9 billion in corrosion prevention and other materials damage, and $0.8 billion in increased property values.

Because studies conducted to quantify the benefits derived from pollution control expenditures have presented widely divergent findings, it has made the public suspicious of benefit claims. In large part, studies are conducted using vastly different assumptions, methodologies, and techniques; hence, the reader should make a careful assessment of any individual study before making inferences on its findings.

Controlling Pollution

Although our approach to controlling air pollution has been through regulation, there are other proposals calling for different approaches. In addition to regulation, three often-mentioned schemes are those of direct payments, effluent fees, and market permits. Emissions trading is a market-based incentive program to control pollution currently being used by the Environmental Protection Agency. At this point, it may be helpful to compare and contrast regulation with other possible approaches.

Direct Regulation. Of the suggested approaches, direct regulation is most often used. Usually minimum acceptable levels of air pollution are established and firms are then required to meet these standards. One advantage of direct regulation is the ease with which it is administered. This results mainly from the fact that noncompliance is easily detected. A second advantage

---

1. Pollution control expenditures are contained in Council on Environmental Quality, *Environmental Quality—1979,* Tenth Annual Report (Washington, D.C.: U.S. Government Printing Office, December 1979).
2. A. Myrick Freeman, III, "The Benefits of Air and Water Pollution Control: A Review and Synthesis of Recent Estimates" (a report prepared for the Council on Environmental Quality, December 1979). This report is recommended to those who wish to examine the benefit claims presented in the text as well as the wide disparity in the results of many studies.

is low cost. Finally, the direct-regulation method is considered equitable, since all firms in an industry face the same standards.

Opponents of direct regulation point to its disadvantages. As they see it, little incentive exists for improving air standards once the minimum standards are satisfied. Also, the system may not be as equitable as it seems. Firms may face the same standards, but each firm is, in fact, different. Individual firms have different cost structures and impose different costs on society. The application of general standards can result in too much control in some cases and insufficient control in others.

Another drawback to the direct-regulation approach is the heavy burden it places on government in both the investigating and the decision-making areas. To place these responsibilities in the hands of government is to encourage strong political lobbying for "proper" regulation.

Direct Payments. This approach entails the granting of tax reductions and subsidies for the acquisition of pollution control equipment. Direct payments may take the form of local property tax exemptions on pollution control equipment, accelerated depreciation of control equipment, or tax credits for investments in control devices. The primary purpose of the direct-payments approach is to lessen the financial burden involved in acquiring control equipment. In this manner, it serves as an incentive to invest. The biggest advantage of the direct-payments plan is its ability to gain legislative approval since firms and industries see this plan as the least painful alternative.

There are several disadvantages associated with the direct-payments approach. For one thing there is little incentive to acquire pollution control equipment, even with a subsidy. Pollution control equipment is inherently unprofitable because it adds nothing to revenue and does not serve to reduce the costs of production. Critics also see great difficulty in determining the amount of subsidy that firms should receive. Once acquired, they argue, there is no incentive to use the pollution control equipment effectively or to maintain it. Another disadvantage is that a direct-payments approach ignores the possibility of other adjustments in the production process, or the product itself, which may prove more beneficial to society. In essence, the direct-payments plan is one in which society pays the producer, thus reducing the indirect costs that the producer imposes upon society.

Effluent Fees. The third approach to pollution control involves the use of effluent fees or taxes. Some estimate of the indirect costs of pollution must be made prior to calculating and levying the appropriate fee. When implemented, the fee is adjusted to reflect the marginal cost to society of pollution-causing activities. Thus, if the fee equals the marginal social cost, the external costs of production will be internalized. Through the effluent fee, the producer is paying society for the indirect costs imposed upon it.

If effluent fees result in higher prices, they are passed on to those for whom the product is produced. This differs markedly from the direct-payments plan, in which the cost of pollution control is passed along to society as a whole.

Advocates of the effluent-fees approach claim that the obvious difficulty of measuring indirect costs is outweighed by several major advantages. First, this approach provides the incentive not only to reduce emissions at a cost less than the effluent fee, but also to develop less costly control equipment. The effluent-fees approach is also a decentralized approach insofar as it places the burden of investigation and decision making on management and not government. This approach also has the advantage of providing revenue to control agencies. Finally, the effluent-fees method is flexible in that fees could be altered according to such things as weather, time of day, and time of year.

Market Permits. A fourth approach to pollution control entails the creation of a marketplace for the buying and selling of pollution rights. After having made a decision on the maximum allowable discharges of a pollutant within the designated market area, a pollution control agency could print and auction off pollution permits. Firms could purchase permits either in the original auction or in a secondary market created by firms and individuals who had previously purchased permits in the original auction. Only after having acquired the right to pollute could a firm discharge pollutants into the atmosphere. Noncompliance would be handled in much the same way as violations of the direct-regulation approach.

The market-permits approach provides an economic incentive factor somewhat similar to that of the effluent-fees approach. Firms that discharge pollutants would have the incentive to seek pollution-free production processes in order to avoid the purchase of expensive pollution permits. Polluting firms would have to compare the costs of pollution control equipment to the cost of permits. In cases in which the cost of reducing or eliminating air pollution is less than the cost of permits, pollution would be reduced. In those instances in which the acquisition of pollution rights is more economical, firms would forsake pollution control expenditures. But, unlike the effluent-fees approach, the total amount of pollution is fixed. Therefore, the prices of pollution permits can fluctuate widely, depending on the supply and demand in the marketplace at a given time.

Like effluent fees, market permits automatically encourage reductions from firms that can inexpensively reduce pollution, minimizing pollution control's social cost. Hence, of the four approaches presented, economists favor the effluent-fees and market-permits approaches because they incorporate economic incentives to reduce pollution. Although recognized as theoretically sound, these approaches have been shunned for the most part in favor of the more simplistic and practical direct-regulation approach.

Emissions Trading. In recent years the Environmental Protection Agency has initiated market-based economic incentives to control pollution. Four different but interrelated incentives-based mechanisms presently comprise the EPA's emissions trading system: bubbles, banking, offsets, and netting. Each technique incorporates the incentives of the marketplace with the "command and control" approach of direct regulation. In adopting market incentives long advocated by economists, the EPA not only seeks to develop innovative and less costly methods of meeting current pollution standards, but also hopes to induce industries to control pollution beyond the letter of the law.

The *bubble* mechanism allows existing plants or groups to be excused from controls in one or more emissions sources in exchange for compensating controls in other, less costly to control, sources. The concept can be viewed as plants covered by an imaginary bubble dome with all pollutants being discharged from a single emissions pipe protruding through the domed roof. As long as pollution escaping from the bubble does not exceed stated limits, companies can choose which pollution source to control and how.

Once a firm reduces emissions through approved programs, it can gain emission reduction credits (ERCs) that can be held aside for the firm's own current or future expansion or sold in the marketplace to another firm in the region. This process of storing ERCs is known as *banking*.

If the company decides to sell its banked credits to another firm, it is said to *offset* its discharge permits. The impetus for creating offsets resulted from the dilemma created by the Clean Air Act, which appeared to prohibit growth in areas already in violation of primary air quality standards. Thus, if a firm seeks to locate a new plant in a particular area already in violation of pollution standards, the firm would have to seek out existing sources of pollution and offer to compensate them for banked credits. Therefore, the offset policy creates a market for pollution rights, with reduction in pollution undertaken by plants that can do so most cheaply.

Netting is similar to offsetting with the difference being that netting applies to firms that must reduce certain pollutants from specific sources in order to expand the use of other sources of the same pollutant. In essence, the firm is trading off pollution rights within itself.

The results of emissions trading are indicative of the workability of combining marketplace incentives with compliance standards. In Bristol, Pennsylvania, a 3M factory used the bubble concept to reduce emissions from one source by 1,000 tons a year more than would have been achieved by direct regulation. Savings to the company were estimated at $3 million in capital costs and over $1 million a year in operating costs.

The General Electric Company's Louisville plant faced a decision to spend $1.5 million to install pollution control equipment to retrofit an old process line or shut it down. Instead, GE opted to lease emissions credits banked by International Harvester, which found compliance relatively easy. The arrangement cost GE $60,000 instead of the estimated $1.5 million. The Narragansett Electric Company in Rhode Island recorded annual savings of $3 million and 600,000 barrels of crude oil by trading ERCs, and Armco Steel saved approximately $15 million in capital costs alone for particulate-abatement equipment.

The EPA estimates that approved bubbles have saved industry an average of some $3 million. In a 1982 report, the Government Accounting Office stated that an open market in air pollution entitlements could in some instances save industry as much as 90 percent in pollution abatement costs, as compared to direct regulation, without sacrificing the benefits of good air quality.

ACID RAIN

In recent years the nation has become aware of another form of pollution threat—acid rain. In fact, acid rain is not a new environmental phenomenon, but it has gained widespread attention because of its adverse impact on lakes, streams, forests, and soil.

Rain and snow are normally slightly acidic, but rain falling on much of the northeastern United States and southeastern Canada is much more acidic than that resulting from natural causes. *Acid rain* is the name given to rain composed of large amounts of sulfuric acid and nitric acid, which are products of reactions involving sulfur dioxide, nitrogen oxides, and water. Sulfur dioxide is a by-product of the burning of coal, gas, or oil in generating plants and industrial boilers, while nitrogen oxides are emitted by automobiles, trucks, and other mobile sources.

Few hard facts are known of the specific causes and effects of acid rain. The technical difficulties in understanding acid rain can be traced to its general nature. First, the sources of acid rain precursor pollutants are widespread and diverse and stem from both man-made and natural activities. In most ecosystems, the relative contributions of each cannot be easily distinguished. Second, unlike localized air pollution, acidic compounds are transported hundreds of miles before being deposited. The complicating effects of local emissions, variable wind patterns, and atmospheric chemical transformations preclude simple correlation between specific sources and specific effects. Third, the gases emitted into the air are not necessarily those that ultimately damage the environment. Sulphur and nitrogen react in the atmosphere with oxidants, sunlight, water, and heavy metals and are transformed into sulfuric and nitric acids. Chemical conversions of this sort complicate regulatory attempts to define how and where emissions should be controlled. Fourth, environmental damage from acid rain may be cumulative over long periods of time and quantifying the extent of acid rain's effect is difficult using short-term data.

The lack of a clear understanding of the acid rain problem does not mean that there is no cause for concern. Scientists point out that at the present time the major demonstrated effects of acid rain are in aquatic ecosystems. Acid rain is killing fish in small high-altitude lakes in the Adirondack region of New York and eastern Canada. These lakes tend to have granite under their drainage basin with little soil cover and are not able to neutralize acid deposits from rainfall.

Little is known of the effects of acid rain on terrestrial systems. Currently no direct evidence has been found to connect forest damage or agricultural crop damage with acid rain. However, preliminary scientific studies appear to indicate that acid rain may contribute to a decrease in forest growth and farm yields.

Although the northeastern part of the country appears to experience the worst acid rainfall, it is also found in the Midwest and in West Coast cities. In the Northeast, the average pH (acidity) of rainfall is now between 4.0 and 4.5, with some rainfalls having a pH of 3.[3] This is approximately equivalent to the acidity of lemon juice or vinegar.

Since coal is considered a major culprit, residents in the Northeast and Canada are blaming Ohio, Illinois, and West Virginia because of their extensive coal-fired pollution activities. Naturally the coal mining and electric utility industries located in these states are quite sensitive to such criticism since they have spent an entire decade faced with tough and costly energy-versus-environment decisions. The thought of massive expenditures and greater managerial uncertainty in the years ahead is not a welcome one.

The problem of acid rain was not included in the Clean Air Act of 1970, nor was any reference made to acid rain in the Act's 1977 amendments. Consequently, the EPA does not have legal authority to mandate regulatory compliance. At some point it undoubtedly will, for acid rain may prove to be the biggest environmental issue of the next decade.

---

3. Normal rainfall has a pH of 5.6. Pure distilled water has a pH of 7, while battery acid has a pH of 1.

IMPACT OF CLEANER AIR

The implementation of the Clean Air Act has entailed major adjustments for both producers and consumers throughout our economy. Overall, despite the fact that aggregate output, income, and employment have not suffered significantly, pollution control has brought about noticeable changes in the composition of output and allocation of resources among industries.

Most manufacturing industries have found the increased burden of pollution control costly but manageable. The time during which antipollution spending accounted for a large share of total investment expenditures was relatively brief for all but a few industries. As compliance with environmental regulations continues, spending for pollution control equipment should decline, on both an absolute and a relative basis of measurement.

However, several large industries have been seriously affected by antipollution requirements. Of particular concern is the extent to which capital investment aimed at meeting pollution control standards is squeezing out investment needed to modernize and expand productive capital stock. For example, it has been estimated that over the 20-year period ending in 1993, our domestic steel industry expects to install pollution control equipment worth more than $8.2 billion in existing production facilities. The steel industry, however, is in dire need of modernization, and profit in recent years has not been sufficient to satisfy both investment needs. Consequently, the size of our domestic steel industry is shrinking as investment capital is being channeled elsewhere. This shift in resource allocation coincides with plant closings, industry unemployment, and conglomerate mergers. Despite increased protective tariffs and quotas on imported steel, the industry continues to suffer.

The electric utility industry is also beset by a number of pressing problems, not the least of which is pollution control. Because of stringent regulations contained in the Clean Air Act concerning the burning of coal of high sulfur content, electric utilities are seeking efficient ways to convert "dirty" coal into "clean" coal. One such way is to install smokestack scrubbers. Between 1980 and 1990, an estimated $127 billion in capital investment will be required for this purpose. As concerns over nuclear power and acid rain intensify, resource allocation within the industry will entail increasingly difficult and costly decisions.

Within industries, individual firms continue to face adjustments. Pollution control requirements have had varying impacts on firms comprising an industry. In large part, the extent of impact is related to the individual firm's market position. If the demand for the firm's product is highly inelastic, a large portion of control costs can be passed along in the form of higher prices. On the other hand, if many substitutes are available, the firm may have to decrease production and settle for lower prices. Over a longer period, the firm with lower profits will experience greater difficulty in acquiring capital for expansion purposes. Hence, its position in the industry will decline.

Small firms operating single plants appear most vulnerable, and a good number have ceased production. For the most part, these firms are inefficient and obsolete; they are marginal enterprises at best. In some cases a plant owned by a multiplant company may be closed because it is inefficient. The expense associated with pollution control equipment provides as good an excuse as any for the company to eliminate one of its older, obsolete production facilities. The result is a loss of jobs and a decreased supply of the commodity.

Pollution control costs are passed on to consumers in the form of higher prices. Higher-income families tend to allocate a larger share of their income for services, which by and large are not affected by pollution control costs. Hence, lower-income families may well be penalized more severely by higher prices caused by increasing pollution control costs than will higher-income families.

CONCLUSION

Prospects for the future are mixed. With increasing population, urbanization, and industrialization, the pollution crisis will worsen unless the goal of preventing and abating air pollution continues to receive the utmost national priority. Despite what has been accomplished by all levels of government in abating air pollution, more needs to be done. Research programs need to be expanded, air quality criteria for selecting air standards must be refined, federal grants must be increased, and greater emphasis must be placed upon multijurisdictional control programs.

A continuation of the shift from direct regulation to various approaches involving market incentives is likely. The increased use of emissions trading is proving to be more cost-effective and equitable than the regulatory approach opted for in the past. Bubbles, banking, offsets, and netting allow greater flexibility and individual decision making on the part of the firms as to how best to achieve state requirements.

Industry must continue to internalize the external costs of air pollution and view them as part of the true costs of production. The result can only be a more efficient allocation of resources.

Public understanding and support will also be necessary since the battle against air pollution will entail painful adjustments in our economy. The costs of many items, such as cars, gasoline, chemicals, paper, electricity, and taxes, have already increased as a result of the costs of pollution abatement. More jobs may also be eliminated as marginal producers, finding it impossible to justify expensive control equipment, close their doors.

The fight against air pollution is proving to be costly, but in the long run it will be cheaper than the costs associated with a policy of limited action. In the final analysis, it all depends upon the price we are willing to pay.

QUESTIONS FOR DISCUSSION

1. In compliance with the Clean Air Act of 1970, automobile manufacturers have made substantial reductions in air pollution emitted by new automobiles. What costs have been associated with these improvements?
2. If need be, should economic growth be sacrificed for pollution control?
3. From an economic point of view, do you believe we overestimate our country's productive capacity when we ignore air pollution in calculating the gross national product? Give specific reasons for your answer.
4. How have rapid advances in technology influenced the air pollution problem?
5. What disadvantages do you see in regional control of air pollution?
6. Does your community suffer from an air pollution problem. If so, what is the nature and cause of such pollution?
7. Must our nation control population to control pollution?

SELECTED READINGS

Annual Report. Washington, D.C.: Council on Environmental Quality, annually.

Crandall, Robert W. *Controlling Industrial Pollution.* Washington, D.C.: The Brookings Institution, 1983.

Crocker, Thomas D., ed. *Economic Perspectives on Acid Deposition Control.* Boston: Butterworth Publishers, 1984.

Freeman, A. Myrick, III. *Air and Water Pollution Control: A Benefit-Cost Assessment.* New York: John Wiley & Sons Inc., 1982.

Tietenberg, T.H. *Emissions Trading: An Exercise in Reforming Pollution Policy.* Washington, D.C.: Resources For The Future Inc., 1985.

Yanarella, Ernest J., and Randal H. Ihara. *The Acid Rain Debate.* Boulder, CO: Westview Press, 1985.

11
CRIME
WHAT ARE THE
COSTS?

In recent years crime has surfaced as one of our nation's greatest concerns. This is not to imply that the problem of crime is new, for crime is as old as humanity itself. What is disconcerting to Americans is the dramatic rise in the frequency and severity of criminal activity since the early 1960s. But despite the gravity of the crime problem, only recently have economists applied their analytical and empirical skills to the analysis of criminal activity. Traditionally, crime and criminal behavior have been largely the subject matter of sociologists and criminologists. Since the economist's treatment of the topic differs markedly from the dominant traditional views espoused by sociologists and criminologists, a lively and healthy controversy has been generated among the professionals of several disciplines.

It should be clear in view of the multifaceted nature of the crime problem that economic analysis alone cannot pretend to provide all of the solutions relating to the causes, prevention, and control of crime. But there is no doubt that an understanding of economics does provide greater insight into the determinants of criminal behavior as well as the steps that might be taken to efficiently allocate our nation's resources to the control of crime. Thus, the major contribution of economic analysis should lie in the development of broad policy guidelines that have the effect of reducing criminal activity.

CRIMINAL ACTIVITY

The most straightforward definition of crime is that it is any act in violation of the law. Since criminal law reflects the fundamental values of a society, the definition of what constitutes a criminal act varies from place to place over time. Although crime is prevalent in all modern societies, especially those that are urban and industrialized, particular forms of crime relate to the manner in which a society is organized. As a democratic, market-oriented nation, the United States protects individual political and economic freedoms, emphasizes the value of materialism, and rewards individual economic incentive. At the same time, these values not only serve to increase the motivation for criminal activity, but also permit the freedom to engage in an illegal act before being confronted with punitive measures.[1]

Although crime can be defined rather easily, the distinction between criminals and noncriminals is more complex. If we simplistically conclude that a criminal is one who has violated a law, then one would reasonably assume that nearly everyone reading this book, as well as its authors, can be classified as criminals. If this seems somewhat far-fetched, we might ask the reader if there has not been an occasion when he or she has driven in excess of posted speed limits, parked illegally, or been guilty of jaywalking. Although in the preponderance of cases we escape detection and citations from police officers, the point becomes clear: We all commit acts such as these that violate the law and thus are criminal. Yet we do not think of ourselves as criminals and, more importantly, neither does society. According to a national survey conducted by the President's Crime Commission, 91 percent of all adult Americans admitted they had committed acts for which they might have received jail sentences.[2] But the labeling of a person as a criminal only occurs when an individual violates a set of social norms that are backed up by strict sanctions. A criminal, therefore, is one who negatively evaluates the serious costs of an act on what society deems to be its best interests. Crimes against people, property, and society as a whole constitute acts that our society does not accept as being part of our conduct norm.

---

1. In a dictatorial society, individuals may be incarcerated on the grounds of being likely to commit a crime.
2. President's Commission on Law Enforcement and Administration of Justice, *The Challenge of Crime in a Free Society* (Washington, D.C.: U.S. Government Printing Office, 1967).

EXTENT OF SERIOUS CRIME

Crime can be classified in numerous ways. The source most often referred to for crime statistics is the FBI's *Crime in the United States*. These annual reports present data on major crimes classified into two major groups: violent crime and property crime. Table 11-1 presents statistics indicating the number and types of serious crimes for selected years between 1967 and 1985. The significance of the economic motivations underlying most serious criminal offenses is seen by the fact that of the 12.4 million serious crimes recorded in 1985, property crimes accounted for 11.1 million, or 89 percent of the total. But even this approach tends to underestimate the economic aspects of serious criminal activity since the commission of recorded violent crimes against people was in many instances incidental to obtaining money, property, or both.

Is crime increasing at epidemic rates as some suggest? Calculations made from Table 11-1 indicate that the number of offenses increased by 115 percent between the years 1967 and 1985. However, despite the magnitude of increasing crime, caution should be exercised prior to drawing conclusions based upon crime statistics.

In the first place, a truer picture of the changing crime rate is one that relates changes in the number of crimes to changes in population over a given time period. The Table 11-1 data show that on the basis of indexed crimes per 100,000 population, crime increased by 74 percent between 1967 and 1985. The approach that measures crime rates per 100,000 inhabitants shows that the extent to which crime has increased is somewhat lessened, nevertheless, the figure is a formidable one.

Secondly, the validity of the statistics depends upon the victims reporting crimes to local police departments and local police departments reporting these crimes to the FBI. In 1968, the number of agencies reporting crimes to the FBI totaled 6,187, but by 1985 a total of 13,616 agencies, representing 97 percent of the nation, submitted crime reports. Consequently, some of the increase in indexed crime is the result of a more extensive reporting base and is not solely due to an increase in criminal activity. In any event, the FBI does not vouch for the reliability of its crime statistics.

Thirdly, it should be noted that the statistics are presented only for the seven recorded crimes and should not be taken as a yardstick for total criminal activity. Excluded from the crime reports of the FBI are crimes that are much more numerous and in some cases just as serious as the seven recorded crimes. Considered serious but excluded from the report are such offenses as vandalism, assaults against family and children, fraud, embezzlement, and many others. However, most of the crimes that are not included are usually classified as *victimless* crimes since such

Table 11-1

Crime and Crime Rates, by Type, Selected Years 1967–1985

Year	Total	Violent Crime					Property Crime			
		Total	Murder	Rape	Robbery	Assault	Total	Burglary	Larceny Theft	Vehicle Theft
Number of offenses (thousands):										
1967	5,903	500	12.2	27.6	203	257	5,404	1,632	3,112	600
1970	8,098	739	16.0	38.0	350	335	7,359	2,205	4,226	928
1973	8,718	876	19.6	51.4	384	421	7,842	2,566	4,348	929
1976	11,350	1,004	18.8	57.1	428	501	10,346	3,109	6,271	966
1979	12,250	1,208	21.5	76.4	481	629	11,042	3,328	6,601	1,113
1982	12,974	1,322	21.0	78.8	553	669	11,652	3,447	7,143	1,062
1985	12,430	1,327	19.0	87.3	498	723	11,103	3,073	6,926	1,103
Rate per 100,000 inhabitants:										
1967	2,990	253	6.2	14.0	103	130	2,737	827	1,576	334
1970	3,985	364	7.9	18.7	172	165	3,621	1,805	2,079	457
1973	4,154	417	9.4	24.5	183	201	3,737	1,223	2,072	443
1976	5,287	468	8.8	26.6	199	233	4,820	1,448	2,921	450
1979	5,566	549	9.7	34.7	218	286	5,017	1,512	2,999	506
1982	5,604	571	9.1	34.0	239	289	5,033	1,489	3,085	459
1985	5,207	556	7.9	36.6	209	303	4,651	1,287	2,901	462

SOURCE: *Statistical Abstract of the United States: 1987* and *Crime in the United States, 1986.*

crimes generally involve some form of illegal behavior rather than criminal action against people or property. Examples of victimless crimes include prostitution, drunkenness, gambling, disorderly conduct, and drug-related offenses. It is estimated that victimless crimes account for approximately 40 percent of all arrests, and of the victimless crimes, drunkenness and disorderly conduct are by far the most prevalent.

A final consideration in drawing conclusions about changing crime rates is the fact that the definition of criminal behavior changes with the passage of time. A criminal act undertaken in one year may not be judged criminal in another, or vice versa. The recent relaxation of marijuana laws in many states is a case in point. Nevertheless, despite the aforementioned reservations relating to statistical interpretations, the general consensus is that crime is a serious problem in the country.

COMPUTER CRIME

Although not classified as a serious crime in the *Crime in the United States* reports, the general area of computer crime deserves attention. Computer crime is not a new phenomenon, for it began with the emergence of computer technology in the 1940s. What is new about computer crime is the widespread and sophisticated methods used to commit crimes previously thought to be extremely difficult, if not impossible. Without a doubt, the growing explosion in personal computers and computer literacy has made the potential costs to society from computer crime enormous.

Unfortunately, the legal system is not clear on computer crime. Historically, legal researchers felt either that the problem was merely a small part of the effect of technology on society or that, because of its complex nature, computer crime should be subordinate to each specific type of crime. The result is that while it is a crime to steal a computer terminal because it is a tangible asset, it is not a crime to steal information stored in the computer since information is intangible. Computers have been used in most types of crime including fraud, theft, larceny, embezzlement, bribery, burglary, sabotage, espionage, and extortion. Computer crime has occurred in military systems as well as engineering, science, and private business systems. Computer crime goes beyond what is thought of as white-collar crime. It sometimes includes violent crime, not only destroying computers and their content, but jeopardizing human life and well-being.

The confusion as to what constitutes computer crime stems from the fact that computer crimes differ from traditional crimes relative to the occupations of perpetrators, environments, methods, forms of assets lost, and geography. In this sense, computer technology has engendered a new kind of crime.

Computer technology has created new occupations that extend the traditional categories of criminals to include computer programmers, operators, tape librarians, and technicians, all of whom function in a specialized environment. New automated methods develop and change rapidly. The targets of computer crime are also new. Assets subject to criminal loss include electronic money as well as warehouse inventories, materials leaving and entering factories, and confidential information stored in computers.

In other criminal acts, time is measured in minutes, hours, days, weeks, months, and years. With the use of computers, however, crimes can be committed in milliseconds. Finally, geography no longer poses the constraint in computer crime that it does in other criminal acts. For example, a telephone with a computer terminal attached to it in one part of the world could be used to engage in a crime in an on-line computer system in any other part of the world.

What is Computer Crime?

Although no legal consensus exists as to what constitutes computer crime, an acceptable definition of computer crime is that it includes any illegal act for which knowledge of computer technology by the criminal is essential. The critical distinction contained in this definition can best be understood through some examples. If an individual steals a computer terminal, the act would not be a computer crime. It would be a simple theft in the same sense as if a television set were stolen. Technical knowledge of the workings of the computer is not necessary to the act any more than it would be in stealing a television. Likewise, if an individual enters a bank and deceives a bank employee into transferring $10 million by computer from someone else's account into his or her own personal bank account in Switzerland, this, too, is not a computer crime. In this case, no fraudulent act was directly related to the computer, and no specific knowledge of computer technology was necessary on the part of the criminal. If, however, the individual gains access to the bank's computer codes and makes the transfer by giving instructions to the computer, a computer crime exists because of the knowledge of computer usage and protocol.

Until recently, computer crime involved mainly illegal access to monetary accounts. With increased computer literacy and popularity of home computers, however, experts are warning of a growing computer crime wave. It appears that the target will be information rather than money. This has become particularly true of high school and college students who see the cracking of computer codes as a challenge rather than a criminal act. Teenagers using home computers have broken into computers belonging to a New York City hospital, the Defense Department, and an electronic-mail service, just to name a few. The total number of such offenses is unknown but undoubtedly large, and verified reports indicate that company secrets are being stolen, motor vehicle records for traffic violations are being erased, and scholastic records are being altered.

Unfortunately, society has looked upon computer information raiders as computer geniuses, not criminals. But, in fact, breaking into someone's computer is little different from breaking into someone's home—an act society deems illegal. Society's passive stance in this matter is reflected in the 1983 movie *War Games,* which popularized the underground culture that views breaking into a computer system as an indoor sport.

Cost of Computer Crime

No one can begin to estimate accurately the cost of computer crime, but one sign of its potential cost is the fact that computer-fraud coverage is the fastest-growing form of insurance policy against criminal acts. In 1986, insurance analysts estimated that more than $100 million in computer-fraud policies were sold. In addition to insurance costs, companies are spending an increased share of their data processing budgets on security by regularly changing passwords, acquiring sophisticated protective software, and hiring additional security personnel. Until high-security systems are widely adopted and society demands stiff penalties for computer crimes, our economy is vulnerable to huge economic and social costs resulting from computer crimes.

CRIME PATTERNS

A general profile of crime across the country indicates that it is mainly an urban phenomenon and that criminals are likely to be young, male, and nonwhite. FBI statistics showing crime rates for violent and property crimes according to city size are presented in Table 11-2. Crime statistics presented in the table refer to offenses known to the police per 100,000 people. In all categories the direct relationship between crime rate and city size is evident. Data presented for selected cities indicate that on the basis of the seven indexed crimes, Detroit is the city with the most crime per 100,000 residents. It is the leader in crimes against persons as well as in crimes against property.

Urban areas contain large numbers of people at both extremes of the income scale: the very rich and the very poor. The obvious wealth of affluent families makes criminal activity appear very lucrative to the large number of poor concentrated in our larger cities and suffering from slum conditions and economic deprivation. Several studies cited by the

President's Commission support the view that most of the serious crime in cities is committed by individuals at the lower end of the income ladder. Research conducted in Philadelphia found that 90 percent or more of the criminal homicide offenders, rape offenders, and robbery offenders were persons ranging on the economic scale from the unemployed to skilled laborers.[3] A separate study in which a comparison was made of the crime rates of male youths from both high- and low-income areas of Philadelphia also supports the contention that there is an inverse relationship between violent crimes and household income.

Age and sex are also important factors in violent crimes. Table 11-3 depicts the numbers of arrests by age and sex for 1985. Approximately 31 percent of all individuals arrested for serious offenses committed in 1985 were "juveniles" (young people under 18 years old). Of the youths arrested for serious crimes, 79 percent were male. Many of those arrested were not yet teenagers. The total number of juvenile arrests in 1985 for serious crimes was about 519,000. Young people under 18 accounted for 8.5 percent of all arrests for killing, 26 percent of those for robbery, 15 percent of those for forcible rape, 38 percent of those for burglary, 36 percent of those for auto theft, and 33 percent of those for larceny theft. Of all the aspects of crime today, what alarms Americans most is the extraordinarily high rate of criminality among the nation's youth.

There is also a much higher crime rate for nonwhites than for whites.[4] But these differences are thought to be primarily the result of living conditions in the urban core. Since the central cities of most major urban areas are becoming increasingly nonwhite and since crime tends to be city oriented, the higher rate of crime among nonwhites is largely a function of environment.

Interestingly, recent evidence indicates that Americans are experiencing a relief from steadily rising crime rates. Two factors largely account for this phenomenon. The first is the change in population mix. As those born in the baby boom years grow older, teenagers comprise a smaller part of the total population, and it is teenagers who are most prone to commit violent crimes. A second factor is that courts have been sending more criminals to prison and for longer stays.

Experts caution, however, that criminal activity is cyclical and may resume its upward trend with the next recession and corresponding increase in unemployment.

---

3. *Final Report of the National Commission on the Causes and Prevention of Violence* (Washington, D.C.: U.S. Government Printing Office, 1969), 22.
4. The crime rate measures the number of arrests per 100,000 whites as well as the number per 100,000 nonwhites.

Table 11-2

Crime Rates, by Type—Population-Size Groups and Selected Cities, 1985
(Offenses known to the police per 100,000 people, as of July 1.)

	Violent Crime					Property Crime			
	Total	Murder	Rape	Robbery	Assault	Total	Burglary	Larceny Theft	Vehicle Theft
Cities with population of—									
250,000 or more	1,344	19.0	78	684	563	7,606	2,156	4,339	1,111
100,000–249,999	802	10.2	52	298	441	7,002	1,949	4,471	582
50,000–99,999	536	5.9	36	186	307	5,499	1,456	3,521	522
25,000–49,999	420	4.7	28	124	263	5,130	1,265	3,480	385
10,000–24,999	307	3.7	20	70	214	4,247	1,000	2,965	283
Under 10,000	272	3.5	17	41	211	3,946	881	2,850	215
Suburban areas	341	4.7	25	88	223	3,883	1,063	2,506	315
Rural counties	177	5.6	19	15	138	1,744	668	967	108
Suburban counties	359	6.0	29	86	237	3,387	1,091	2,002	293

Selected cities:

Baltimore, MD	2,010	27.6	77	1,008	898	6,585	1,799	3,985	781
Chicago, IL	1,949	22.2	80	897	970	7,297	1,756	4,044	1,497
Dallas, TX	1,440	30.2	115	607	688	11,542	3,154	7,359	1,030
Detroit, MI	2,375	58.2	144	1,538	635	11,375	3,703	4,219	3,452
Indianapolis, IN	980	12.5	73	381	514	5,307	1,740	2,953	613
Los Angeles, CA	1,658	24.4	73	877	684	7,581	2,007	3,953	1,621
Memphis, TN	1,488	18.6	138	766	565	7,673	2,427	3,452	1,794
New York, NY	1,881	19.3	54	1,107	701	6,491	1,738	3,648	1,106
Philadelphia, PA	988	16.6	62	572	337	4,113	1,159	2,184	770
Phoenix, AZ	844	10.0	71	272	491	8,420	2,621	5,248	552
San Antonio, TX	625	20.9	95	311	198	9,062	2,843	5,350	870
San Diego, CA	632	9.7	34	310	279	6,237	1,656	3,564	1,018
San Francisco, CA	1,296	11.6	70	697	517	6,693	1,603	4,281	809
Washington, DC	1,625	23.5	54	835	712	6,374	1,598	3,973	803

SOURCE: *Crime in the United States, 1986.*

Table 11-3

Persons Arrested by Charge, Sex, and Age, 1985
(In thousands)

Charge	Male Total	Male Under 18	Female Total	Female Under 18
Serious crimes	1,327.7	417.3	360.7	101.9
Murder	11.3	1.0	1.5	0.1
Forcible rape	25.2	3.8	0.3	0.1
Robbery	96.8	25.0	8.0	1.8
Aggravated assault	181.2	25.0	28.5	4.6
Burglary	275.8	105.2	22.9	8.4
Larceny-theft	641.8	221.2	289.1	82.5
Motor vehicle theft	84.2	31.1	8.6	3.9
Arson	11.4	4.9	1.8	0.5

SOURCES: *Statistical Abstract of the United States: 1987* and *Crime in the United States, 1986.*

COSTS OF CRIME

The economic costs of crime are not easily measured, but estimates run well into the billions of dollars. A major problem in developing cost data is the uncertainty about what should be included in the cost of crime and how to arrive at dollar estimates. It is evident that different measurement approaches are required for the various types of crime. For 1965, the President's Crime Commission estimates the costs of all crime at $20.9 billion. More recent data are unavailable, but cost estimates would be appreciably greater than those for 1965 because not only has the number of crimes increased, but the economic loss involved in each crime has risen due to inflation.

The economic costs of crimes perpetrated against persons are difficult to measure because opportunity cost estimates are necessary. The costs resulting from crimes against persons would correctly include (1) the present value of future income losses in the case of death or permanent disability that prevents future employment, (2) loss of earnings if disability is temporary, (3) medical expenses, and (4) the costs of pain and suffering to victims and their families. It is because of such measurement problems that the cost of such crimes as murder and rape are so difficult to grasp. The commission calculated the cost of crimes against persons to be $815 million for the year 1965. However, this figure excluded dollar estimates for the costs of pain and suffering.

The measurement of economic loss in cases of crimes against property is somewhat less difficult since the costs of such crimes can be measured by the value of the lost or destroyed property. This is particularly true in cases of vandalism, sabotage, and arson, where there is physical destruction of property and a reduction of national wealth. From the standpoint of society, however, a complication arises in the case of theft since the value of stolen property does not change—only its ownership does. Although costly to the victim, theft essentially involves a redistribution of income from the victim to the thief, resulting in a net economic loss to society of zero. The commission listed the economic loss of crimes against property as $3.9 billion for 1965.

But the real economic cost of crimes against property is much greater than the value of diminished or destroyed property. With rising crime rates, there has been an increase in self-imposed costs to society for protection and deterrence of crime. To avoid being victimized by crime, people spend billions of dollars for such things as burglar alarms, property insurance, legal expenses, security personnel, and avoidance of high-crime areas. Crimes against property also impose hidden costs on nonvictims in the form of higher prices for goods and services resulting from such activities as shoplifting, hijacking, robbery, and burglary.

Finally, the costs of crime imposed upon society include expenditures associated with public law enforcement and criminal justice. Some of these services would be required even if crime could be eliminated. Police services such as traffic control, crowd control, and the tracing of lost persons would continue without crime. But the bulk of the costs of law enforcement, justice, and corrections are linked to crime. The governmental costs of crime prevention and control are borne primarily by local governments.

OPTIMIZING CRIME LEVELS

As a starting point, it is safe to say that in a civilized society people prefer less crime to more crime and that the amount of crime is inversely related to the amount of resources allocated to crime control. Because crime control is costly, society must compare the economic loss resulting from criminal activity to the cost of providing better crime prevention and control. Theoretically, it would be possible to achieve a zero level of crime by devoting nearly all of our resources to law enforcement agencies. However, the opportunity costs of forgone goods and services would be so overwhelming that it would be irrational to seek a crime-free society. Beyond some point the cost associated with crime reduction exceeds the benefits. Instead, society prefers to opt for a tolerable level of criminal activity that optimizes resource allocation. In this sense, the annual budget appropriations of federal, state, and local law enforcement agencies reflect the amount of crime society will accept in a given year, as well as the number of crimes it is willing to control. The optimal amount of resources can be examined by means of a cost-benefit analysis, as presented in Figure 11-1.

The economic damage resulting from additional acts of crime is measured along the marginal cost of crime curve, whereas the increased cost of crime control is measured along the marginal cost of crime control curve. The horizontal axis in Figure 11-1 represents both the amount of crime committed in a community (read from left to right) and the amount of crime control (read from right to left). The marginal cost of additional crime begins at point T, which reflects the maximum number of crimes committed with no control expenditures. At first this curve increases slowly, since in the early stages small increases in control expenditures can significantly reduce criminal activity. However, beyond some point the marginal cost of controlling additional crime is likely to increase more rapidly, indicating the extremely high extra cost in reducing crime as the number of crimes approaches zero.

The socially optimal level of crime is determined by the intersection of the two marginal functions and occurs at point X. At this point, a dollar expended to prevent an additional criminal act is equal to a reduction in

Figure 11-1

Optimizing Crime Expenditures

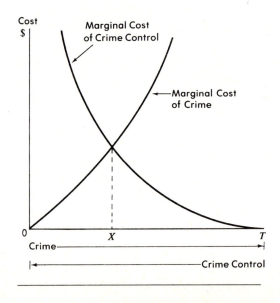

crime costs of one dollar. At crime levels greater than *OX,* the marginal benefits from crime reduction exceed the marginal costs of crime control. The result is that for every dollar society spends on law enforcement, society will save more than a dollar in damages. Thus, it is economically efficient to reduce crime by the amount *TX* and to tolerate *OX* number of crimes. To reduce criminal activity below point *X,* however, is to act irrationally because the marginal cost of crime control exceeds the marginal dollar amount of damage due to crime.

Deciding how many resources to devote to crime prevention and control is not the only major allocative decision required. Once the total budget is determined, law enforcement officials must then allocate resources among the various departments. Again cost-benefit analysis serves as a guideline. If, for example, an increase of $1.00 spent on controlling crimes such as murder and homicide yields a return of $3.00 in the form of reduced social cost, while the same $1.00 channeled into the vice squad benefits society by an amount of $1.50, then the dollar should be properly allocated to the homicide division.

The public position taken by some police officials that police departments do not enact laws but only enforce them is not sufficient grounds for employing resources as though all criminal offenses were of the same approximate severity. Police departments that seek to maximize the total number of arrests, particularly in the easier cases involving victimless crimes, in order to present the image of an efficient police force are guilty of inefficiently allocating scarce economic resources. As long as economic resources for crime prevention and control are limited, then priorities must be established according to the costs of certain criminal acts and the benefits accruing to society from their reduction. Although the use of cost-benefit analysis in determining the social costs and social benefits associated with the various amounts and types of criminal offenses is difficult, in most cases reasonable estimates can be made.

CRIMINAL BEHAVIOR

Before considering possible public policy approaches for reducing the extent of criminal activity, an economic analysis can provide insight into several motivating factors that determine criminal behavior. It is this particular facet of crime analysis that has triggered the greatest controversy.

Rational Behavior

Economics is concerned with choices and assumes that as individuals we possess the freedom to exercise our choices in the marketplace. Faced with numerous constraints that limit our freedoms, we nevertheless choose the best option, given the many choices available. To consistently act in such a way as to maximize the returns available is to behave rationally in the economic sense. This principle of behavior serves as a motivating factor for both consumers and producers.

Criminal activity may simply be another example of rational behavior since engaging in criminal acts is a matter of choice. The fact that an individual violates the law and risks a monetary fine or prison term does not necessarily imply irrational behavior, for every choice involves some costly risks. If costs associated with risk taking constituted irrational behavior, then many financial investors, entrepreneurs, steeplejacks, high-rise window washers, and Hollywood stunt performers would fail the rationality test. However, we usually do not think of their actions as being irrational since it is assumed that each of them arrived at the decision to engage in such activity after having examined the costs and benefits entailed in the next best choices. The fact that some criminals are apprehended and punished is not proof of irrationality any more than the fact that some steeplejacks fall off the Golden Gate Bridge. It only indicates that rationality need not assume perfect knowledge of future events. The implementation of a rational decision can entail mistakes, but unlike the steeplejack, the criminal usually has the opportunity of increasing professionalism by eliminating such mistakes in the future.

Benefit-Cost Approach

If rational behavior can be assumed, then economically motivated crime can be examined within the framework of benefits and costs. Consider first the expected benefits from a criminal act. Although in some cases nonmonetary returns in the form of prestige, revenge, or thrilling experiences serve as motivating forces, the gains from crime are usually of a monetary nature. Thus, for most participants crime is an income-producing activity, and it is anticipated income that constitutes the major benefit to the criminal.

But, as in other occupations, a person considering criminal activity must consider the costs of attaining the expected benefits, for it is net income, not gross income, that is important. One major cost to be considered is the opportunity cost of giving up the chance to use one's skills and abilities in legal employment. Forgone income from legitimate sources serves as the appropriate measure of income loss for the criminal. However, since most criminals are lacking in skills and education, their forgone income from legal employment opportunities can be quite low. This has the effect of reducing the cost of criminal activity, since crime may be viewed as the best available alternative.

In addition to forgone earnings from legal endeavors, the criminal must also weigh the costs of being arrested, convicted, and imprisoned. These are considered the "occupational" costs associated with crime. In essence, these are the probabilities and costs of failure.

The probability of being arrested for the crime is largely a function of the type of crime involved, the skills of the criminal, and a certain element of luck. The probability of being convicted depends upon the evidence presented and the skills of the defense attorney. A third probability estimate must be made for the likelihood of receiving a prison sentence if convicted, as well as the length of time to be served. Factors affecting the outcome include the severity of the crime, the criminal's past record, and the judge's view of appropriate punishment.

The cost of failure varies greatly among individuals. The nonmonetary cost of arrest, conviction, and imprisonment is the social stigma of having a criminal record and having "served time." For repeaters, however, these costs may be relatively low since their criminal records have been well established, and if they have been incarcerated previously, they are more easily able to adjust to prison life. But for many, the cost of imprisonment is largely the loss of income that results. Obviously, the opportunity cost of imprisonment will be greater for highly educated and skilled people. The uneducated and unskilled may experience little income loss, particularly when free room and board are provided at public expense. The significant nonmonetary cost is the individual's loss of freedom, but again this cost is nonmeasurable and differs widely among those removed from society.

A hypothetical example of the rational decision-making process is illustrated in Table 11-4. Assume that two individuals, *A* and *B,* are independently considering a robbery and the target is a local retail store that is open for business until 10 p.m. Routine surveillance plus inside information indicate that on Friday nights the store holds an estimated $15,000 in cash at closing time. Thus, $15,000 constitutes the marginal private benefit resulting from the robbery. Note that this sum represents only the anticipated value since the individuals cannot be positive that on a given Friday night the store won't be light on cash. Having reasonably estimated the marginal private benefits from the robbery, each individual must calculate marginal private costs, for in order to perpetrate the robbery, anticipated benefits must exceed anticipated costs.

Cost calculations on the part of individuals *A* and *B* would require the multiplying of expected costs associated with crime by the probability of arrest, conviction, and imprisonment. For the sake of simplicity, assume that both *A* and *B* estimate the probability of being arrested at 0.4 and the probability of conviction at 0.5. By multiplying the two probability figures, the combined probability of arrest and conviction is 0.2. If *A* and *B* expect

Table 11-4

Expected Benefits and Costs of Robbery

	A	B
Private benefits	$15,000	$15,000
Cash from retail store		
Private costs		
Probability of being arrested	0.4	0.4
Probability of being convicted	0.5	0.5
Expected length of sentence	10 years	10 years
Annual income loss from legal employment	$3,000	$10,000
Expected private costs	$6,000	$20,000
Net private benefits	+$9,000	−$5,000

a prison sentence of 10 years if arrested and convicted, then the expected sentence term is reduced to only 2 years (0.2 × 10) since there is only a 0.2 probability of arrest and conviction.

The next step is to approximate the forgone earnings from legitimate employment and multiply the estimated income loss by the probability of arrest and conviction. If A calculates possible income loss of $3,000 per year, then A's total forgone earnings would be $6,000 ($3,000 × 2 years). Assuming B estimates forgone annual earnings at $10,000, B's total income loss becomes $20,000 ($10,000 × 2 years).

To arrive at net private benefits, costs must be deducted from benefits. The net gain from committing the robbery is $9,000 for A ($15,000 less $6,000) but is a negative $5,000 for B ($15,000 less $20,000). Therefore, A has a positive economic incentive to carry out the crime, while for B the act would be economically irrational. The deciding factor influencing the result is the greater opportunity costs of B's legitimate earnings loss, since both individuals had identical anticipated benefits and anticipated risks of arrest and conviction. On an annual income basis, B's income loss was over three times that of A.[5]

In general, it can be concluded that anticipated future income is a function of human capital value. Therefore, individuals possessing higher

---

5. Readers familiar with the process of discounting future values to arrive at present worth recognize that the net benefits presented in Table 11-4 are undiscounted figures. Since A's stock of human capital is lower than B's, the discount rate applied to A's loss of future income would be higher, making the crime even more profitable to A than to B.

stocks of human capital from investment in education or job skills are less likely to commit economically motivated crimes. Those with little in the way of acquired investment in human capital would expect a smaller flow of future income and would have far less to lose from committing criminal acts.

Thus, the motivating factors contributing to criminal behavior are much the same as those that serve to guide decision making in everyday life. As long as individual behavior is the result of subjectively weighing benefits and costs and acting accordingly, economists consider such behavior rational. Since the calculated benefits and costs of criminal activity vary among individuals, depending upon their attitudinal makeup and perceived income opportunities, different crime levels are to be expected from different types of people. For many individuals, crime is rational because it may be the best available economic opportunity.

ECONOMIC POLICY

The preceding theoretical approach to explaining criminal activity is not meant to imply that all criminals behave rationally in the economic sense of the word. It is true that some criminals are irrational or "sick" in that they do not consider the costs of their actions. But the economic approach appears applicable for any crime for which it can be presumed that the individual is not wholly irrational and is therefore responsive to changes in the costs and benefits of crime. It follows that in order to be effective, public policy must make crime less attractive by increasing its costs.

Increasing Legitimate Income Opportunities

A positive approach to deterring crime is to create more legitimate income opportunities, particularly for minority youth. Sufficient evidence exists that indicates a direct relationship between unemployment and crime. As long as the unemployment rate for urban youths consistently remains above 20 percent, the opportunity cost of crime will remain quite low. By making legal employment more attractive through an upgrading of economic opportunities, the cost of crime will increase and the quantity of crime should fall.

Law enforcement officials repeatedly claim that in periods of prosperity, with the economy nearing full employment, crime rates decline. But nagging unemployment and reduced economic growth rates have characterized the American economy in recent years. An economy that boasts full employment is difficult to achieve and impossible to sustain as long as the private market sector is subject to the business cycles. Even if full employment was achieved and many urban youths were on private payrolls, cyclical prosperity might only serve as a short-term palliative. With the downturn, those with low job skills or education would be the first to be released from employment and the last hired back when prosperity returns. Thus, legal opportunities for employment may be viewed by many as only temporary and lacking a steady long-term income flow. Crime may seem to be a more rewarding activity to these individuals.

An expanding economy is an important factor in reducing crime, but a greater investment in human capital is also required. Increasing an individual's stock of human capital through education and vocational training is a form of investment since it increases the economic value of human services and results in a greater flow of future income. The beneficial long-term effects of increasing economic opportunities have long been recognized by the government. This is evidenced by the various federally financed training programs as well as educational loan guarantee programs.

Unfortunately, upgrading economic opportunities is easier said than done. Thus far, training programs have been expensive failures. But continued experimentation, albeit on a smaller scale, with programs aiming to increase the value of human capital may yet prove successful. If and when that occurs, the cost of crime will increase for large numbers of the poor. But even with greater economic opportunities and income stability, there will always be wage earners along any point on the income scale who are at the threshold of viewing crime as a more rewarding alternative to legitimate employment, or perhaps as an attractive source of supplementary income. This is true for individuals earning very high incomes as well, for no matter how high the costs of crime, there will always be some individuals at the margin. However, the greater the costs of crime, the greater the rationing effect and, consequently, the fewer the number of crimes.

Increasing the Cost of Punishment

Because of the dismal experiences with programs geared to increasing economic opportunities, many economists place greater emphasis on the role of punishment in deterring criminal behavior. Unlike the former approach, which is generally supported by sociologists, urbanologists, and criminologists, the punitive approach to crime reduction is met with stiff opposition from members of these groups. The root of the controversy lies in the economist's assumption of rational criminal behavior.

Traditional View. The traditional view held by many sociologists and criminologists is that criminals are sick or abnormal individuals who are totally unresponsive to the costs of crime, or they are socioeconomically deprived because of environmental factors. According to this view, the criminal mind is not rational. It functions irrespective of the expected costs and benefits of its actions. Individual behavior is assumed to be independent of past experiences and the experiences of others. Increased costs of punishment are not only ineffective, they also reflect a cold and somewhat cruel attitude toward the sick and environmentally deprived. The function of criminal justice, according to this view, is to detain criminals until such time as they are rehabilitated and ready to resume their proper place in society.

Economist's View. The economist's view of criminal behavior rests on the assumption of rational behavior, as developed earlier in this chapter. Acceptance of criminal behavior as rational logically leads to a policy recommendation of increased punishment as a deterrent to crime. But it must be emphasized that many crimes may not be prevented by increasing the costs of punishment. Included in this category are such crimes as drunkenness and narcotic addiction, which are really illnesses rather than crimes, as well as crimes of passion and crimes committed by the criminally insane. Many of these people are better treated by social programs that include education and physical and mental health care.

But for the majority, economists maintain that punishment deters crime. Rehabilitation of prisoners is not likely to be successful in changing behavior if the crime was a rational act in the first place. Behavioral modification rests on the assumption that the criminal was socially deviant and in need of reform. However, economists agree that prisoner rehabilitation is likely to be successful if it increases the chances of employment upon release from prison, for it will increase the individual's opportunity cost of committing future crimes. Unfortunately, however, relatively few rehabilitation programs are successful in that regard.

If one gives any merit to the saying "a person can be judged by the company he keeps," it is not likely that a person sent to prison for a serious crime will come out any better than when initially imprisoned. Consequently, many people convicted and imprisoned for criminal offenses return to the life of crime almost immediately upon release from confinement. To these people, crime still pays. Many repeaters were considered model prisoners and were paroled for good behavior. But good behavior within the confines of prison walls does not necessarily result in good behavior in a free society where economic choices concerning income must be made. Ex-convicts must face the same decisions concerning legal versus criminal income opportunities as they did prior to imprisonment. In a good number of cases, the costs of criminal activity have changed little despite imprisonment. This is particularly true where judges are lenient in sentencing criminals and parole boards are lenient in releasing them.

Economists hold that crime can be reduced by increasing the probabilities of arrest, conviction, and imprisonment as well as increasing the

length of time served in prison. Increasing the probabilities of being apprehended and punished deters crime because, unlike the traditional view, the
economist's view is that criminal behavior is affected by past experiences
and the experiences of others. Stiffer punishments also serve to reduce
crime, but care must be exercised in relating punishment to the severity of
the crime. For example, if the crime of armed robbery carries with it the
death penalty, then the armed robber faces no additional costs in committing crimes of murder, kidnapping, hijacking, or treason. To be effective,
penalties must be gradationally implemented to ensure marginal deterrence.
In practice, this has become increasingly difficult as the number and types
of illegal activities increase.

Supply and Demand Analysis. The deterrent effect of punishment on criminal activity is
shown graphically in Figure 11-2 by the use of market supply and demand
curves. Assume the crime is auto theft and participants in the market include
the buyers of stolen cars as well as the car thieves. Thus, the demanders of
stolen cars and the suppliers of stolen cars are assumed to be two different
groups of people. Since most stolen cars are quickly sold by the thief, this
represents a realistic view of the marketplace.

The price of stolen cars is measured on the vertical axis and the quantity
of stolen cars is measured on the horizontal axis. The demand curve for
stolen cars slopes downward and to the right as does the demand curve for
any normal and legal commodity. Buyers of illegal goods are also willing

Figure 11-2

Supply and Demand for Stolen Automobiles

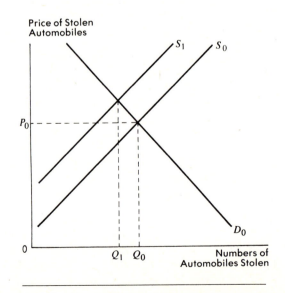

to purchase a greater quantity at a lower price than at a higher price. As usual, demand at any given price is determined by such factors as tastes, incomes, price expectations, and the prices of related goods.

The supply curve, on the other hand, slopes upward and to the right, indicating that the quantity of automobiles that thieves are willing to steal and offer to demanders is directly related to the price received in the market. At any given price, the supply of illegal goods is a function of the costs of production (special tools, labor, etc.), available technology, distribution costs, and the costs of punishment.

Given the supply and demand functions (S_0 and D_0) in Figure 11-2, market equilibrium is reached at a price of P_0 and a quantity of Q_0. If a lower level of market equilibrium is desired by society, this can be achieved by increasing the costs to suppliers. Society can increase supply costs by increasing the probability of being apprehended and punished for the crime. The effect of higher supply costs is to shift the market supply to the left (S_1), resulting in fewer automobiles stolen (Q_1). If, however, criminals are "irrational" and therefore insensitive to changes in the expected costs of punishment, the number of cars stolen would not diminish but would remain at Q_0 despite the greater likelihood of arrest and punishment.[6]

Policy Toward Victims

In recent years, many states have introduced compensation programs for the victims of crime. Although such programs are new and their economic impact is uncertain at this time, they do reflect a policy change in the area of criminal justice. Compensatory programs recognize that crimes of violence involve victims and that for every victim there is a personal loss. Essentially, compensation programs involve some type of public insurance against the personal costs of victimization. Premiums are collected in the form of tax dollars, and claims are paid to those subjected to criminal attack.

Public pressure for compensatory programs is founded on the view that our present system has ignored the victims of crime. The criminal chooses to commit the criminal act and selects the person to be victimized. Even if apprehended and sentenced, there are cases in which the criminal is left better off than the victim. From the time of arrest, the criminal receives medical attention, an attorney if the criminal cannot afford one,

---

6. Supply curve S_0 is assumed to contain some expected costs of punishment. If the market were comprised of irrational criminals, the market equilibrium would lie to the right of Q_0 and would not change with higher punishment costs.

room and board, and, if sentenced, opportunities to increase educational and vocational skills—all at public expense. By contrast, the victim may have to pay medical bills, replace property losses, miss work, and absorb the cost of trial proceedings. In addition, the victim may be dissatisfied with the results of the trial and live in fear of retaliation.

Although Congress has not passed a national compensation program for the victims of crime, it is currently studying the various state programs to determine its feasibility. The state of Pennsylvania, for one, has had a compensation program in effect since 1979. Since that time, the Crime Victim's Compensation Board has paid out more than $10 million to individuals victimized by crime.

Ohio initiated a compensation program in late 1976. Under Ohio law, innocent victims of crime are eligible for awards to compensate for lost earnings due to absence from work and any expenses incurred as a result of the crime for which the victim is not reimbursed. The victim may receive up to $50,000 as compensation. Financing for the program is collected by means of a surcharge on all criminal court costs throughout the state.

Should a national program to compensate crime victims be implemented, it will more than likely be one in which the federal government shares the funding costs with the states. It is also likely that any federal program will include some type of "means test" whereby victims must show financial hardship. The prevailing opinion is that there is little justification for providing public insurance against crime to those who have the financial means to acquire private insurance. Thus, federal support will probably be limited to the poor, and it is the poor who face the highest probability of criminal injury.

Policy Toward Victimless Crimes

One area of criminal activity that is becoming an increasingly controversial policy issue is that of victimless crime. Various estimates indicate that victimless crimes account for approximately 40 percent of all arrests and nearly 50 percent of the total economic costs of crime. The latter figure is based on the revenue received by the criminal sector from the sale of illegal goods and services. Although the list of laws prohibiting certain types of behavior is extensive and differs among communities according to citizen attitudes, activities presently commanding the greatest attention are those of prostitution, gambling, and narcotics. There appears to be growing support in this country for legalizing and decriminalizing a number of victimless crimes.

The first argument for legalizing victimless crimes is based on economic costs. The relevant factor is not whether such activities entail large costs, but whether the costs are borne by willing participants. The direct costs of purchasing drugs, acquiring the services of prostitutes, and absorbing gambling losses constitute a large share of the private costs to the participant. Total private costs would also include the possibility of experiencing deteriorating physical and mental health from venereal disease, drug addiction, or gambling addiction. Although private costs can be substantial, the economic approach to victimless crimes is based on whether negative externalities or social costs result from such activities, for if the costs of victimless crimes spill over to nonparticipants, then legal prohibitions or restrictions can be justified. Unless social costs result, making such activities illegal directly interferes with the rights of individuals to exercise their consumer freedoms.

But it appears that most laws pertaining to victimless crimes are based on moral and not economic principles. Although members of the community may feel morally victimized by "sinful" activities, this does not constitute grounds for making them illegal; such laws are the expression of value judgments based on personal beliefs as to what is right and wrong. Nor is the fact that participants may be engaging in self-destructive activity a justification to protect individuals from the consequences of their own behavior. The deleterious mental and physical effects that can result are the costs stemming from the lack of complete information. Rather than interfere with consumer sovereignty, a more suitable alternative in some cases may be regulation or public information concerning possible costs to potential demanders.

A second argument concerns the role of organized crime in monopolizing markets for illegal goods and services. The greater the number of goods and services made illegal, the greater is the incentive for organized crime to assume the role of supplier. Outlawing supply does not make the demand for an illegal good or service disappear. It merely makes its acquisition more expensive and channels business away from the legitimate private sector to the criminal sector. Thus, many crime experts argue that since people are going to gamble, buy drugs, and hire prostitutes anyway, why force them to patronize organized crime? Legalization of these activities could provide two benefits: a reduction in organized crime activity and an increase in tax revenues, which could be used to provide sorely needed public services.

In recent years there has been a dramatic shift in public policy toward gambling. Many states have instituted legalized gambling in the form of lotteries, numbers games, and off-track betting. Only Nevada and Atlantic City have legalized casino gambling. Although considered successful in some quarters, it is still questionable whether revenue raising and the undercutting of organized crime are compatible. It is also questionable whether these operations are completely independent of organized crime.

Some experts doubt if legalized gambling can effectively compete with the gambling odds and tax-free status of syndicate games.

A third argument in support of legalizing victimless crimes is that it will permit police, prosecutors, and courts to devote greater resources to preventing and controlling serious crime. The extent to which this is true depends on the amount of resources individual communities allocate to enforcing victimless crime. Overall, however, the benefits of relieving an overburdened criminal justice system by progressively decriminalizing victimless crime could be worth literally billions of dollars in crime-control funds and could bring about a substantial reduction in the number of serious offenses.

CONCLUSION

The information presented in this chapter should leave no doubt that the crime problem is a very serious one. However, crime can be reduced to a more tolerable level by a redirection of public policy aimed at making the payoff from crime less attractive. Crime cannot be totally eliminated any more than pollution can be eliminated, regardless of public policy. Economists believe that the public must realize that the traditional approaches emphasizing environmental factors and irrational behavior as causes of crime and prisoner rehabilitation as the cure have been largely unsuccessful in achieving reduced crime levels. Public policy would be more effective in emphasizing the rational behavior of many criminals and the deterrent effect of punishment on many crimes.

In many states it appears that such a change in public policy is taking place. In at least twelve states, discretionary sentences have been replaced by fixed-term mandatory sentences for various offenses, such as committing a crime with a gun, kidnapping, and arson. A majority of states have passed new death penalty provisions since 1972. During the years 1977–1982, the number of executions resulting from the death penalty totaled only six. In 1985, however, eighteen people were executed. Prison sentences also appear to be lengthening, and there is a growing movement to apply stiffer penalties to juveniles committing serious crimes. Gun control legislation continues to be a controversial issue in many state legislatures.

Whether taxpayers are willing to bear the increased costs of a criminal justice system that emphasizes deterrence remains to be seen. Nearly all observers agree that the public will have to commit substantial additional resources in order for prisons, courts, and prosecutors to reduce crime levels. Increased crime prevention and control is going to be costly, but less so than crime itself.

QUESTIONS FOR DISCUSSION

1. In your opinion, which kinds of crime are likely to be rational in the economic sense and which are likely to be irrational?
2. Should public policy be geared more to increased punishment as a means of deterring crime? If so, is the certainty of punishment or the severity of punishment more important?
3. What should the role of public policy be in so-called victimless crimes such as prostitution, purchasing drugs, and gambling? Consider as possibilities complete prohibition, decriminalization, and legalization, with and without regulation.
4. What role should the death penalty play in the area of criminal justice?
5. White-collar crimes are usually committed by individuals with relatively high stocks of human capital. How can cost-benefit analysis explain such criminal behavior?
6. A public policy emphasizing the deterrent effects of punishment will require substantially more resources than are presently allocated to the criminal justice system. What costs and benefits would this entail for your own community?
7. Do you believe that prison rehabilitation programs can succeed? What changes would you propose in present rehabilitation programs?

SELECTED READINGS

Hagan, John. *Modern Criminology: Crime, Criminal Behavior and Its Control*. New York: McGraw-Hill, 1985.

Hellman, Daryl A. *The Economics of Crime*. New York: St. Martin's Press, 1980.

Parker, Donn B. *Crime by Computer*. New York: Scribner, 1976.

Rogers, A. T., III. *The Economics of Crime*. Hinsdale, IL: The Dryden Press, 1973.

Rottenberg, Simon, ed. *The Economics of Crime and Punishment*. Washington D.C.: American Enterprise Institute, 1973.

Schmidt, Peter, and Ann D. White. *An Economic Analysis of Crime and Justice: Theory, Methods and Applications*. Orlando, FL: Academic Press, 1984.

Stigler, George. "The Optimum Enforcement of Laws." *Journal of Political Economy* (May 1970).

Wilson, James, and Richard J. Herinstein. *Crime and Human Nature*. New York: Simon and Schuster, 1985.

12

NATURAL RESOURCES
COULD MARKETS
HELP?

Economics is defined as the study of the allocation of scarce resources. Thus, economics is very relevant to the investigation of issues of scarce *natural* resources. Although one can learn much about natural resources by looking at their geology, ecology, or physics, economics is particularly suited to evaluating alternative natural resource policies.

In the past decade people have become more and more concerned about natural resources. Not only does society worry that we are depleting resources, but different groups care both deeply and differently about how resources are used. As with any scarce good, conflicts arise over how natural resources should be used. In the case of publicly owned land, for example, some believe that the wilderness areas should be extended, while others believe that public lands should be sold to private interests for mining or industrial use. Although conflict over use is common to all scarce goods, there are two respects in which the conflict over scarce natural resources is unique.

First, since many natural resources are nonrenewable, users of natural resources compete not only with present users, but also with future users. If mankind is going to survive into the future, it is clear that some system must exist to effectively allocate the use of these scarce resources between the present and the future.

Secondly, many natural resources have only recently become scarce goods, generating conflicts among competing users. In the past, goods such as land, water, and fishing grounds were often so abundant relative to demands that they were essentially of no interest to economists. This is no longer true.

This chapter consists of three parts. The first part discusses the history of natural resource scarcity. The second part asks directly whether free markets can adequately distribute natural resources among generations. Is the visible hand of government necessary to complement the invisible hand of the market? The third part investigates the cases of water resources and government land management as examples of how economics is relevant to specific natural resource issues. Each of these issues is of particular interest to those who wish both to use the resources today and to conserve some for future generations.

THE HISTORY OF THE USE OF NATURAL RESOURCES

With the current emphasis on energy and other resource crises, it might seem surprising that the past thousand years are replete with periods in which resources ran critically low. Table 12-1 lists several.

The Resource Problem in England: 1200–1900

The expansion of English agriculture during the thirteenth century eliminated most accessible forests. This depleted the major energy source of the period, wood. The wood shortage led to the substitution of coal for wood (known at the time as "that awfule sea coale"—"awfule" because even then it created urban pollution; "sea coale" because the known deposits were found on pieces of land that extended into the North Sea). Similar circumstances in the Netherlands led the Dutch to reclaim land from the sea for the first time and led to unrealized plans in England to drain the Fens, swamps northeast of London. Ironically, the Black Plague of the fourteenth century saved the English from this crisis by applying a positive check on population. The population declined by as much as 50 percent, agricultural lands were abandoned, and reforestation occurred.

It was not until the sixteenth and seventeen centuries that the problem of scarce wood was again encountered. Again coal replaced wood; and, although coal continued to pollute, the development of the steam engine allowed miners to pump water out of coal mines inexpensively. As a result, coal proved far more abundant and far less expensive than it had been previously.

Table 12-1

Resources Shortages

Date	Location	Resource	Probable Causes
c.500 B.C.	Babylon	Land	Excessive accumulation of salt in the irrigated land of the Tigris-Euphrates basin
1300 A.D.	England	Wood	Harvesting of forests
1200 to the present	Netherlands	Land	Population expansion
1600	England	Wood	Reharvestation of forests
1800s	U.S.A.	Forests	Population expansion, Monopolistic interests
1960s	U.S.A.	General environmental deterioration	Technological expansion
1970s	World	Energy	Overuse of resources, Forming of OPEC
1970s	World	Resources in general	No rational planning
1970s	Developing nations	Food	Population expansion
1980s	Western U.S.A.	Land and water	Government ownership, Growth of western agriculture

The Reverend Thomas Robert Malthus (1766–1834) was the first important commentator on finite natural resources. Although Malthus is best known for his pessimistic analysis of human population (see Chapter 15), the same analysis is also pessimistic about the possibility for rapid expansion of food production.

In the nineteenth century, English economists continued to be concerned about resources. Both David Ricardo and John Stuart Mill asserted the relevancy of the law of diminishing returns according to which the productivity of additional resources that were used would decline. However, mechanization at least postponed the inevitability of the law of diminishing returns, and the social problems of the industrial worker overshadowed resource problems. Later, William Stanley Jevons forecast in *The Coal Question* (1865) that England had no more than 20 more years of coal resources at its then current level of output.

Resources in America: The Conservation Movement

As the U.S. economy expanded geographically and economically through the nineteenth century, and as the frontier remained open, the notion of resource limits was of little concern. But by 1890 the last Indian treaties were signed symbolizing the closing of the frontier. The psychology of the generation changed from the belief that our nation was a boundless cornucopia to the belief that the United States was an unusually well-endowed but, nonetheless, finitely endowed country. At the 1873 meeting of the American Academy for the Advancement of Science the view was expressed that our resources needed to be rationally husbanded. By the first decade of the twentieth century, President Theodore Roosevelt's chief conservationist, Gifford Pinchot, forecast that if the current trend continued the United States would run out of timber by 1940 and out of coal by the end of the twentieth century.

During the post–Civil War era big business expanded rapidly, so it should not be surprising that conservationists blamed free-market capitalism, and big business in particular, for the irresponsible exploitation of natural resources. It seemed clear to observers of the time, having just seen the virtual extinction of the buffalo and the rapid disappearance of forest land, that the unregulated profit motive overexploited resources in general. Timber resources in particular were overexploited, endangering the public interest both in the present and in the future.

The solution was clear. The government must assume the role of steward of this stock of resources, either by assuming ownership of large tracts of natural resources or by establishing control of the rate at which private enterprise could remove natural resources. The movement culminated in 1908 when President Theodore Roosevelt called a conference of those governors who agreed with him on the goals of resource conservation. Not only did the governors agree that public ownership was desirable, but the meeting also led to an expanded scale of federally owned land in national parks, national monuments, and national forests.

Resource Problems after World War II

Resource concern returned to the public eye on three different fronts following World War II. First, President Truman appointed a Presidential Commission to study materials and resource requirements for the deepening conflict with the USSR during the 1950s. This commission reported that resources were a problem in that their costs were likely to rise over the next decades but that extinction of resources was not likely to be a serious consideration. Secondly, Rachel Carson's book *Silent Spring* received widespread attention. Carson argued that our use of insecticides was doing irreparable harm to the natural environment. Third, books appeared expressing deep concern that the current excessive population growth would create catastrophe.[1] At the same time, technologists predicted a new abundance based on new technology.[2] Others, expressing mixed optimism, argued that fossil fuels were indeed very finite but that some form of energy, typically nuclear power, would ultimately replace them.[3]

At the beginning of the 1970s, in spite of some technologists who proclaimed the beginning of a new age of plenty, many people became increasingly convinced that the limits of our natural resources had been reached. The clearest warnings came in energy resources, where the output of natural gas and oil in the United States had begun to decline after peaking in 1970, as shown in Table 12-2.

As smog alerts became routine in cities, people began to recognize that the air itself was scarce. Famine occurred successively in Biafra, Bangladesh, and then the Sahel of sub-Saharan West Africa. Evidence grew that water, particularly in the American West, was being overused and that the water table was being drawn down. Congestion developed in the wilderness. But what was particularly distressing was that all these things happened at once.

Economist Kenneth Boulding coined the term "Spaceship Earth," suggesting that the entire planet is a finite interdependent system. The Club of Rome sponsored work that resulted in the 1972 publication *The Limits to Growth*. This book used a computer model to demonstrate that mankind was simultaneously outstripping resource limits in fossil fuels, land, water, air, and nonmineral resources. As seen in Figure 12-1, the study projected a catastrophic collapse of the world's food and industrial production sometime in the twenty-first century. As if it were foreshadowing the prophecy, the price of oil rose 400 percent from 1972 to 1974. As a result, most of the world was left with a lower living standard in 1974 than had existed in 1972.

ECONOMIC ANALYSIS OF FINITE RESOURCES—WILL THE FREE MARKET CONSERVE FINITE NATURAL RESOURCES?

Beginning with the early conservationists but continuing through today, most noneconomist natural resource experts have generally believed that

---

1. William Vogt, *Road to Survival* (New York: William Sloan Associates, 1948); Fairfield Osborn, *Our Plundered Planet* (Boston: Little, Brown, 1948).
2. John Von Neuman, "Can We Survive Technology?" *The Fabulous Future* (New York: Dutton and Company, 1955).
3. Palmer Putnam, *Energy in the Future* (New York: Van Nostrand, 1953); Hans Thirring, *Energy for Man* (New York: Harper and Row, 1976).

Table 12-2

U.S. Oil and Natural Gas Production and Reserves

Year	Oil Production (Billions of barrels)	Oil Reserves (Billions of barrels)	Gas Production (Thousand cubic feet)	Gas Reserves (Thousand cubic feet)
1950	1.974	26	6.28	185
1960	2.575	32	12.77	262
1970	3.517	38[a]	21.92	290
1971	3.453	37	22.49	280
1972	3.455	36	22.53	265
1973	3.361	35	21.73	250
1974	3.202	34	21.60	237
1975	3.057	33	19.24	228
1976	2.976	31	19.10	215
1977	3.009	32	19.16	208
1978	3.178	31	19.12	208
1979	3.121	30	19.66	201
1980	3.146	30	19.40	199
1981	3.129	29	19.18	202
1982	3.157	28	17.76	202
1983	3.171	28	16.03	200
1984	3.250	28	17.23	197

[a]The Alaskan North Slope oil discovery occurred this year.

SOURCES: *Statistical Abstract of the United States: 1982–1983,* and *1987.* Sam Schurr *et al.,* *Energy in America's Future* (Baltimore: Johns Hopkins University Press, 1979).

unregulated free markets will lead to overuse of scarce natural resources. Many argue that leaving natural resources to the free market would be like leaving the management of the chicken coop to the discretion of a fox! Most economists have greater faith that the free market will be able to sensibly allocate natural resources over time. The economic theory of exhaustible resources suggests that free markets will induce the incentive to conserve. The following suggests how it works.

Suppose one owns an oil reservoir of 100,000 barrels. Should the oil all be sold today or should some of it be conserved? Assuming sufficient demand exists to sell all of the oil, the only reason not to sell it all today is that the owner chooses to forgo the revenue he could earn this year by waiting to sell it next year. In the language of economics, the forgone revenue is the *opportunity cost* of selling oil today. Suppose that the profit on oil after subtracting the costs of extraction is $30 this year, and

Figure 12-1

Club of Rome Analysis of Relative Quantities of
Crucial Socioeconomic Variables for the World

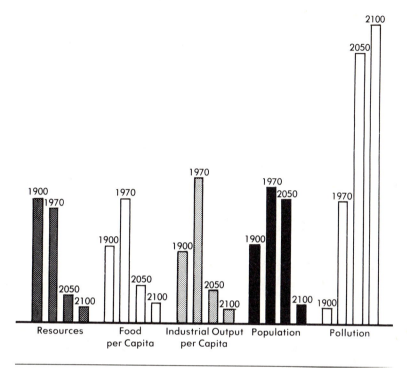

SOURCE: Meadows *et al., The Limits to Growth* (New York: Universe
Books, 1972).

suppose the owner expects the price of oil to be the same next year. Now, what should the owner, as a profit-maximizing producer, do? The only responsible thing that the producer could do is to sell all of his oil today. Waiting until next year to sell the same oil for a $30 profit is the same as putting the money in the bank for a year and earning no interest on it. The producer would certainly expect to do better than this since alternative uses for the money exist that would earn a positive return.

On the other hand, if the owner expected the profit to be $60 next year, he or she would be foolish to sell any oil this year because by waiting just one year, the value of the oil would double. That would be the same as putting money in a bank and earning 100 percent interest.

Economic theory reasons that the profit (or what is known technically as the economic rent) on the oil or any other natural resource will neither double nor remain constant. Instead, it will rise by the rate of return on alternative investments. Competition will force this to occur. This results because the increase in the economic rent (or profit) of a natural resource is the only return the owner receives on the resource. Unlike a savings account, natural resources do not earn interest. If the rent is currently not increasing at a level equivalent to the interest rate on alternatives, people will sell all of their resources now rather than wait until next year. The result is that the supply next year will fall, causing the rent next year to rise until the rent increase between this year and next year is the same as the return on competing alternatives. On the other hand, if the return on natural resources is greater than the return on alternatives, people will try to buy natural resources immediately, causing a rise in the price of the natural resources this year and creating a lower increase in rent between this year and next. But whatever the starting point, the theory of natural resources concludes that competition will make the rent of natural resources rise by an amount equal to the rate of return on alternatives.

The fact that in a free market the rent on natural resources tends to rise by the rate of return on alternatives naturally induces conservation. As the rent on natural resources increases, natural resource prices increase, so the quantity demanded decreases. The fall in quantity demanded necessarily means resource owners increase their conservation.

Not only does the price increase encourage users to conserve the resource by reducing the quantity demanded, but it also encourages resource users to substitute other more abundant, less expensive alternatives. Resource suppliers are encouraged to look for more costly deposits of the resource to extract.

Concerning alternatives, natural scientists have indicated that there exist abundant substitutes for all current mineral resources with the possible exception of phosphorous.[4] Clearly the market leads to some conservation, but does it lead to the correct amount of conservation? To economists the term *correct amount* has a very specific meaning. The correct amount means that amount which puts resources to their highest valued use. Conservation is thought of as the use of not being used at all. This notion of correct use is what economists call *economic* or *allocative efficiency*. Economists provide two reasons why the amount of conservation may be too little.

Common Property Resources

Assume there exists a reservoir of oil under the ground to which anyone has the right and ability to drop a well and pump out oil. In this case,

---

4. H.E. Goeller and A.M. Weinberg, "The Age of Substitutability," *Science* (February 20, 1976), pp. 683–689.

drillers have no reason to conserve, because competitors can always enter and extract any oil that the driller conserved. The cost of pumping out oil today is only the extraction cost. There is essentially no opportunity cost because if one doesn't pump out the oil today, rivals will. The result is inadequate conservation.

Economists observe this problem not only among the owners of oil wells, but also among fishermen who overfish fishing grounds, users of groundwater in the High Plains, ranchers who overuse common rangeland, and campers who previously overused Yosemite National Park. The same problem leads to pollution of our air and to congestion of our urban highways. The problem is known as the problem of common property resources. The cause is simply that decision makers have, for whatever reason, no incentive to include the complete costs in their private calculations of what is profitable. This problem typically arises when there is no clear owner of the resource.

Future Consciousness

The second possible reason that conservation will be inadequate is that decision makers will be insufficiently conscious of the future. People may have an innately inadequate facility for looking out for future concerns, or decision-making institutions may be organized in such a way that the emphasis is irrationally on short-term rather than on long-term goals. The second argument has been made both for political decision makers who prefer results during their term in office and for corporations in which the decision maker's emphasis may be centered on present profit at the cost of future welfare.

How valid these arguments are has been debated often, and there is surely something to them. Yet, the opposite argument has also been made. Other economists argue that with rapid technological progress, the bias of our present versus future allocation is, perhaps surprisingly, toward the future rather than the present. They suggest that with the incredible progress of the past 100 years, it might have been better to use more resources in the nineteenth century to bring the living standard at that time closer to our present living standard.

POLICY RESPONSES

At the urging of conservationists as early as the start of the century, policymakers have responded to resource problems with government action. Many of these government actions are controversial. There are innumerable ways in which the government is involved with natural resources. Two of the most important are in the areas of water and land resources policy. The problems associated with these two issues will be discussed in some depth. This is to give a sense of the issues involved and to illustrate the place of economic analysis in understanding resource problems and alternative solutions to them.

THE WATER RESOURCE PROBLEM

As with most natural resource issues, the water resource problem is simultaneously a problem of natural science, law, politics, and economics. As the United States moved from being a very sparsely populated, primarily rural nation at the beginning of the nineteenth century to a fairly densely populated, primarily urban population today, water has turned into an economic good. Today, in order to obtain water, other things of value must be sacrificed.

Although the United States has water resources sufficient to provide for our total water needs, the nation suffers two very serious regional water problems. First, there are regions with an insufficient quantity of water resources. This is the problem in most of the western United States and in certain major eastern metropolitan areas, particularly New York City. Secondly, there are regions where there is a problem of inadequate water quality due to pollution. This is the problem in most of the eastern United States. The first of these problems, that of insufficient quantity, will be discussed here.

Water Use

It is hard to define what is meant by water use because water is recyclable. In particular, one must distinguish between *water withdrawals* and *water consumption*. Withdrawals are any diversion of water from nature irrespective of what happens to it after withdrawal. Consumption is any use of water after which the water is not returned directly to groundwater or to surface water. Water that is consumed cannot be used again until it runs back through the hydrological cycle. So, if an electric generating plant uses stream water for cooling after which the water is returned to the stream, it is considered a withdrawal, not consumption. Water used to irrigate land after which the water transpires or evaporates into the atmosphere is both a withdrawal and a consumption use. The notion of consumption will be emphasized here.

More water in the United States is consumed for agricultural irrigation than for any other use. Table 12-3 shows total water consumption by use in the United States.

Table 12-3

U.S. Consumption of Water
(Billions of gallons per day)

Year	Total Consumption	Irrigation	Public Water Supply	Rural Domestic	Industrial Use
1960	61	52	3.5	2.8	3.2
1970	88	73	5.9	3.4	4.9
1975	96	80	6.7	3.4	6.1
1980	100	83	7.1	4.0	8.3

SOURCE: *Statistical Abstract of the United States: 1987.*

Specific Water Use Problems and Policies

In the following sections two of the most important policy responses to the problem of insufficient water quantities are discussed. Typically, the responses to inadequate water quantity have been either to utilize groundwater resources or to engineer a technologically sophisticated water transit system to move water from one area to another.

New York City. Ever since the nineteenth century, the New York City region has consumed far more water than is available within the city. As long ago as 1842, water was impounded in the Croton reservoir north of the city and then transferred to New York. Over the next century, the Croton system was expanded, water from Catskill mountain streams was transferred, and finally, water was transferred from the Delaware River Basin. The last transfer introduced complex legal issues because the Delaware River Basin provides water not only for New York State but also for New Jersey and Pennsylvania.

The High Plains. The High Plains, the region of the United States that lies east of the Rocky Mountains and west of the 100 degree meridian running through Texas, Oklahoma, Kansas, Nebraska, and the Dakotas receives less than 20 inches of rain per year. For most agriculture to be viable, supplementary water is necessary. In the west Texas portion of the High Plains, farmers have developed a very successful cotton-growing industry with the aid of irrigation. The water was provided by "mining" the Ogallala aquifer that lies underneath the land. The aquifer is simply an enormous deposit of subterranean water. Withdrawals from this formation have greatly exceeded the local recharging of the aquifer through water runoff and rainfall.

California. The San Joaquin and Imperial valleys of central and southeast California contain climates and soil capable of great agricultural productivity.

Unfortunately, the rainfall, particularly in the Imperial Valley and in the southern portion of the San Joaquin Valley, provides too little water for successful agriculture. Added to this agricultural demand is the dramatic growth of the southern California urban area, which stretches down the coast from Santa Barbara to San Diego and inland to San Bernardino and Riverside. From 1940 to 1980, this area expanded in population from about 3 million people to 13 million. As with the California valleys, rainfall is inadequate to provide for water needs.

Just before World War I, this problem was addressed by transferring water into Los Angeles through an aqueduct built from the Owens Valley on the east side of the Sierra Nevada mountains to Los Angeles. As urban growth outstripped these water resources, it became necessary to transfer additional water. This was accomplished by moving water from the Lower Colorado River to the Los Angeles Basin. In the 1930s, the Central Valley Project constructed dams and canals to transfer water from the Sacramento River Basin and to store the melting snow pack on the western side of the Sierra Nevada for timely use in the San Joaquin Valley and the San Francisco metropolitan area. Additional water was diverted to the Imperial Valley from the Lower Colorado during the 1940s, this time through the American Canal. Finally, work was begun in the 1960s on the largest water diversion project to date, the California State Water Project. This project transports water from the Feather River, a tributary of the Sacramento River, both to the Southern San Joaquin Valley and across the Tehachapi mountains to the Los Angeles Basin.

Other Transfer Plans. There are many other water transfer proposals. These range in ambition from small-scale plans to transfer water, such as a plan to redistribute water in Wyoming, to moderate-sized projects such as the Central Arizona Project that also withdraws water from the Colorado River.

In 1964, the Ralph W. Parsons Company proposed and designed a plan on an unheard-of scale to transport up to 300 billion gallons of water per day from Alaska and northwest Canada to 7 provinces of Canada, 33 U.S. states, and 3 Mexican states. This made it roughly 100 times as large as the California State Water Project. The project was never carried out.

Conflicts over Water Resources

At the heart of each of these water problems is the basic economic problem. Water has become a scarce resource in precisely the sense that economists use the term. To get additional water, someone must give up something that has value. The particular resource that must be given up varies depending on circumstance, but, invariably, giving up the something means that someone will bear a cost, and that someone is not necessarily the same person who gains the water. This gives the water problem a political as well as an economic dimension. There are several important examples of this that can be drawn from the previously mentioned problems and policies.

The transfer of water from one basin to another means that the water is transferred from one group of users to another. The problem is more difficult when the transfer is across state lines as in the case of the Delaware River Basin. Still more difficult are the problems brought about by transferring water out of the Colorado River, because the Colorado flows from southern California across an international boundary into Mexico. Therefore, the water that is used by Californians comes at the expense of Mexicans. The problem is further complicated because the Navajo Indian Nation also has claims on the water of the Colorado River. These issues can lead to deep discord.

The construction of physical transfer systems is typically very expensive. In California, not only do northern Californians resent that southern Californians are taking their surplus water, but they are further upset at the prospect that they must share in paying for it either through higher taxes or lower alternative public services.

There are also environmental costs. The storage systems build dams, often in very scenic locations. To even out the flow of the Lower Colorado, there was a plan in the 1960s to dam the lower reaches of the Grand Canyon. The Glen Canyon, just upstream from the Grand Canyon, was dammed up in the late 1950s, turning a unique environment into Lake Powell. Water transfers will change the ecological balance in ways that many fear are typically for the worse. This is done by lowering water levels in streams that are water sources, by building reservoirs out of rivers, and by adding to water levels in locations that receive the water.

Irrigation can have deleterious long-term effects on the very land irrigated by building up the saline content of the land. This happens because even fresh water has a slight salt content; and as the water evaporates, the sediment of salt remains. This has become a serious problem in the Imperial Valley. The current solution has been to build outflow canals that flush the salt back into the Colorado River. But as a result, the Lower Colorado's saline content has increased 30 percent in the last 20 years. This has jeopardized the Mexican cotton industry, which uses the reinjected water for agriculture. This competition for fresh water creates political conflicts between the United States and Mexico.

There is a cost to future generations. "Mining" groundwater reduces the supply of easily available water in those regions. As a result, future residents of those regions must either do without water or use less accessible water resources. This is a particular problem in west Texas and in California.

Economic Analysis of Water Use Policies

Economic analysis cannot tell what should be done; it can only determine, at best, what the effects of alternative actions will be. Most economists do believe that, all other things being equal, allocative efficiency is desirable. Much economic analysis tries to establish whether a policy is allocatively efficient.

Economic Analysis of Groundwater Use. Economists accept that using groundwater may be a good way of obtaining water on the High Plains. It can be extracted at comparatively low cost, and it is generally free of pollutants. It might even be sensible to use more groundwater than is recharged into the aquifer. However, economists do question whether the quantity of groundwater used is the allocatively efficient amount. The reason goes back to the common property resource problem. As discussed earlier, if there is no clear owner of a resource, there is inadequate incentive to conserve because the full costs of use are not borne by the user.

The basic debate concerns what policy response is appropriate. Some favor government ownership of groundwater. The government would then be the authentic owner of the resource, and to that extent it would have the motive to consider all the costs and benefits in the use of the water. The federal government rather than state governments would have to manage the aquifers because aquifer boundaries are not coincident with state lines. While the government could institute a strategy that would efficiently manage the aquifers, economists do note two potential problems. First, the government, being influenced by political interests, might allocate the water on a political basis. As a result, the water might not end up going to the users of highest value. Secondly, even if the government tries to allocate the resources efficiently, it may find it very difficult to determine the uses of highest value. This occurs because it is extraordinarily difficult for a centralized decision maker to know the details of use and cost of a resource that has many different uses and sources.

Alternatively, the government could devise rational methods that employ markets to allocate the use of the groundwater. These methods are likely to be more efficient than centralized management. Two options exist. First, the government could decide each year how much water it wished to allow to be pumped out and then allow potential users to bid competitively for the right to use that water. Of course there is no guarantee that the quantity of water used would be the efficient quantity, but this method would ensure that the quantity of water that the government decides should be used will be used by the highest bidders. If implemented, this plan would lead to considerable reallocation from agricultural uses to industrial and municipal uses. Reallocation would result because agricultural users generally attach lower value to incremental supplies of water than do industrial and municipal users.

Secondly, the government could establish clear private ownership of the water, but it could impose a fee or tax for using the water. The fee or tax would encourage the private owners to internalize the opportunity cost of using the water today versus in the future. In this way, the owners would consider the full costs and benefits of water use. In principle, this solution works very much like the water rights scheme just mentioned. The most important difference is that the government sets the price rather than the amount of water it wants used. It is not clear which could be more efficient.

Economic Analysis of Interbasin Transfer Schemes. To evaluate large capital expenditure plans by the government, economists have developed a technique known as cost-benefit analysis. This method compares the economic benefits, which in water projects are multiple, to the economic costs, which are also multiple. For example, in the California State Water Project, the benefits include the value of the added water in southern California, the value of the recreational lakes created in northern California, the value of the additional electrical energy, and the value of the improved flood control. The costs include the value of the forgone water in northern California, the value of the forgone resources used to construct and operate the project, and the value of the wild rivers destroyed in the north.

The cost-benefit comparison is often addressed in the form of a question. But rather than asking whether the benefits are greater than the costs, economists ask: (1) At what price would the transferred water have to be sold to provide benefits that cover the costs? and (2) Is this price roughly comparable to the value of the water to the users?

In the case of the California State Water Project it turns out that the charge for water that would be necessary to pay for the project is considerably greater than the value of the added water to users. It has been estimated that for Los Angeles in 1960 the cost of a unit of water from the project was, depending on assumptions, between three and eight times the then-current price of water.[5] This means that the California State Water Project was very likely allocatively inefficient. It was an example of taking resources of high value and putting them to low value uses.

What is of particular interest to economists is that there is an alternative way of "solving" the water problem that is less expensive. It involves making the markets for water work better.

---

5. Jack Hirschleifer, James De Haven, and Jerome Milliman, *Water Supply* (Chicago: University of Chicago Press, 1969), p. 969.

An Alternative Solution to the Water Problem. The first part of the solution is to price water at its marginal value. Invariably water is priced well below its marginal value. If water were priced at its marginal value, which in the case of Los Angeles would mean at about twice its present price, a new "source" of water would appear. This new source would be water conservation. Again, according to the law of demand, price increases will encourage people to reduce the quantity of water demanded. Evidence suggests that for every 10 percent rise in water price, the demand among agricultural users will fall by about 6.5 percent.

The second part of the solution is to price all uses of water at the same rate relative to the cost of providing the water to different users. Currently, the price charged for agricultural use of water is considerably lower compared to the cost of providing it than the prices charged for home and industrial use are compared to their costs. This means that there is water that is being used for agriculture that has higher value uses in industry and urban use. Since agricultural use of water is such a large percentage of total water use, a fairly small reduction in agricultural use would lead to an enormous increase in the supply available for nonagricultural use.

The third part of the solution is to price water differently depending on the available supply. Just as it makes sense to charge higher subway tolls at hours of peak use, it is also efficient to charge more for water at times of peak use during the year. When the quantity supplied is less than the quantity demanded (as it often is in periods of peak use), the quantity of water supplied can be increased by charging higher prices.

The fourth part of the solution is to clarify rights of ownership. The ownership of water is an enormously complex issue. Water resources would be better allocated if it were possible to sell water rights on a free market. Often the water right goes with the property, so in order to buy the water right you must buy the entire property. It would be helpful if the water could be sold independently of the rest of the property. In the western United States, traditional law has it that water becomes the property of the first user. Ownership difficulties can result in excessive transactions costs, which can prevent the transfer of water to its highest valued use.

While these policy proposals would probably lead to a more efficient allocation of water resources, they would not necessarily lead to a better use of water resources. Pricing water at its marginal cost would redistribute wealth among groups. If free-market pricing had been used rather than the California Water Project, the price of water would be considerably higher in southern California. Agriculture in the region would be less profitable since water would be considerably more expensive. Those in the north would have more water at a lower cost. Such conflicts are serious and involve more than the economic concept of allocative efficiency. Nevertheless, one would hope that those who do resolve such issues in the future will consider economic analysis.

THE FEDERAL LANDS

Approximately one-third of the land area of the United States is owned by the various agencies and departments of the U.S. government. The pattern of land ownership is geographically uneven. At the extremes, the federal government owns over 80 percent of the land areas of both Alaska and Nevada but owns less than 1 percent of the land areas of Connecticut, Iowa, and Maine. The Bureau of Land Management of the Department of the Interior and the Forest Service of the Department of Agriculture are the two major land owners within the federal government. Most of the rest of federal lands are owned by the National Parks Service, the Fish and Wildlife Service, the Department of Defense, and the Department of Energy.

Many different groups use or desire to use these lands. Table 12-4 shows some of the important resources of federal lands. Ranchers use them for grazing animals; mineral companies use them for mining; recreational users use them for hunting, fishing, camping, and hiking; timber companies use them for forest supplies; and environmental preservationists use them in the sense that they restrict the use of them. Conflicting claims by at least some of these parties abound.

There are two particular areas in which economics can significantly contribute to our thinking about these public lands. First, economics can help decide how one should use these lands. For example, on what basis should certain lands be used for oil exploration? On what basis should

Table 12-4

Resources Found on Federal Lands

Resources	Amount on Federal Lands
Undiscovered oil and gas	50% of total
Coal reserves	33% of total
Uranium	35% of total
Softwood timber inventory	54% of total
Recreational use	6.3 billion visitor hours in 1985
Rangeland	28% of total

SOURCES: *Statistical Abstract of the United States: 1987,* and *The Annual Report of the Council on Environmental Quality: 1982*

certain lands be committed to wilderness? Secondly, economics can help decide among various ownership options. In the past two years, two alternatives to federal government ownership have been suggested. The first alternative, which arose out of discontent among western ranchers at federal land management policies, advocates that the ownership be transferred from the federal to the state governments. The political expression of this viewpoint is known as the "sagebrush revolution." The second alternative to federal ownership rejects government ownership at any level of government and advocates selling off the land to private individuals or organizations. This alternative is known as "privatization." Former Secretary of the Interior James Watt and President Reagan have been identified as advocates of both of these alternatives.

Economic Analysis of Land Use

The economist begins thinking about land use by suggesting that it is efficient to allocate land to its highest valued use. The difficulty with applying this criterion to actual land use decisions is that it is very hard and controversial to determine what the value of various land uses actually is. It is tempting to think that the value of a land use is simply related to the dollar profit that can be made from it. While this is certainly relevant, it is too narrow to be used as a general framework. As a practical matter, what economics can do is organize one's thinking by carefully considering both the benefits and the costs of various land use alternatives. It will be helpful to look at some examples of land use to see how economics can be used.

Northeastern Wyoming. In certain cases the conflict of use involves cases in which both alternatives involve conventional monetary returns. An example of this is the decision about whether to use federal lands in northeastern Wyoming as rangeland or for extensive strip mining of coal. Economists would counsel the decision maker to compare the values of these two alternatives. This is done by adding up the benefits minus the costs of rangeland use and then comparing that to the benefits minus the costs of the strip-mining alternative. Special attention would need to be drawn to scarce water used by strip mining and to the number of years that strip mining will preclude range use of the land, but these can be measured fairly accurately by using market prices. In addition, if the strip mine is fairly sizable, new communities might need to be created, and the ranchers of the area will be diminished in number. Together, these would change the culture of the area. While these effects are not easily measured in dollars, many observers, although not all, would agree that these effects are comparatively small, albeit intensely felt by those affected.

Rainey Wildlife Sanctuary. There are some cases in which two uses do not conflict with one another. For instance, there is the case of the Rainey Wildlife Sanctuary in the Louisiana marshlands, which is owned by the Audubon Society. The society has leased oil and gas exploration rights in the sanctuary to Mobil Oil at a price that sufficiently compensates the Audubon Society for whatever detrimental effects on the wildlife refuge there might be through the exploration.

Grand Canyon. In some cases the conflict involves a unique amenity such as the Grand Canyon. Such a case developed several years ago, when it was proposed that a dam be built to form a reservoir along part of the Colorado River as it runs through the Grand Canyon. In this case it is impossible to assign the full dollar value to the amenity. We could measure the dollar amount that visitors to the Grand Canyon actually pay to visit it. However, this underestimates the value that visitors place on the Grand Canyon, because the amount they actually pay will typically be less than the amount they would be willing to pay. But the important reason that this does not fully measure the value of the Grand Canyon is that there are other people who gain from its unspoiled existence. These parties include both future generations who will value the Canyon and people who, although they never visit the Grand Canyon, nonetheless derive substantial pride from its unspoiled existence.

    Not only do economists ask whether the Grand Canyon assumes greater value in its original form, but they further ask whether the benefits of expanding the size of the park by one square mile exceed the cost. By the same logic they would also ask whether the benefits exceed the cost of reducing the size of the park by one square mile. In the language of economics, economists are interested in incremental or marginal decision making. Economists ask the same sort of question about any use of land.

Use of National Parks. Given that one has areas like national parks, the additional question arises, how should one decide who should get to use them? Both in Yosemite Valley and at the base of the Grand Canyon, congestion would be extreme were there no mechanism to control use. Many economists prefer the setting of a monetary toll for the scarce resources like national parks, because that way only users who value the use of the resource above that toll would use it. However, others argue that a toll discriminates in favor of the wealthy and against the poor, both of whom should have equal access to our natural amenities. As a result other methods are used to reduce demand. These include such techniques as restricting automobile traffic into Yosemite, requiring inconvenient procedures for making campsite reservations, or simply requiring people to suffer through frustrating traffic jams in the Great Smoky Mountains.

Wilderness Areas. Some land areas have been declared wilderness areas. This means that no structures may be constructed; individuals may enter only with permits; and natural changes, be they through forest fire, flood, or disease, will not be hindered. It is nearly impossible to quantify the costs and benefits of setting aside such areas in comparison to using them in other ways. However, it is instructive to inspect the arguments in favor of creating such areas. Generally, the benefits result from ecological diversification in much the same way that an investor gains benefits by diversifying his or her portfolio.

Two specific arguments are made in favor of creating wilderness areas. First, benefits derive from maintaining a diversity of species controlled by natural selection rather than by artificial selection. Specifically, this diversity may provide species that may furnish unique vaccines for man at some unknown time in the future. Secondly, benefits derive from maintaining diversity within a species. The argument is simply that maintaining a variety of subspecies might well provide subspecies that would be unaffected by catastrophic blights or plagues. As a result, rapid recovery by the species from such unforeseeable events is ensured. These arguments are cogent, but economists would propose the use of economic analysis before decisions are made to create wilderness areas.

The economist suggests looking at benefits of alternative allocations of those same resources. How much wilderness area is desirable? In other words, what are the marginal benefits and marginal costs of added wilderness land? Is there some optimal amount of wilderness? Are there other possible ways to achieve the desired objectives at lesser opportunity cost?

The Issue of Land Ownership

The federal government has owned a great percentage of the land in the western states ever since they were acquired in the nineteenth century. As a condition of statehood, many of the territorial governments renounced all claims to federally owned lands. As population and development have proceeded over the past 100 years, the issue of federal ownership has been increasingly contested.

The Sagebrush Revolution. In 1979 the Nevada State Legislature passed a bill declaring the state's right to 49 million acres of land within Nevada that is owned by the U.S. Bureau of Land Management. This acreage is over 70 percent of the land area of Nevada. The Sagebrush Revolution was born with this symbolic legislation.

The explicit purpose of the Sagebrush Revolution is to transfer the ownership of federal lands to the state government. Some change in land use toward greater commercial and industrial development would no doubt result from such a change in ownership. The western states are generally among the most rapidly growing states, and their state legislatures generally favor more development than the federal government does. Still, land use decisions would continue to be made by a political body, in this case state governments rather than the federal government, and state government would also be subject to the pull of political interests rather than market interests.

As with land use decisions, land management would resemble the old patterns. This is likely because, just like the federal government, state governments would have the very difficult problem of centrally managing massive properties, each with its own idiosyncrasies. Many economists, although not all, fear that the states would be unable to do this with any greater expertise than the federal government.

Privatization. Several arguments have been made by economists who favor selling most federal lands (the process called privatization). First, the sale of the lands could pay off a considerable share of the national debt. Second, the sale of the lands would allocate the use of land in patterns that are more economically efficient. Third, sale would decentralize the control of the lands. It would put land management into the hands of those who know more fully the details of the individual circumstances of particular parcels of land.

The sale of half of the 700 million acres of federally owned land would raise as much as $500 billion according to one estimate.[6] With this amount, it would be possible to buy back almost 25 percent of the current national debt. While this would make many people feel more comfortable, economists point out that such a policy would have dramatic effects on the private bond market because the demand for loanable funds would decrease significantly. It might well lower interest rates, which would have positive effects on private investment but at the same time might have negative effects on private consumption since, as resources move into producing interest-sensitive capital goods, resources must move out of the production of consumer goods. The net effects might be very great, and they would be very hard to forecast. Economists also point out that the impact of this debt reduction on the health of the government is less clear than might appear at first. For example, when a homeowner sells his or her home to retire the mortgage, the owner also no longer owns the asset of the home. In the same manner, when the federal government reduces its indebtedness by

---

6. "The Think Tank Where Jim Watt's Ideas Live On," *Business Week* (October 24, 1983), p. 165.

retiring the debt, it also reduces its holding of assets. Whether this change is desirable or undesirable is difficult to answer.

Privatization may lead to a more efficient use of resources. Private ownership of land creates a strong incentive to use its resources most profitably. To the extent that highest profit equates to the highest-value use to society, private ownership is a benefit. Economists argue that this relationship between profit and social value is often the case. For example, the free-market pricing system probably does a better job of determining whether a parcel of land is better used for grazing or for agricultural crops.

However, as already discussed above, economists argue that there are certain cases in which free markets may fail to provide for the allocation of resources to their highest social value. First, private owners might tend to look too much to present use and not enough to future uses. However, present use tends to push future prices upward, which creates an incentive for conservation. Secondly, some uses create external social costs or social benefits. External social costs encourage private use contrary to the social interest. For example, strip miners of coal have no reason to prevent their wastes from running off as pollutants into public waterways because they do not bear the cost of the deterioration created by the runoff. Economists who are particularly supportive of free markets point out that the solution to the problem of external social costs lies in private bargaining between the polluting coal miners and those who must bear the pollution. Others point out that such a solution is infeasible when there are many miners and many concerned individuals. Each of them might have a slightly different view of the problem, making any agreement impossible from a practical point of view.

Proponents of the free-market view point out that the effects of privatization should not be compared to the ideal, but rather to the actual alternative of public ownership. They point out that government decision makers, just like private decision makers, are driven by incentives that may lead to failures of public decisions to be the socially optimal decisions. The primary specific argument is that special interest groups are disproportionately successful in influencing the political decision-making process because they can trade votes with others to attain objectives that are of great benefit to their constituents. Opponents of privatization would counter by saying that political decision making gives greater weight to less wealthy interests by relying on voting that is proportional to population rather than on free-market "voting" that is proportional to wealth.

Privatization may lead to a more detailed understanding of issues. Many economists argue that private ownership will lead to decisions being made by those who have an intimate understanding of the alternative uses for a piece of land. They point out that decisions made by even the most beneficent bureaucracy suffer from being distant from the actual properties that are to be managed. Examples abound in which well-meaning government decision makers have required land management procedures that are generally wise but, in specific cases, irrational.

One example is the practice of rest rotation. The idea is to divide grazing areas into four parts and allow cattle to use one part each year. Rest rotation is rational for parcels of land that have adequate rainfall and gentle terrain. But to simplify management, the Bureau of Land Management requires the procedure under a variety of land and climatic conditions. Unfortunately, it is very costly when rainfall is inadequate or terrain is rugged. When rainfall is low, rest rotation can led to excessive erosion, soil compaction, and deterioration of stream banks. When the terrain is rugged, the costs of building the necessary fencing do not justify the resulting benefits.

On the other hand, it is pointed out that many private landowners are large companies that are themselves so bureaucratized that they are no closer to the managed properties than the government itself. Supporters of free markets accept this but point out that with market decision making as opposed to government decision making, there is a competitive process that penalizes bad managerial decisions. Finally, those who oppose privatization point out that as a practical fact the quality of personnel in the public land management agencies is often higher than in private firms. Proponents of privatization agree, but they argue that were privatization to occur, this circumstance would change.

Generally, most authorities agree that privatization, like the Sagebrush Revolution, is likely to lead to greater development of western lands. Furthermore, many economists would suspect that such a change in land use probably would lead to moving resources toward higher-valued uses as measured by dollar values. It is also very clear that there are important interests, including possibly those of future generations and certainly those of environmental preservationists, that would be compromised by such a policy. As usual, economic analysis cannot say that such a change is desirable, though economists may feel that it promotes economic efficiency. Whether one regards economic efficiency to be desirable, given conflicting claims, is an issue of values.

CONCLUSION

There is no simple answer to the question: Can the free market work? It does not appear that it will lead to a rush of dramatic overexploitation of natural resources. On the other hand, in the case of common property resources there is a tendency for free markets to fail to achieve the most effective use of the resources. We have seen that too much conservation of natural resources is no better than too much use of them. There are costs and benefits to both paths. Economists find the idea of allocative efficiency useful in defining what a reasonable rate of use of natural resources is. Stated differently, economists recommend that natural resources be used now as long as the benefits of using additional units exceed the costs of using the additional units. Yet economics can only be used to advise policymakers since decisions must involve the weighing of ethical values.

The ideas of prices and markets have roles in deciding how to use natural resources. Groundwater resources cannot be effectively allocated using free markets, but the idea of bidding for rights to use groundwater or introducing a tax to allow users to recognize the full costs of their use is a partial use of markets. These policies have redistributive consequences to which some will object. Interbasin transfers are expensive, and many believe that their costs often outweigh their benefits. Alternative policies in areas where water is scarce center around raising the price of water. Higher prices will encourage conservation and a shift of resources to higher-valued uses. On the other hand, it has redistributive consequences both between regions and between groups within the region.

Land uses can be efficiently allocated using market prices, but that may redistribute wealth in ways that some will find undesirable. Land ownership changes will likely change the pattern of land use. In cases in which present demand is crucial, privatization will probably lead to a more effective pattern of use. When future preservation is essential or when common property resource problems dominate, privatization will probably reduce the effectiveness of patterns of use.

It is very hard to know what policies are most in our interest or are most in the interest of our grandchildren or great-grandchildren. What is very clear is that natural resources have become economic goods in that their use involves clear costs. While the best policy cannot be determined solely by using economic analysis, it is assuredly a step in the right direction to recognize that the use of our natural resources involves opportunity costs in the same sense as our use of anything of value.

QUESTIONS FOR DISCUSSION

1. Do you think society has conserved too many resources from the past? Too few? About the right number? What are the implications of this for the economic analysis of finite resources?

2. Suppose Yosemite National Park were sold to a private company like Walt Disney Enterprises. What would be the impact of this?
3. Who would be helped and who would be hurt by pricing water at the efficient level? What do you think the consequences of this would be not only economically but politically and socially as well?
4. Discuss whether fishing in publicly owned water would lead to the same sort of problem as using aquifers in the High Plains of west Texas. What solution(s) do you think is(are) possible if a similar problem arises?
5. Is there a natural resource whose use is of particular concern to you? Do you think that markets could be used to help the allocation of this resource, or do you think they would hurt the allocation? Explain.
6. Suppose that solar energy is of infinite supply and can be substituted for finite natural gas, but presently solar energy is much more expensive than natural gas. How would this affect society's use of gas through time? The use of solar energy?
7. Food is a natural resource. Would you favor using markets to allocate its use in a country on the borderline of malnutrition such as Ethiopia or Bangladesh?
8. Suppose that before the State Water Project had been built, California had a price for water that varied seasonally according to the monthly availability of water. How would that have changed the need for the new water supply system?

SELECTED READINGS

Baden, John A. "The Case for Private Property Rights and Market Allocation." *Center Magazine* (January–February 1981).

Environmental Quality: The Annual Report of the Council on Environmental Quality: 1982. Washington, D.C.: U.S. Government Printing Office, 1982.

Goeller, H.E., and A.M. Weinberg. "The Age of Substitutability." *Science* (February 20, 1976).

Hirschleifer, Jack, James De Haven, and Jerome Milliman. *Water Supply.* Chicago: University of Chicago Press, 1969.

Howe, Charles W., and K. William Easter. *Interbasin Transfers of Water.* Baltimore: Johns Hopkins Press, 1971.

Meadows, Donella, *et al. The Limits to Growth.* New York: Universe Books, 1972.

National Water Commission. *Water Policies for the Future.* Port Washington, NY: Water Information Center, 1973.

Owen, Oliver S. *Natural Resource Conservation*. 3d ed. New York: Macmillan, 1980.

Reppetto, Robert. *World Enough and Time: Successful Strategies for Resource Management*. New Haven, CN: Yale University Press, 1986.

———, ed. *The Global Possible: Resources, Development, and the New Century*. New Haven, CN: Yale University Press, 1986.

Schurr, Sam, *et al. Energy in America's Future*. Baltimore: John Hopkins Press, 1979.

Zaslowsky, Dyan. "Does the West Have a Death Wish?" *American Heritage* (June–July 1982).

13

THE U.S. DOLLAR
IS IT
TOO STRONG?

During most of the past 50 years the U.S. dollar has been the strongest currency in the world. Americans and foreigners found its purchasing power in the United States and abroad to be strong in relation to other currencies. It became the world standard against which to measure other currencies and served as the prime unit for settling international payment balances. For a time the U.S. dollar was considered more valuable and useful than gold itself.

Throughout the 1970s the purchasing power of the U.S. dollar dwindled, it could no longer be redeemed for gold, and its value compared to that of some other key currencies depreciated. Other currencies were sometimes sought in preference to the dollar, and some international transactions were being determined in terms other than the U.S. dollar.

Although the U.S. dollar was buffeted severely in the tempestuous gyrations of international finance, it stayed afloat. Measures were undertaken to prevent it from capsizing and to restore it to its premier position among world currencies. In the 1980s the dollar rebounded and again is the strongest currency in the world—perhaps too strong.

What knocked the dollar off its pedestal and tarnished its image? How does it stand in relation to gold and other major currencies in the world today? What was done to restore the prestige of the U.S. dollar? Why is it considered by some to be too strong? To better understand this changing value of the U.S. dollar, it is necessary to trace its history and its relation to gold and other world currencies.

HISTORY OF THE U.S. DOLLAR

Gold has been the elixir of humanity since the early days. It has been exchanged, hoarded, coined, and used for both industrial and decorative purposes. It has been instrumental in winning wars, bringing wealth and culture to nations, and causing jealousy and disaster. People have worked for it and fought for it. In addition to decorative purposes, its greatest uses have been as a medium of exchange and as a store of value. It has been coined since ancient times.

In spite of attempts by various governments in the past to limit its use as a money supply, gold still holds a prominent role in determining the value of money. Even when not used as currency, it has frequently been used to give backing and value to the money supply. Its wide use as money, or backing for money, results from its general acceptance throughout the world. Its limited total supply, its relatively stable annual production, and its durability add to this acceptance. It is an important determinant of exchange value between currencies of different nations. It is an ultimate means of settling international balances of payment. Consequently, the amount of gold that a nation may have for the support of its money supply has been of major, if not critical, economic importance. Thus when the United States established its monetary system, it gave gold an integral role.

The first coinage act in the United States in 1792 set up a bimetallic money standard. It defined the dollar in terms of both gold and silver. A dollar was defined as 371.25 grains of pure silver or 24.75 of pure gold. Thus, the mint ratio of the two metals was 15 to 1. When the market ratio subsequently deviated from the mint ratio, a shortage of gold coin developed. The mint ratio was changed to 16 to 1 when the gold content of the dollar was reduced to 23.22 grains by the Currency Act of 1834. Unfortunately, the new ratio overvalued silver at the mint, which resulted in a shortage of silver coin. The virtual disappearance of silver from the monetary system was further hastened by the Subsidiary Coinage Act of 1853, by which the government stopped the free coinage of fractional silver. Subsequent demonetization of silver took place when the Coinage Act of 1873 practically eliminated the coinage of silver dollars. In 1879 the United States abandoned the inconvertible paper standard adopted at the outbreak of the Civil War and established a gold standard by giving the dollar its prewar gold content of 23.22 grains and making the dollar convertible into gold.

The Gold Standard

It was not until the turn of the century, however, that Congress established a singular gold standard for our money supply with the passage of the Gold Standard Act of 1900. This ushered in the so-called golden age of the gold standard, which lasted until the outbreak of World War I. During this period, not only the United States but also most other major nations were on the gold standard. The most important of these nations were on the gold-coin standard, while others were on a gold-bullion or a gold-exchange standard. The heyday of the gold standard, however, was interrupted by the outbreak of World War I. As the war commenced, all belligerent nations abandoned the gold standard by refusing to redeem their currencies in gold and by prohibiting gold exports. In many cases, gold coin and bullion were called in by the government or the central bank. This was done to prevent the nation's gold supply from falling into the hands of the enemy, to conserve the gold supply for the purchase of essential war material, to continue operations in the foreign exchange markets, and to maintain enough gold reserve to preserve confidence in the nation's money supply.

In spite of various problems involved, most nations returned to some form of gold standard during the 1920s, with the United States leading the way in 1919 by re-establishing the prewar gold content of the dollar. Britain, Switzerland, France, Germany, and other nations followed some years later. Gold coin, however, virtually disappeared (except in the United States) as most countries adopted gold-bullion or gold-exchange standards. In general, these postwar standards were managed to a greater extent than were the pre–World War I gold standards. High tariffs and other restrictions, heavy war debts, an unstable flow of international lending, and other disturbances made the operation of the gold standard more difficult, especially in serving its function of settling the international balance of payments.

With the beginning of the Great Depression in the early 1930s, countries abandoned the gold standard in great numbers. The United States and France were the only major nations left on the gold standard at the beginning of 1933, and the United States went off it later that year. The general abandonment of the gold standard was precipitated in large measure by an international financial panic caused by foreign creditor demands for repayment in gold of short-term liabilities, such as bank deposits, and short-term government and commercial obligations. In most cases, the total credit demands exceeded the gold stock held by individual nations. Since they did not have the ability to redeem these obligations in gold, many nations were naturally forced to go off the gold standard.

Devaluation — 1933

With a 50 percent drop in GNP between 1929 and 1933, an increase of unemployment from 1.5 million to 12.8 million, widespread bank failures, and a rash of commercial bankruptcies taking place, foreigners began large-scale withdrawals of short-term liabilities from the United States. Although its gold supply, then about 40 percent of the world total, remained above $4 billion, the United States lost over $270 million in early 1933. Under these conditions, the United States abandoned the gold standard on March 6, 1933, when President Roosevelt placed an embargo on gold exports. Subsequently, the country returned to a gold standard, but it was feared that re-establishment of the dollar at the old gold content with full convertibility would lead to such an extensive gold drain that it would seriously affect the abilities of the banks to issue credit and would jeopardize the opportunity for domestic economic recovery.

After calling in practically all of the nation's gold held by citizens, businesses, banks, and other organizations, the United States returned to the gold standard on January 30, 1934. The new gold standard, however, differed substantially from that in operation prior to March, 1933.

1. According to the Gold Reserve Act of 1934, the value of the dollar in terms of gold was reduced. The gold content of the new dollar was 13.71 grains compared to the previous content of 23.22 grains, a reduction of approximately 40 percent. Consequently, the price of gold was increased from $20.67 to $35.00 per ounce. This meant that the gold supply, then worth nearly $4.2 billion, increased in dollar value to $7 billion.
2. Gold was, in effect, nationalized with some minor exceptions, and no currency in the United States was to be redeemed in gold.
3. The coinage of gold ceased, and all existing gold coin in the hands of the Treasury was converted into bullion.
4. Individuals and firms were prohibited from holding, transporting, exporting, or otherwise dealing with gold except under regulations specified by the Secretary of the Treasury.

Although a number of other nations subsequently returned to the gold standard, each had devalued its currency and introduced considerable management into its monetary system. By the outbreak of World War II, gold had lost much of its appeal, and most of the gold standards were of a gold-bullion or a gold-exchange type. No major nation had returned to a gold-coin standard, and most nations of the world were on an inconvertible paper standard. Of course, during World War II the entire world went off the gold standard.

The Gold Avalanche

An avalanche of gold hit the world, especially the United States, in the latter part of the 1930s. The causes for this increase in gold supply were manifold. First, the increase in the price of gold encouraged its production. During the decade of the 1930s, the annual physical output of gold more than doubled. Second, trade increased with the Orient, which paid for its imports in large part with gold. The gold supply was further increased by the melting of scrap gold, which brought a greater monetary reward to the seller.

While the world was experiencing a substantial increase in the gold supply, the United States was enjoying an even greater influx of the precious metal. The amount of gold held in the United States rose from approximately $7 billion immediately after devaluation of the dollar in 1934 to $22 billion at the end of the decade. It continued to increase in the 1940s, peaking at $34.5 billion in 1949. The bulk of the increase, more than $16 billion, resulted from gold imports. This was due not only to a favorable balance of trade, but more so to the flight of capital that took place in Europe because of unsettled economic and political conditions preceding World War II. By 1949 the United States held 69 percent of the world's gold supply.

Bretton Woods Agreement (1944)

After World War II several nations returned to a form of gold standard under the auspices of the newly organized International Monetary Fund (IMF). Most of these gold standards were limited. With the exception of the United States, redeemability of currency for gold, even for international purposes, was severely restricted by most nations throughout the world.

Since the U.S. dollar was the one major currency redeemable in gold, other nations agreed to define their currencies in relationship to the U.S. dollar. Moreover, they agreed to hold their exchange rates to within a plus or minus 1 percent range of the exchange rate established at Bretton Woods, New Hampshire. If the exchange rate of a nation tended to deviate beyond the established range, that nation's monetary authorities had an obligation to buy or sell U.S. dollars or its own currency to keep the exchange rate in line. The IMF agreed to make loans of gold, dollars, and other currencies to help nations keep their exchange rates stable.

Dollar Shortage

Due to the scarcity of goods and services throughout the world after World War II, especially in the war-damaged nations, there was a great demand for American dollars as foreigners sought dollars to buy American products and to settle their international balances of payments. During this time, there was a greater demand for American dollars than there was for gold. The Marshall Plan and reconstruction loans made by the United States to other nations at that time were helpful in alleviating to some extent the dollar shortage.

Moreover, a number of foreign nations began to accumulate dollar reserves for future use in purchasing goods and settling trade deficits. Since the dollar could be converted readily into gold by foreign governments and central banks, it was as good as gold. In addition, American dollars could earn interest income when held in deposit at banks.

Surplus Dollars and the Decline in the U.S. Gold Stock

Circumstances favorable to the U.S. dollar changed, however, in the 1950s. As war-torn nations rebuilt their economies and increased their production of goods and services, as U.S. prices rose making its products more expensive to foreigners, and as the U.S. encountered more competition in world trade markets, the American favorable international balance of trade became less favorable. As the United States spent more money to maintain military installations overseas and granted more foreign aid and as Americans increased their investments abroad, the dollar shortage was eased, U.S. international balance of payments became negative, and many foreigners began converting American dollars into gold. This caused a gold outflow from the United States that continued until 1971. By the middle of 1971 its total gold reserve had dwindled from a peak of $24.5 billion in 1949 to $10.5 billion. Changes in American gold holdings are shown in Table 13-1.

This table indicates that the world's gold supply increased from $35.4 billion to $41.3 billion during the period 1949–1970. During that time, however, the gold holdings of the United States dwindled by almost 50 percent, and the share of the world's gold supply it held decreased from 69 to 27 percent. Increased production and shifts in gold holdings resulted in sizable amounts of gold flowing into such nations as Belgium, France, West Germany, Italy, the Netherlands, Switzerland, and even the United Kingdom.

Between 1965 and 1970, gold reserves of central banks and governments dropped by almost $2 billion. This was due to the sale of gold by central banks and governments to private sources, especially through the

Table 13-1

Estimated Gold Reserves of Central Banks and Governments, 1947–1986
(Millions of dollars)

Year	Estimated World Total[a]	United States	International Monetary Fund	Rest of World	United States (as a % of total)
1949	$35,410	$24,563	$1,451	$ 9,396	69
1950	35,820	22,820	1,495	11,505	64
1955	37,730	21,753	1,808	14,170	58
1960	40,525	17,804	2,439	20,280	44
1965	43,230	13,806	1,869	27,285	32
1970	41,275	11,072	4,339	25,865	27
1975	49,555	11,559	6,446	31,550	23
1980	47,750	11,172	4,369	32,209	23
1985	44,026	11,120	4,380	28,562	26
1986 (June)	44,415	11,084	4,380	28,951	25

[a]Excludes USSR, other Eastern European countries, and the People's Republic of China.

SOURCE: *Federal Reserve Bulletin* (January 1987) and *International Financial Statistics* (October 1986).

London and other gold exchanges. Speculation regarding possible devaluation of the currencies of some major nations forced the price of gold upward in the free markets. The drain of gold out of central banks and into the free markets was arrested for several months early in 1968 when the ten major nations of the world agreed not to supply the free market with gold. This was done in connection with the devaluation of the British pound. The gold outflow from the United States resumed, however, and continued until President Nixon placed a ban on gold exports in August, 1971.

BALANCE OF PAYMENTS

The balance of payments was an integral factor in the gold outflow, and it is an important determinant today of the dollar's value in world markets.

There are several ways to show U.S. international transactions. Much will depend on the items included and the categorization of each item. A typical presentation for 1985 is shown in Table 13-2. It indicates that the United States imported $124.5 billion more in merchandise than it

Table 13-2

United States International Transactions, 1985
(Billions of dollars)

U.S. Balance of Payments		
Merchandise trade balance	− 124.5	
Exports	+ 214.4	
Imports	− 338.9	
Net military transactions		− 2.9
Net investment income		+ 25.2
Receipts	+ 90.0	
Payments	− 64.8	
Net travel and transportation expenditures		− 11.1
Other services		+ 10.6
Balance of trade on goods and services		− 102.7
Remittances, pensions, and other unilateral transfers	− 15.0	
Balance on current accounts		− 117.7
Changes in U.S. and Foreign Assets		
Net U.S. assets abroad (−)		− 32.4
U.S. official reserve	− 3.9	
U.S. government assets	− 2.8	
U.S. private assets	− 25.7	
Net foreign assets in the U.S. (+)		+ 127.1
Foreign official assets	− 1.3	
Other foreign assets	+ 128.4	
Statistical discrepancy		+ 23.0
Net change		+ 119.7
U.S. Official Reserve Assets (Net)		
Gold stock		11.1
SDRs		7.2
Reserve position in IMF		12.0
Foreign currencies		12.9
Total		43.2

SOURCE: *Economic Indicators* (October 1986) and *Federal Reserve Bulletin* (October 1986).

exported. This deficit was due largely to U.S. imports of oil, autos, steel and machinery. American tourists spent $11.1 billion more abroad than foreign tourists spent in America. We also sent $15 billion in pensions and other unilateral payments to persons living abroad. On the plus side, the $25.2 billion in net investment income from abroad offset some of our deficit in the balance of payments for 1985. In total, the United States ended up owing foreigners $117.7 billion as a result of the deficit current account.

Settlement of International Payments

Until 1971 gold was the ultimate means by which international balances of payments were settled. If there was a balance of payments due to the United States, a foreign nation could settle it in one of three ways: (1) by using its dollars previously obtained and held with American banks to pay the difference; (2) by selling financial assets to Americans, which amounted to borrowing U.S. dollars; or (3) by selling gold to the United States. On the other hand, if the United States owed a balance to foreigners, it settled by paying out dollars from its various accounts in foreign or American banks, by selling American securities in foreign markets and using the money to pay trade debts, or by selling (paying) gold to foreigners.

Whether a foreign nation holding a balance of payments against the United States would increase its dollar holdings, increase its holdings of American financial assets, or require payment in gold depended on many circumstances, such as the country's need for dollars, its current holdings of gold, and its desire to hold short-term financial assets for the purpose of obtaining interest income. The practice followed in each country depended in large part on its reserves. Such countries as West Germany, France, Switzerland, Italy, and Belgium had a tendency to build up gold reserves. Consequently, any substantial increase in their dollar claims could easily result in a gold outflow from the United States. Other nations, however, preferred to hold part of their reserves in dollars and part in gold, while still others tended to hold most of their claims in U.S. short-term assets.

Not only was the gold flow affected by the action of individual nations, but it was also affected by the desire of foreign persons or firms to increase or decrease their dollar holdings. If they decided to decrease their dollar holdings, their dollars, through proper channels, could be exchanged for gold, resulting in an aggravation of the gold outflow from the United States. Even though they did not convert their dollars directly into gold, the dollars or dollar claims could be exchanged for domestic currency at banks, which in turn converted the dollars into gold.

Settlement of international balances of payments by the sale of gold from one nation to another could take place without any physical movement of gold stock. A foreign nation, for example, could purchase gold from the United States to resolve a deficit U.S. balance. Instead of having the gold shipped, the foreign nation could simply have its newly purchased gold earmarked or segregated for its gold account at the Federal Reserve Bank of New York. This same nation could settle an adverse balance of payments with the United States by selling gold from its earmarked account to the United States. In July, 1986, the Federal Reserve was still holding over $14 billion in earmarked gold for foreign and international accounts. This was in addition to the $11.1 billion gold reserve held by the United States.

Moreover, there were $145 billion in U.S. government securities and $233 million in deposits being held at Federal Reserve Banks for foreigners.

Changes in Gold Reserves

Prior to August, 1971, the United States settled its deficit balance of payments in part through the payment of gold in exchange for dollars held by foreign central banks and governments. Since August, 1971, however, payments by the United States to foreigners have not been made in gold.

Between 1949 and 1971 the United States had numerous, and sometimes sizable, deficits in its balance of payments; and foreigners increased their dollar claims against the United States by a substantial degree. This caused a sizable increase in foreign holdings of short-term liquid assets and in some years a drain of gold exceeding $1 billion from the United States. As the United States lost gold despite the overall increase in the total world's supply, other nations increased their holdings of gold for a variety of reasons, including the desire by the United Staes that they make their currencies freely convertible.

Noticeable gains in the gold holdings of certain nations since 1949 can be observed in Table 13-3.

Table 13-3

Reported Gold Reserves of Central Banks and Governments,
1949 and 1986

	1949 (Millions)[a]	1986 (Millions)[b]	Percentage Increase
Belgium	$698	$1,443	107%
Canada	496	849	72
France	523	3,456	560
International Monetary Fund	1,451	4,380	202
Italy	256	2,814	999
Netherlands	195	1,855	851
Organization of Petroleum Exporting Countries	—	1,848	—
Switzerland	1,504	3,516	134
United Kingdom	1,688	803	−55
United States	24,563	11,083	−54
West Germany	—	4,018	—

[a]Valued at $35.00 per ounce.
[b]Valued at $42.22 per ounce.

SOURCE: *International Financial Statistics* (September 1986).

THE DOLLAR AND THE GOLD POSITION

The dollar problem is brought into clearer focus when the amount of gold is compared to formerly required reserves and a study is made of the causes of the gold outflow.

U.S. Free Gold Reserves

In respect to total gold holdings, in 1964, for example, the United States held $15.5 billion of gold, of which $13.4 billion was required as reserve behind Federal Reserve notes and deposit liabilities. This left about $2 billion in so-called free gold available for other uses, including the settlement of deficit balances of payments. This was quite a change from 1949, when the United States had over $12 billion in free gold reserves. Its free reserve position was made more precarious by the fact that foreigners also held $24 billion in dollar deposits and short-term American securities that could easily and quickly be converted into a demand for gold. Therefore, it was imperative that foreigners continue to hold dollars and short-term securities and settle future deficit balances against the United States by the use of dollar deposits and short-term securities if the U.S. gold supply was to remain free from jeopardy. This situation was eased considerably in 1965, however, when Congress voted to remove the gold reserve requirement behind Federal Reserve Bank deposits. This action freed an additional $4 billion of the total gold reserve. Finally, in 1968, when free gold reserves had dwindled to less than $1.5 billion, Congress removed the 25 percent gold cover on Federal Reserve notes. This freed all of the U.S. gold supply for international use.

Liquid dollar balances, rather than gold, were being held by foreigners for a number of reasons. Some were held for the purpose of receiving interest income, some as emergency reserves, and some as a working balance to carry on daily transactions of international trade. In addition to government securities, foreigners also held several billion dollars in U.S. corporate and state and local government securities, which could have been sold for dollars and eventually converted into gold.

Another possible discharge of the gold supply could result if there were a loss of confidence in the dollar. If, for example, Americans suspected that the dollar was to be devalued, it would behoove them to convert dollars to foreign currencies with the idea of converting the foreign currency back to dollars after the devaluation. In the interim, however, the American dollars held by foreigners could be converted into gold by foreign monetary authorities.

Role of International Monetary Fund

Nations that are short of foreign currency or gold to meet a temporary deficit balance of payments can obtain help from the IMF. The fund, established in 1944, has resources consisting of gold and the currencies of the 150 member nations, which are paid in by each nation on a quota basis. The quota for each country is determined by the relative importance of the country in regard to international trade, national income, and population. In 1986 the IMF held gold and foreign currency subscriptions in value equivalent to more than $100 billion (in American dollars). Of this total, about $4 billion was in gold.

A nation desiring to borrow foreign currency or gold from the fund can do so by depositing a like amount of its own currency with the fund. Drawings from the fund for an amount equal to 25 percent of the borrowing nation's own quota are almost automatic. Approval of additional borrowing is dependent upon the borrowing nation's effort to take steps to eliminate its imbalance of payments. The borrowing nation has an obligation to repay its loan within a period not to exceed three to five years. Repayment is generally made by the borrower by repurchasing its own currency from the fund with gold or convertible currencies.

Cause of Gold Outflow and the Weakening of the Dollar

One common way often used to determine the reason for our gold outflow was to cite specific categories of international payments that accounted for sizeable portions of our deficit balance of payments. In 1974, for example, when our deficit balance of payments was $5.1 billion, some analysts cited our $4.9 billion in military expenditures as the major cause of our deficit balance of payments. Others noted that we sent abroad over $8.1 billion in unilateral transfers, remittances, and pensions. On the other hand, some mentioned as inadequate certain categories on the credit side of the balance of payments, such as exports or the investment of foreign capital in the United States. In fact, many times such factors as high wage rates, high prices, or the Common Market were cited as causes limiting our export of goods and services and stimulating imports. Others, using a chronological approach, blamed the deficit on the last major phenomenon or two that occurred before or while we were shifting from a positive to a negative balance of payments. As the supply of U.S. dollars or dollar claims held by foreigners increased compared to the demand, the underlying strength of the dollar began to weaken. The one thing maintaining its strength was its convertibility into gold.

REMEDIES FOR GOLD OUTFLOW

Possible solutions for arresting the gold outflow and strengthening the position of the dollar have centered on remedying the U.S. deficit balance of payments. But even without altering the balance of payments, the gold outflow could have been abated by inducing foreigners to hold more dollars and short- and long-term American securities. This latter method, however, would not have removed the potentially dangerous situation of having excessive dollar claims that could be converted readily into gold. Some of the corrective measures suggested involved a greater degree of control or management over exports, imports, capital flows, the gold supply, and domestic economic measures. Here the United States would have to measure the advantage of a desirable balance of payments and a stable gold supply against the disadvantage of a certain amount of economic restriction. It also would have to weigh the merits of a favorable balance of payments against the possible adverse effect on the domestic economy of those measures adopted to obtain an improvement in the balance of payments.

Included among the measures suggested and/or adopted in the 1960s to arrest the gold outflow were the following:

1. Restrict imports.
2. Promote exports.
3. Reduce government spending abroad.
4. Reduce military spending overseas.
5. Reduce or eliminate foreign aid by
   a. using tying contracts, or
   b. encouraging other nations to help with aid.
6. Reverse capital outflows
   a. through higher interest rates,
   b. by paying higher interest rates on foreign deposits,
   c. by adopting an interest equalization tax, or
   d. by restraining credit on foreign investment.
7. Reduce required gold reserves.
8. Devalue the dollar.
9. Revalue other currencies.
10. Abandon fixed exchange rates.
11. Increase IMF lending ability.

During the throes of the gold outflow from the mid-1950s to 1971, the United States used each of the preceding measures. It restricted imports by the use of tariffs, quotas, and limited duty-free foreign purchases by

American tourists. It promoted exports through trade shows. It found ways to reduce government spending abroad and cut back on overseas military spending, except the spending for the war in Vietnam. In addition, it reduced foreign grants and loans by sizeable amounts and encouraged other nations to help with aid to the developing nations. It tried to reduce U.S. capital flows with higher interest rates, paid premiums on foreign deposits in the United States, and imposed an interest equalization tax. In addition, the banks, through the Federal Reserve, voluntarily restrained loans for foreign spending. The United States also freed more gold for foreign payments by eliminating the gold reserve requirement behind Federal Reserve Bank deposits and Federal Reserve notes. Other nations, such as West Germany and Japan, revalued their currencies. In addition to these measures, the International Monetary Fund created Special Drawing Rights (SDRs) to expand its lending ability, and rigid, fixed exchange rates were generally abandoned. When these measures failed to correct the deficit balance of payments and arrest the gold outflow, the United States finally devalued the dollar and stopped the convertability of dollars into gold.

THE DOLLAR PRICE OF GOLD

Gold is bought and sold not only by governments and central banks throughout the world, but also in private markets. Until August, 1971, the world monetary price of gold was well established, especially by the U.S. Treasury, at $35 per ounce. In the private gold markets, the largest of which is London, the price of gold is free to fluctuate with changes in supply and demand.

From the time the London Gold Market reopened in 1954 (after closing during World War II) until 1960, the free-market price of gold stayed close to the official monetary price of $35 per ounce. The supply of gold for the free market was obtained from three major sources: (1) new production, especially from the Union of South Africa, which mines three-fourths of the West's gold; (2) gold sales from Communist nations; and (3) monetary gold of central banks, whenever the free-market price of gold rose above its monetary price.

The Gold Pool

With increased industrial use of gold, limited supplies coming into the market, and the speculation in 1960 that the United States might change its policy and devalue the dollar, the heavy demand for gold pushed up the price of gold in the free market to more than $40 per ounce. At that time eight major nations, including the United States and Great Britain, formed a Gold Pool for the purpose of stabilizing the free-market price of gold near the monetary price of $35 per ounce.[1] Technically, the Gold Pool consortium agreed to sell gold when the price rose above $35 per ounce and agreed to buy gold when the price fell below the $35 monetary value.

With the perennial heavy demand for gold for industrial uses, a modest increase in supply, and the continuous belief that the United States, Great Britain, France, or some other major nation might devalue its currency, the Gold Pool was a net seller of gold, with its gold supplies coming primarily from the United States.

The Two-Tier System

As a result of its continuous balance of payments deficits, the United Kingdom, with the approval of the IMF, devalued the pound sterling by 14.3 percent, from $2.80 to $2.40, in November, 1967. At that time a wave of gold buying hit the free market. Gold hoarders speculating that the United Kingdom would be forced to devalue again or that the United States or some other major nation would subsequently devalue its currency drove the price of gold well beyond $40 per ounce. In the next five months the Gold Pool nations sold about $3.5 billion worth of gold, primarily from the United States, to the London market in an attempt to keep the market price from rising too greatly. Faced with a continuous rise in the free-market price of gold in contrast to dwindling free gold reserves, the ability of the Gold Pool nations to supply the free market with gold was threatened. At that time, which was prior to removal of the gold cover on its money supply, the United States had free gold reserves amounting to $1.5 billion. Consequently, on March 17, 1968, the Gold Pool nations agreed to end the operation of the Gold Pool and endorsed the establishment of a two-tier gold price. By this agreement, the Gold Pool nations ceased to supply gold to the free market. At the same time, they agreed to freeze world monetary reserves and stop buying gold for monetary purposes.

This action permitted the toleration of a two-tier system in which the free-market price could deviate from the monetary price of gold without bringing about stabilizing measures by the major nations. It was hoped, however, that the freezing of world monetary gold reserves would force the producers to sell all their gold in the free market, thus driving the market

---

1. Other Gold Pool members besides the United States and Great Britain were Belgium, West Germany, France, Italy, the Netherlands, and Switzerland. France, however, ceased its participation in June, 1967, shortly before the pool's heaviest gold losses began.

price downward toward the monetary price of $35 per ounce. The Union of South Africa, however, contended that it should be able to sell in both the monetary market and the free market. In reaction to the agreement, South Africa for a year or more sold very little gold to either the private market or the central banks. From March, 1967 until June, 1969, the price of gold on the free market fluctuated between $37.00 and $42.40 per ounce. The scheme eventually worked as planned, however, because by December, 1969 the free-market price of gold was down to $35 an ounce.

Special Drawing Rights (SDRs)

During the 1960s, world trade increased by more than 10 percent annually. In contrast, the total stock of international reserves, consisting primarily of gold, dollars, and pounds, increased by only 3 percent annually. Much of this relatively small increase in world reserves was accounted for by dollars, since the United States, through its perennial balance of payments deficits, contributed to world international liquidity in the form of short-term liabilities to foreigners.

To overcome the world shortage of international liquidity, the Group of Ten worked and deliberated over a period of four years on a system of SDRs as a means of increasing world reserves.[2] In September, 1968, member nations of the International Monetary Fund unanimously agreed to the concept of an SDR plan.

Debtor nations, such as the United States and the United Kingdom, were strongly in favor of the SDRs. Some of the creditor nations, however, were not so enthusiastic about the plan. Some nations, such as the United States, thought that the SDR amount should be a permanent addition to world reserves. Other nations saw the proposal only as a temporary expedient. In July, 1969, a compromise was reached when it was agreed at a meeting of the Group of Ten that the equivalent of $9.5 billion in SDRs, called "Paper Gold," would be created over a three-year period. The new reserve asset became available January 1, 1970.

The final agreement called for the creation of supplementary reserves in the form of SDRs established through the IMF. The SDRs were subsequently distributed to the member nations of the IMF according to each country's respective contribution quota. Thus, the United States received nearly one-fourth of the total SDRs created. The total issue of SDRs was

---

2. The Group of Ten consisted of ten world trade leaders, including Canada, Belgium, France, Italy, Japan, the Netherlands, Sweden, West Germany, the United Kingdom, and the United States.

subsequently increased, and by 1986 the total allocation of SDRs amounted to $21.4 billion.

Member nations cannot buy SDRs from the fund but can use their assigned SDR reserves to settle balance of payments deficits. All member countries are obligated to accept SDRs from debtor nations. No creditor nation, however, can be compelled to accept SDRs in excess of three times its allocation. Since SDRs carry an interest rate paid by the debtor nation, a creditor nation may desire to hold a greater volume than that required.

SDRs can be issued for balance of payments purposes and to protect a nation's reserve position. Upon mutual agreement, one nation can use its SDRs to buy currency of another nation. To encourage nations to accumulate other forms of reserves and not rely solely on their accumulation of SDRs, a provision in the agreement indicated that the average use of SDRs by a member nation must not exceed 70 percent of that nation's average accumulation over a five-year period.

SDRs can be used between nations either on the basis of bilateral agreement or at the designation of the IMF. In the latter case, the country desiring to use SDRs can request the IMF to designate certain countries to receive (accept) SDRs and the amount to be received.

The United States has been active in using SDRs in its exchange transaction for both payments and receipts. In 1985 the United States was holding $8.0 billion in SDRs; other large holders were Japan and West Germany.

Although SDRs cannot be converted directly into gold, the unit value of an SDR was initially equivalent to the gold content of the U.S. dollar. Consequently, when the U.S. dollar was devalued and the price of gold was raised from $35 per ounce to $38 per ounce, the value of an SDR was likewise increased to $1.09 per unit. Again, when the United States raised the price of gold to $42.22 in February, 1973, the transaction value of an SDR increased to $1.21 per unit ($42.22 ÷ $35.00 = $1.21). To improve the transferability of SDRs and move away from the exclusive reliance on the dollar as a means of determining the value of SDRs, the IMF in 1974 widened the base of valuation by including 15 other currencies besides the U.S. dollar in the "currency basket." For simplification this was reduced to 5 currencies on January 1, 1981. In 1987 the U.S. dollar was weighted 42 percent and the German mark 19 percent. The Japanese yen 15 percent, and the French franc and British pound had weights of 12 percent each.

The Demise of Gold

The various currency adjustments, along with the creation of SDRs, tight money, and higher interest rates at home, took some of the burden off the

U.S. dollar as a supplier of international liquidity. But the improvement of the U.S. balance of payments position was short-lived. Early in 1971 the United States experienced a substantial deterioration of its trade position and faced a deficit balance of trade for the first time in nearly a century. Speculation emerged that the dollar and/or some other currencies might subsequently be devalued. Consequently, the market price of gold began to rise substantially. By mid-1971, even nations that for years had retained "excess" U.S. dollars and refrained from redeeming them for gold in the interest of international financial stability began to question the wisdom of holding dollars as opposed to gold. By the end of July, 1971, it was estimated that foreign official institutions held between $40 and $50 billion in liquid claims against the United States. This was double the amount of a year earlier. On the other hand, the total U.S. gold stock was only $10.4 billion.

In the spring of 1971, many dollars were being sold in exchange for other currencies, particularly West German deutsche marks. Speculation was that the deutsche mark would be revalued and thus worth more dollars and that the exchange rate between the dollar and the deutsche mark would be moved away from the official monetary exchange rate (27.5 cents) in favor of the deutsche mark. The sale of dollars became so heavy that the foreign exchange markets in West Germany, Switzerland, Belgium, Austria, and the Netherlands were closed in early May, 1971.. When the markets reopened, changes were made. Effective May 9, 1971, the Austrian schilling was revalued. The next day the West German government announced that the deutsche mark would be allowed to "float"—that is, the rate of exchange was unpegged—to permit the market to seek and establish new exchange rates between the deutsche mark and other currencies, particularly the U.S. dollar. At the same time, the Netherlands' guilder also was floated and the Swiss franc was revalued to 4.08 per U.S. dollar (24.5 cents). Within a short period the average appreciation of world currencies vis-à-vis the dollar was 6–8 percent, and among major U.S. trading partners it was 10–12 percent. By July, 1971, the deutsche mark had floated upward from 27.5 cents to a market value of 31.6 cents.

Devaluation of the Dollar—1972

Under these circumstances President Nixon, on August 15, 1971, announced his New Economic Policy. In addition to imposing a wage-price freeze for the domestic economy, the New Economic Policy established a 10 percent surcharge on imports and suspended the convertibility of dollars for gold.

It was evident that the U.S. dollar was overvalued in world markets and that currencies of those nations with substantial balance of payments surpluses, such as Japan and West Germany, were undervalued. It was also evident that the United States could not continue forever as a major supplier of international liquidity for the world. In spite of this, the action of President Nixon on August 15, 1971 startled the financial world and brought about serious repercussions.

The United States subsequently used the 10 percent surcharge as a club to encourage various nations to adjust their currencies and to take other steps for the improvement of world trade. In December, 1971, after numerous meetings, the Group of Ten, in cooperation with the IMF, agreed to the so-called Smithsonian Accord, by which they pledged to work for an "effective" realignment of important world currencies. Subsequently, in early 1972 the U.S. import surcharge was modified and the U.S. dollar was devalued by 8.57 percent when Congress officially raised the price of gold to $38 per ounce. As a part of the international accord, the Japanese agreed to revalue the yen, and the deutsche mark and the guilder were to continue to float before new exchange values were set for them. The French franc and the British pound were to hold their previous par values.

The U.S. balance of payments failed to improve to any substantial degree in 1972. Moreover, the international monetary authorities failed to come up with any further solutions to the world monetary problems. The relationship of the American dollar to foreign currency, especially the deutsche mark and the yen, continued to deteriorate. In addition, the free-market price of gold was driven up over $60 per ounce, and it was evident that speculators were still unconvinced that the price of gold had been settled. Consequently, in February, 1973, the United States again devalued the dollar. This time the value of the dollar was decreased by 10 percent and the price of gold was raised to $42.22 per ounce.

A prolonged meeting of the Committee of Twenty (an enlargement of the original Group of Ten) in 1973 developed an outline of reforms and set a target date of July 31, 1974 to implement the reforms. The meeting, however, failed to produce any substantial remedies for the world monetary situation. In addition, the Arab oil embargo in late 1973 and early 1974 and the subsequent heavy increases in the price of oil had an adverse effect on the economies and balances of payments of the United States, Japan, West Germany, France, the United Kingdom, and other industrial nations. Those nations that belonged to the Organization of Petroleum Exporting Countries (OPEC) began having more influence on world financial matters as a result of their accumulation of dollars and other currencies due to selling oil at the new higher prices. This brought a new monetary problem to the international scene as the question now became how to invest or recycle these dollars back into the world money markets.

As world financial conditions became more uncertain, speculators bid the price of gold to more than $175 per ounce by the summer of 1974.

With the announcement that the United States would permit its citizens to purchase and hold gold beginning December 31, 1974, the price of gold in the free market reached $200 per ounce by Christmas. When gold sales in America proved to be less vigorous than anticipated, however, the market price of gold dropped back below $200 per ounce in early 1975.

WEAKENING OF THE U.S. DOLLAR AFTER DEVALUATION

Since the United States no longer converts dollars into gold for foreign treasuries and central banks and foreign nations no longer fix their currencies in terms of gold or U.S. dollars, supply and demand now play a more dominant role in determining foreign exchange rates. Meetings under the sponsorship of the IMF, the World Bank, and the Group of Ten countries in the past few years have made little progress in the correction of disorderly exchange markets or in the establishment of a new international monetary system.

In the absence of a new international monetary order, the currencies of many countries have been allowed to float. This means that the supply and demand for a particular currency affects it exchange rates. An increase in demand for a particular currency or a shortage of supply will strengthen the value of the currency and cause its price to rise in terms of other currencies. On the other hand, a decrease in demand for a particular currency or an increase in supply will weaken a currency and its value will fall in terms of other currencies.

The United States has experienced substantial deficits in its balances of trade and its balances of payments, as shown in Figure 13-1, and the supply of dollars abroad, or dollar claims against the United States, increased substantially during the 1970s. This caused the value of the dollar to weaken vis-à-vis many other currencies of the world, as shown in Table 13-4. Although the U.S. dollar strengthened in relation to a few currencies, it weakened relative to numerous major currencies. The weighted average exchange value of the U.S. dollar against currencies of the other Group of Ten countries plus Switzerland, for example, declined 15.4 percent between 1970 and 1980.

A falling value of the U.S. dollar, of course, acts as a corrective factor for the U.S. trade deficits. As the dollar falls in value, U.S. dollars cost less in terms of foreign currencies that are rising in value relative to the dollar. Therefore, foreigners pay less for American goods, and Americans pay more for foreign goods and services. This eventually will cause U.S. imports to decrease and U.S. exports to rise in the absence of offsetting changes. Many other factors could also help reverse trade deficits, such as

Figure 13-1

U.S. International Transactions, 1978–1986

Billions of Dollars★

Balance on Goods and Services

Balance on Current Account

Merchandise Trade Balance

★Seasonally Adjusted

SOURCE: *Economic Indicators* (January, 1987).

a more rapid expansion in the economies of other major nations that may encourage them to import more goods and services, revaluation of strong or undervalued foreign currencies, a slowdown in the U.S. inflationary rate compared to other nations, and the establishment of a new international monetary system. There are also dozens of minor factors that could be used to limit imports and dollar claims against the United States.

Eurodollars

One of the most mystifying forces affecting the value of the dollar, the price of gold, and the status of international liquidity in the past two decades has been the Eurodollar market. By simple definition, a *Eurodollar* is a U.S. dollar on deposit in a foreign commercial bank. It can come into existence when someone transfers a dollar deposit from a U.S. bank to a foreign bank or when someone buys U.S. dollars in exchange for other currencies and then deposits the dollars in a foreign bank. Although there is a difference

Table 13-4

Percentage Changes in Foreign Exchange Rates, 1970–1986
(U.S. cents per unit of foreign currency, except percents)

Country	Currency Unit	1970	1980	1985	1986[a]	% Change[b] 1970–1980	% Change[b] 1980–1985	% Change[b] 1985–1986
Australia	Dollar	111.36	115.85	70.03	62.91	+4.0	−39.6	−10.2
Austria	Shilling	3.87	8.06	4.84	6.62	+108.2	−60.0	+36.8
Belgium	Franc	2.01	3.58	1.68	2.26	+78.1	−36.9	+34.5
Canada	Dollar	95.80	86.78	72.99	72.46	−9.4	−15.9	−0.8
Denmark	Krone	13.33	18.49	9.44	12.40	+38.7	−49.0	+31.3
France	Franc	18.09	24.66	11.14	14.43	+36.3	−54.9	+26.6
Germany	Deutsche mark	27.42	57.25	34.01	46.51	+108.8	−40.6	+36.8
India	Rupee	13.23	12.88	8.11	7.99	−2.6	−37.1	−01.5
Ireland	Pound	239.59	214.74	106.62	139.00	−10.4	50.4	+30.4
Italy	Lira	0.16	0.12	0.05	0.07	−25.0	−56.7	+40.0

Country	Currency						
Japan	Yen	0.28	0.45	0.42	0.63	+60.7	−6.7	+50.0
Netherlands	Guilder	27.65	52.34	30.12	41.05	+89.3	−42.5	+36.3
Norway	Krone	13.99	20.76	11.64	13.37	+48.4	−44.0	+14.9
Portugal	Escudo	3.50	2.05	0.58	0.67	−41.1	−71.8	+15.5
South Africa	Rand	139.24	130.79	45.57	39.04	−6.1	−56.1	−14.3
Spain	Peseta	1.43	1.41	0.69	0.72	−1.4	−57.2	+22.0
Sweden	Krona	19.28	24.24	11.63	14.14	+25.7	−42.1	+21.6
Switzerland	Franc	23.20	62.20	40.73	57.32	+168.1	−34.6	+40.7
United Kingdom	Pound	239.59	237.32	129.74	150.71	−1.0	−45.3	+16.2
Memo								
United States	Dollar[c]	100.00	84.65	143.01	110.38	−15.4	+68.9	−22.8

[a] As of July, 1986.
[b] A positive change designates a gain in value of foreign currency in relation to the U.S. dollar. A minus change designates a loss in value of foreign currency in relation to the U.S. dollar.
[c] Index of weighted average exchange value of U.S. dollar against currencies of other gold countries plus Switzerland. March, 1973 = 100.

SOURCE: Federal Reserve Bulletin (June 1971 and October 1986).

of opinion, Professor Milton Friedman maintains that Eurodollars are also created in the same manner that dollar demand deposits are created in regular domestic banking practice. More mystifying is the fact that most of the actual dollars underlying Eurodollar deposits never leave the United States.

According to the Federal Reserve, "A bank accepting a Eurodollar deposit receives, in settlement of the transaction, a dollar balance with a bank in the U.S. A bank making a Eurodollar deposit or loan . . . completes the transaction from its U.S. bank balances."[3] Consequently, as Eurodollars are transferred around the world, the dollar balances supporting them are merely changed from one bank account to another within the American domestic banking system or even within an individual U.S. bank.

Although the Eurodollar market is worldwide, London is the center of the market. Furthermore, large U.S. banks are the major holders of and dealers in Eurodollars. The Eurodollar market operates under the supervision of the Bank for International Settlements (BIS); but to date there is little control or regulation over the Eurodollar market.

Petrodollars

In the 1970s, also, there was a growing concern about the effect on international payments of the dramatic rise in dollar and other currency holdings of OPEC. The dollar holdings, often referred to as petrodollars, rose sharply as a result of increases in the price of oil exported to the United States and other industrial nations. In a peak year, for example, the gross transfer of purchasing power realized through import receipts of OPEC countries was estimated to be the equivalent of $100 billion, of which $95 billion was from the export of oil. After subtracting imports and grants to developing nations and making a few other adjustments, the OPEC nations were left with a $60 billion surplus in their balance of payments. Estimates indicated that of the $60 billion, OPEC nations invested $11 billion directly in the United States. About one-half of the investments made in the United States were in the form of marketable government securities, $1 billion was used to buy real estate, and the remainder was placed in banking and money market liquid assets, such as large negotiable certificates of deposit.

The high price of oil caused substantial shifts in balances of payments, especially of the oil-exporting and oil-importing nations. In addition, the "recycling" of petrocurrencies has had a substantial impact on various economies depending on the flow and direction of OPEC spending, lending, and investment.

---

3. *Federal Reserve Bulletin* (April 1984).

European Monetary System

In March, 1979, the European Economic Community, in an effort to help bring about stability in the international exchange markets, launched the European Monetary System (EMS). Early in the 1970s an attempt had been made to do something of the same nature through an arrangement called the "snake," in which Common Market nations agreed to limit currency fluctuations against each other to no more than 2.5 percent so that their currencies would move up and down in unison against the U.S. dollar and the Japanese yen. It was sometimes referred to as the "snake in a tunnel" because the average value of the currencies was to fluctuate no more than 2.5 percent against the U.S. dollar. Stability was to be accomplished through the buying and selling of dollars to prevent currency deviations beyond the targets. The original snake was short-lived. Britain, France, and Italy withdrew because of high domestic inflation rates and an inability to keep on par with the soaring value of the West German deutsche mark.

The current EMS is an improved three-part arrangement:

1. It links ten European currencies (those of the Common Market nations) by permitting their exchange rates to fluctuate no more than plus or minus 2.25 percent against each other.
2. It established a new currency, the European Currency Unit (ECU), as a reserve asset to alleviate the burden on the U.S. dollar. The ECU is defined as a basket of European Community currencies and is equivalent in value to an SDR.
3. It established a European Monetary Cooperation Fund (EMCF) into which member nations contributed part of their gold, U.S. dollars, and domestic currencies. The original fund created was equivalent to $65 billion, or 50 billion ECUs.

In the new system, banks and government treasuries endeavor to strengthen their currencies buying and selling Common Market currencies and ECUs instead of U.S. dollars. This, of course, avoids the side effects on the U.S. dollar that arise from the sale and purchase of U.S. dollars.

The EMS survived its early years in spite of the skeptics. It did not encounter any serious problem or face any major tests. By 1986 the fund had accumulated the equivalent of $57 billion in U.S. dollars, other foreign currencies, SDRs, and gold, against which it issued $42.5 billion worth of ECUs to contributing nations.

U.S. External Liabilities

As a result of continuous and sizeable deficits in the U.S. balance of payments in the 1960s and 1970s, foreigners accumulated large dollar claims against the United States and many of them enhanced their reserve positions. Table 13-5 shows the size of these claims in 1978 and 1986.

Additional information indicates that considerable portions of the U.S. external liabilities to industrial countries were owed to Japan and West Germany. In addition, OPEC nations held a total of $75 billion in official reserves, compared to $12 billion in 1973 prior to the oil embargo and oil price explosion. Since the value of the U.S. dollar in foreign exchange markets is determined primarily by supply and demand, it is not difficult to see why the dollar weakened in the 1970s vis-à-vis some foreign currencies.

THE 1980s — THE DOLLAR REBOUNDS

The U.S. dollar fell substantially against most major currencies in the 1970s. By mid-1980 it had fallen 15.4 percent against other gold country currencies (G-10 nations) plus Switzerland, as shown in Table 13-4, and in 1980 it took $1.30 to equal 1 SDR ($1.00 = SDR 0.769). After 1980, however, the value of the U.S. dollar rebounded remarkably. By mid-1985, it had gained against every major country of the world and had risen 68.9 percent against G-10 currencies.

Many factors added to the strength of the dollar in the past few years. In the fall of 1979, for example, the Federal Reserve announced its plan to switch from a policy of stressing the use of interest rates in preference to emphasis on the money supply as a means of combating inflation. The resulting high interest rates ranging up to 20 plus percent encouraged a flow of foreign investment into the United States and discouraged foreign borrowing from the United States. This flow was accelerated by the rapid decline of inflation in the United States since 1982. The United States has once again become a prime area for investment with high interest rates, low inflation, and a stable political environment. It was estimated by French authorities that $150 billion of European capital was transferred to the United States in 1983 and that it would rise to $300 billion by 1988. Some of this transfer or investment of funds was no doubt by speculators, who would move back into their own currencies if and when the dollar fell in value.

As the dollar strengthened, it became easier for Americans to purchase foreign goods and services and more difficult for foreigners to purchase U.S. goods and services or to borrow U.S. dollars. This, of course, aggravated the U.S. balance of payments. The relationship between the strength of the dollar and the U.S. merchandise trade balance can be observed in Figure 13-2.

At the annual economic summit of the seven major industrial nations held in London in June, 1984, a common major complaint was aired about

Table 13-5

U.S. External Liabilities and Claims, March 1986 and 1978
(Billions of dollars)

	1986			1978		
U.S. External Liabilities			$625.25			$244.37
Central banks and governments		$174.38			$156.85	
Industrial countries	$97.94			$124.80		
OPEC countries	42.64			24.30		
Other countries	33.80			7.75		
Other banks and foreigners		431.81			79.76	
International agencies		19.06			7.76	
U.S. External Claims			440.87			130.05
Net Liabilities			$184.38			$114.32

SOURCE: *International Financial Statistics* (September 1986).

Figure 13-2

Comparison of Trade Deficits and Value of the Dollar

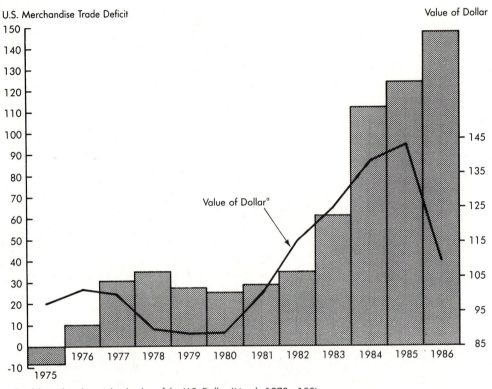

U.S. Merchandise Trade Deficit

Value of Dollar

°Multilateral trade-weighted value of the U.S. Dollar (March, 1973 = 100).

SOURCE: *Economic Report of the President,* 1987 and *Economic Indicators* (January 1987).

high U.S. interest rates. Some nations suggested that the United States take action, including lowering interest rates, easing its money supply, reducing its fiscal deficit, and intervening in foreign exchange markets in order to depress the value of the dollar. Some even suggested that the United States should moderate (slow down) its rate of disinflation. Early in 1984 the Organization of Economic Cooperation and Development (OECD) urged the United States to reduce its deficit and lower interest rates. At the same time, political factions in France and West Germany were suggesting the imposition of a domestic tax penalty on any of their currencies invested in the United States.

352   Chapter 13

IS THE U.S. DOLLAR TOO STRONG?

Not only do foreigners complain about the strong dollar, but many Americans do likewise. Since the strong dollar makes American goods and services and U.S. travel more expensive to foreigners, some American economists and international financial banks blame part of our growing balance of payments deficit on an overvalued dollar.

In addition, many U.S. firms and industries indicate that the strong dollar adds to their loss of markets here and abroad. Union leaders and others blame the strong dollar for some of the U.S. unemployment. U.S. trade representative William Brock stated that the dollar's strength would force more companies to establish plants overseas in order to compete in U.S. and world markets. Ford Motor Company executives maintain that $750 of the $2,000 cost advantage that the average car made in Japan has over a similar car produced in the United States is due to currency misalignment.

John Williamson, senior fellow at the Institute for International Economists, calculated the dollar to be overvalued by 24 percent in 1983 against major currencies and calculated the Japanese yen and West German deutsche mark to be undervalued by 5 percent. Ford Motor Company officials suggested that the dollar valued at 233 yen in Japan should really be valued at 180 to 190 yen.

Although a strong U.S. dollar is popular with Americans traveling abroad in Europe, Asia, and Latin America and with those who buy imports, it spells disaster and unemployment to U.S. exporters. It is blamed in large part for the record U.S. balance-of-payments deficits, markets lost by U.S. firms, decreased agricultural sales, increased unemployment, and moves by U.S. firms to locate plants overseas.

As a final result of the adverse effects of the strong dollar, U.S. finance and trade officials in the mid-1980s were encouraging other industrial nations, especially Japan and West Germany, to stimulate their economies. It was hoped that more growth in these economies would lead to increased imports from the United States and lessen the U.S. deficit balance of payments. In 1985 and 1986, for example, the U.S. deficit balance of payments exceeded $100 billion annually and the United States was a net importer of agricultural products for the first time in recent history. Moreover, a strong protectionist movement was growing in Congress.

At a meeting in the fall of 1985, finance ministers from five leading nations (France, West Germany, Japan, the United Kingdom and the United States) agreed to stimulate their economies. Agreement was reached also to reduce currency imbalances among the countries. This, in effect, called for strengthening other currencies vis-à-vis the U.S. dollar through exchange market intervention. Agreement was reached also to coordinate reductions of interest rates to stimulate the economies of various nations. These aims were reinforced at the economic summit meeting of the heads of state in May, 1986.

As a result of subsequent economic changes including the declining inflation rate, lower interest rates, continued reduction of oil prices, and the enactment of the Balanced Budget and Emergency Deficit Control Act (Gramm-Rudman-Hollings), the U.S. dollar's strength was diminished. By the fall of 1986 it had declined by 22.8 percent to a value of 110.38 against currencies of other leading nations, compared to a high of 143.00 in 1985 (March, 1973 = 100) as shown in Table 13-4 and Figure 13-3. In addition, the dollar value dropped from $1.00 to $1.20 per SDR. Although the U.S.

Figure 13-3

Exchange Rate Indices of Five Major Currencies 1970–1985
(Monthly nominal effective exchange rates)

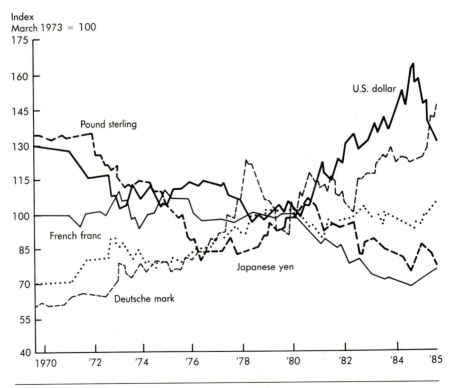

SOURCE: *Finance & Development* (June 1986).

balance of payments had improved a little by that time, it was still running in excess of $100 billion annually.

At that time some U.S. officials, such as Federal Reserve Board Chairman Paul Volcker, thought that the dollar had fallen far enough. Others, such as Treasury Secretary James Baker, suggested that it might fall a bit more in order to improve matters.

CONCLUSION

The decline in the value of the U.S. dollar during the 1970s was attributable in large part to continuing large deficits in the U.S. balances of trade and payments. Numerous factors contributed to the U.S. trade deficits. One was our continued reliance on large-scale imports of foreign oil; a second was our continuous high rate of inflation. A third factor was the increased demand by Americans for foreign products, such as autos, steel, shoes, electronic equipment, and machine tools. A fourth cause was the pace of U.S. economic expansion compared to other major industrial nations of the world. U.S. economic recovery from the 1974–1975 recession preceded and was more vigorous than that of the other industrial nations. In addition, the U.S. money supply grew at a brisk clip. Therefore, Americans purchased more goods from foreigners than foreigners purchased from the United States.

Our large trade deficits aggravated the fall in the value of the dollar in foreign exchange markets. Complicating the situation has been the rise of the Eurodollar and petrodollar markets and the growth of U.S. external liabilities. At recent international monetary meetings, the United States and some others have suggested that political leaders in such nations as Japan and West Germany, which have had sizeable surplus trade balances against the United States, should stimulate their economies more vigorously. This, it was hoped, would encourage more U.S. exports, reduce the U.S. trade deficit, and stabilize the value of the dollar in foreign exchange markets.

At times a government may intervene directly in the foreign exchange market to shore up its currency. In the summer and the fall of 1977, for example, as the dollar continued to fall in value, the Federal Reserve entered the foreign exchange markets. It sold deutsche marks and bought sizeable amounts of dollars to help stabilize trading in the major monetary centers. Again, in the fall of 1978, in response to a particularly severe decline in the exchange value of the dollar, the United States announced a dramatic program to bolster the position of the dollar. This program included borrowing SDRs from the IMF to purchase foreign currencies, borrowing from foreign central banks, selling gold from the U.S. Treasury to buy dollars, issuing U.S. government securities in foreign currency denominations, and raising the member bank discount rate by a full percentage point.

A turnaround in the value of the dollar came in 1981. The continuation of high interest rates, tight money, disinflation, growing investment in the United States by foreigners, and stable political conditions improved the strength of the U.S. dollar. Between 1980 and June, 1985, the U.S. dollar gained against every major currency in the world and improved 69 percent against the currencies of the G-10 countries.

Although the strengthening of the dollar was favorable in many respects, especially to American travelers abroad and U.S. importers of foreign goods, it has had adverse effects. It has aggravated U.S. unemployment, caused a loss of markets here and abroad for some U.S. firms, accelerated the movement of U.S. producers to overseas locations, and aggravated the U.S. balance-of-payments deficit.

In the mid-1980s, the situation was reversed from the 1970s. Instead of concern about a weak dollar, the big question was "Is the U.S. dollar too strong?" Consequently, by 1986 measures were being taken to reduce the strength of the dollar.

QUESTIONS FOR DISCUSSION

1. Should the United States be concerned about the rise and fall in the foreign exchange value of the dollar?
2. Do you think the United States should lower U.S. interest rates?
3. Would you recommend that the United States intervene in the foreign exchange market to lower the value of the dollar?
4. Do you favor fixed or flexible exchange rates?
5. Suggest a program for reducing or eliminating the U.S. deficit balance of payments.
6. Do you think the United States should restore the gold reserve requirement (cover) for Federal Reserve notes?
7. Do you see any serious economic consequences from the use of petrodollars or foreign currencies to purchase U.S. real estate and real assets, such as banks and manufacturing firms?
8. What is your position regarding the creation of additional SDRs as a means of increasing international liquidity?

9. Do you see any serious threat to the U.S. dollar resulting from the creation of the European Monetary System?
10. What do you think is the answer to foreigners' complaints about the strength of the U.S. dollar?

SELECTED READINGS

Anderson, Gerald H., and Owen F. Humpage. "Exchange Rates and U.S. Prices." *Economic Commentary*. Federal Reserve Bank of Cleveland (April 18, 1983).

Carluzzi, Nicholas. "Exchange Rate Volatility: Is Intervention the Answer?" *Business Review*. Federal Reserve Bank of Philadelphia (November–December 1983).

Finance & Development. International Monetary Fund, monthly.

"Foreign Exchange Market: U.S. Dollar Up Still More." *Business Review*. Bank of Montreal (August 1983).

Hervey, Jack L. "The Internationalization of Uncle Sam." *Economic Prospectives*. (May/June 1986).

International Economic Indicators. Washington D.C.: U.S. Department of Commerce, annually in January.

International Financial Statistics. International Monetary Fund, annually in January.

Pearce, Douglas K. "Alternative Views of Exchange-Rate Determination." *Economic Review*. Federal Reserve Bank of Kansas City (February 1983).

Rosenweig, Jeffrey A. "A New Dollar Index: Capturing a More Global Perspective." *Economic Review*. Federal Reserve Bank of Atlanta (June/July 1986).

"World Economy in Transition: Exchange Rate Indices of Five Major Currencies, 1970–1985." *Finance & Development* (June 1986).

14

THIRD WORLD
DEBT
IS A BAILOUT
THE ANSWER?

The current external debt problem of less developed countries (LDCs) is
and will remain one of the most serious challenges of the next decade.
Many of these countries are on the verge of default and bankruptcy. They
simply do not have the ability to repay their debts or, in some cases, even
meet their interest payments.

   Lending to LDCs by private banks located in wealthier nations is not a
recent phenomenon. Governments, and government-owned financial institutions, also have a long history of lending to governments and businesses
within developing countries. Infrastructure development and the building
of industrial bases require huge amounts of loanable funds. Traditionally,
LDCs, because of their limited production capability, have found it difficult
to forgo current consumption in order to devote resources to the growth of
heavy industry and the facilities necessary to the development and support
of an industrial base. Often, a significant portion of the machinery, materials, and expertise necessary to support the developmental process must
be imported from already developed countries. This usually requires large
loans to provide the huge amounts of needed funds.

   With the exception of England, most of the developed nations
depended heavily upon foreign borrowing to support their internal growth.

The United States, for example, was a major borrower of funds from foreign banks during the industrialization period of 1840–1880. The inflow of funds allowed a higher rate of economic growth and the ability to finance a balance-of-payments deficit until U.S. businesses were able to repay interest and principal out of profits.

Although there is a long history of lending to developing countries, the last decade has seen unprecedented growth in such loans. According to *World Economic Outlook,* released by the International Monetary Fund (IMF) in 1986, developing countries now have external debts of $990.5 billion, as shown in Table 14-1. About 60 percent of this debt is owed to commercial banks and other private creditors. The cost of servicing this debt (interest plus amortization) is nearly $140 billion annually. Some private sources, however, estimate the total debt of LDCs to be nearly $1.5 trillion.[1]

NEED FOR LOANS BY LESS DEVELOPED COUNTRIES

A tremendous increase in borrowing by the LDCs occurred following the oil embargo of 1973–1974. Prior to the OPEC oil embargo, the LDCs had never depended heavily upon external borrowing to finance balance-of-payments deficits. However, the tremendous increase in the price of crude oil following the embargo of 1973–1974 and the large price increases for oil again in 1979–1980, stimulated by worldwide inflation, caused problems in the current accounts of the balance of payments for most oil-importing countries. The burden on the balances of payments for the developed countries was alleviated by the inflow of funds through their capital accounts. This was a result of the oil-exporting countries' investments of their surplus funds in the commercial banks of developed countries.

The non-oil-producing LDCs did not offer the type of investment opportunities that the oil-exporting countries sought. Most of the surplus funds during this time were generated by a small number of Middle Eastern countries with sparse populations, such as Saudi Arabia and Kuwait, that were unable to increase their merchandise imports in proportion to the growth in revenue from their oil exports.[2] The IMF estimated that the oil-exporting countries generated cash surpluses of approximately $475 billion

1. Lipson, Charles. "The International Organization of Third World Debt," *International Political Economy* (New York: St. Martin's Press, 1987).
2. Kuwait, Iraq, the Libyan Arab Jamahirija, Qater, Saudi Arabia, and the United Arab Emirates have had surpluses since 1970.

Table 14-1

Capital Importing Developing Countries: External Debt, by Class of Creditor, and Debt Service Payments, 1978–1987[1]
(In billions of U.S. dollars)

	1978	1979	1980	1981	1982	1983	1984	1985	1986	1987
Total debt	399.1	475.9	567.8	662.0	751.6	798.4	840.7	888.3	943.1	990.5
Short-term	71.8	82.4	112.9	135.6	158.0	132.7	132.7	120.4	117.6	120.9
Long-term	327.3	393.5	454.8	526.4	593.6	665.8	707.9	767.9	825.5	869.6
Unguaranteed[2]	58.2	69.7	82.6	101.7	110.3	106.6	104.0	102.8	98.1	98.3
Guaranteed[2]	269.1	323.8	372.3	424.7	483.3	559.2	603.9	665.1	727.4	771.3
To official creditors	130.4	149.8	174.5	197.8	220.7	247.9	270.3	294.8	323.9	351.1
To financial institutions	102.8	135.8	157.7	184.8	214.3	258.6	280.4	314.3	343.9	356.9
To other private creditors	35.9	38.2	40.1	42.1	48.3	52.7	53.2	56.0	59.6	63.2
Value of debt service payments	57.2	75.3	87.9	110.2	119.5	110.7	126.2	131.4	137.1	139.4
Interest payments	21.5	31.5	44.1	60.5	68.6	65.2	72.2	71.8	73.4	74.7
Amortization payments[1]	35.7	43.8	43.8	49.7	50.9	45.6	53.9	59.5	63.8	64.8
Debt service ratio[2]	19.0	19.1	17.1	20.5	23.6	22.0	22.9	24.1	24.2	22.6
Interest payments ratio	7.1	8.0	8.6	11.2	13.6	13.0	13.1	13.2	13.0	12.1
Amortization ratio[1]	11.9	11.1	8.5	9.2	10.1	9.1	8.8	10.9	11.3	10.5

[1]Excludes debt owed to the Fund.
[2]By an official agency of the debtor country.

SOURCE: *World Economic Outlook*, IMF (April 1986) 243 and 250.

from 1973 through 1981. A significant percentage of the cash surpluses was placed in banks in the United States and western Europe in the form of bank accounts and other short-term investments. U.S. banks were major recipients of those short-term deposits through their home offices and their offshore branches.

Those banks that received the large deposits from the oil-exporting countries needed and sought customers willing to borrow. The LDCs were eager customers because they needed funds to finance their continually increasing costs of crude oil imports. Funds were also needed to finance their often ambitious economic development programs. Many of the LDC economies were experiencing solid growth in the 1970s, and the banks were willing to provide loans to them.

The pace of borrowing by LDCs increased rapidly in 1979 because of the new round of oil price increases caused by the Iranian crisis and surging inflation. Between 1979 and 1982 the long-term external debt of the countries increased by 52 percent. It rose another 46 percent by 1987. The cost of servicing the external debt became an ever-increasing problem. As interest rates on new loans rose to record levels, the average interest rate on total outstanding debt of LDCs increased significantly, as can be seen in Figure 14-1. Their debt service ratio (interest and amortization payments as a percentage of exports of goods and services) rose to more than 20 percent.

Accumulation of short-term debt to meet interest payments was adding still more to the debt burden. Additionally, the burden was increasing much more than the average in a few countries.[3]

The larger developing countries, such as Brazil and Argentina, had been borrowing heavily in order to diversify their economic bases. Historically, LDCs have depended on exports of raw material to finance their imports of manufactured products from the industrialized nations. The worldwide recession that began in 1980 and lasted through 1982 had a major impact on the demand for raw materials. Demand from the industrialized countries fell as their output of manufactured goods was reduced. Therefore, the LDCs' volume of exports declined substantially. In an effort to maintain sales, prices of raw materials were reduced rapidly. The decline in both quantities and prices of raw material exports caused the *terms of trade*[4] to worsen for the developing countries. The decline in commodity

---

3. "External Debt—The Continuing Problem," *Finance & Development* (International Monetary Fund, March, 1983), 22.
4. *Terms of trade* refers to the price a country pays foreigners for import goods compared to the price it charges foreigners for its export goods. Sometimes it is expressed as an exchange rate between imports and exports.

Figure 14-1

Developing Countries: Growth in Total Debt
and Interest Rates, 1971–1985[1]

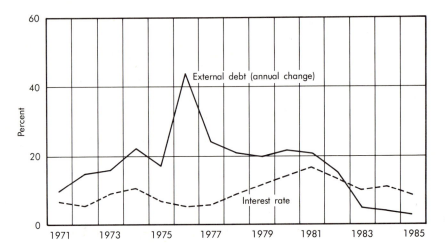

[1]U.S. dollar values of total short-term and long-term external debt, and London interbank offered rate on six-month U.S. dollar deposits.

SOURCE: *World Economic Outlook,* 1986.

prices is shown in Figure 14-2. This forced the LDCs into even more borrowing to finance their current account deficits, including interest payments on their loans.

Under these conditions the banks in the developed countries became more concerned about the ability of the LDCs to repay their large external debts. Banks offered fewer long-term fixed-interest loans. Developing countries were forced to borrow short-term money with floating interest rates based upon the prime or federal funds rates in the United States or the London Interbank Offer Rate (LIBOR) in England. These rates climbed steeply in the early 1980s, as shown in Figure 14-3, creating an even greater strain on the LDCs' balances of payments and their ability to meet loan and interest payments.

BORROWING COUNTRIES

Although estimates vary as to the total amount of external debt of the LDCs, the amount was approximately $1 trillion by 1987, as shown earlier

Figure 14-2

Developing Countries: Non-Oil Commodity Prices, 1980–1986
(Indices expressed in terms of U.S. dollars)

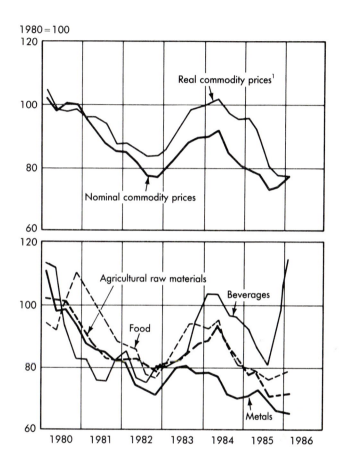

[1]Nominal commodity prices deflated by the index of prices of manufactured exports of developed countries.

SOURCE: *World Economic Outlook,* 1986.

in Table 14-1. New loans are continually being negotiated with private banks and public institutions, but the rate of new loan creation has declined since 1983. However, the need for external sources of funds is expected to remain large into the foreseeable future. According to the IMF, the 11 countries that owe the largest external debt are those shown in Table 14-2.

Figure 14-3

LIBOR and Federal Funds Rates

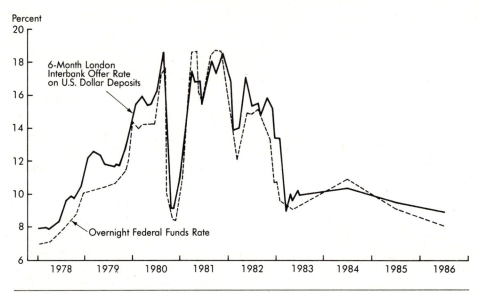

SOURCE: *Weekly Letter,* Federal Reserve Bank of San Francisco (September 23, 1983) and *International Financial Statistics* (September 1986).

Most of the LDCs appear to be capable of servicing their external debt. Those nations that avoided high rates of inflation by pursuing prudent fiscal and monetary policies are not facing severe problems. South Korea, Taiwan, and other Asian nations have been most successful in restraining domestic demand and holding down prices of their exports. They have been

Table 14-2

Major Third World Debtors, 1985
(In billions of U.S. dollars)

Country	Amount	Country	Amount
Brazil	$104.4	Philippines	$32.5
Mexico	96.3	Nigeria	31.0
Argentina	48.6	Chile	19.9
Venezuela	34.9	Yugoslavia	19.0
Indonesia	32.5	Peru	13.4
		Ecuador	7.6

SOURCE: Institute of International Finance, 1985; and Morgan Guaranty Trust Company, 1985.

able to maintain their market shares of exports to other nations. Moreover, they were poised to resume high rates of real growth in 1983 when recovery from the worldwide recession of the early 1980s began.

Unfortunately, not all countries managed to restrain growth in their domestic demand. Some permitted rapid growth in their money supplies and expansionary fiscal policies. The result was a deterioration in their competitive position in the export market, while their import demand remained strong. Many of these nations have recently resorted to extremely restrictive import policies accompanied by devaluations of their currencies.

The countries with the largest external debt and rapid inflation are concentrated in the Western Hemisphere. For example, two of the largest debtors, Brazil and Argentina, have had inflation rates in excess of 200 percent in some recent years. The external debts of the Latin American and Caribbean nations total $330 billion. Of that, two-thirds is owed to international commercial banks. The remainder is owed to various public financial institutions.

Brazil and Mexico have been the largest borrowers. Both nations have large and rapidly growing populations. Each nation has been attempting to diversify its economic base by developing a manufacturing sector and exploiting natural resource wealth more effectively. Domestic inflation and the unexpectedly deep and extended worldwide recession wrecked havoc with their developmental goals, while creating especially acute problems for their many foreign creditors.

It is estimated that Mexico, Brazil, and Argentina together have a private, external debt of approximately $250 billion. All three countries have negotiated, or are currently negotiating, restructured programs with private commercial banks and the IMF in order to delay principal repayment and to acquire additional borrowed funds. The agreements have involved commitments to reduce imports, lower domestic rates of inflation, and promote export sales in an effort to generate more foreign exchange.

SOURCES OF LOANABLE FUNDS

As indicated, the external borrowing of the LDCs currently exceeds $990 billion. Lenders of the external debt are divided rather equally between government and international financial institutions and privately owned commercial banks. The distribution of public and private borrowing among the LDCs has been quite uneven. The large nations in the Western Hemisphere and other rapidly industrializing countries such as South Korea have relied heavily upon private banks for most of their funds.

The smaller LDCs have relied more heavily upon governments and public financial institutions, especially the IMF and the World Bank Group, for their external borrowing. Domestic demand management through restrictive monetary and fiscal policies often is a requirement for loans from those institutions. These constraints may have resulted in better financial management and less ambitious growth plans on the part of the borrowing countries. As a result, it is hoped that they may be in a better position to manage their external debt without having to resort to the emergency measures facing some of the larger developing countries.

Commercial Banks

The large European and U.S. banks were the major recipients of surplus funds (petrodollars) poured into the multinational financial markets by the OPEC countries. Competition among the banks for profitable investments offered exceptional borrowing opportunities for those LDCs believed to have the best growth potential. The LDCs found banks eager to provide long-term loans at fixed rates of interest, and short-term loans were readily available at flexible rates of interest tied to the U.S. prime rate, federal funds rate, or the London LIBOR rate. Additionally, nonbanking multinational businesses, particularly manufacturers, were continually attempting to expand their operations. The result was an excellent borrowing environment for those LDCs whose growth potential was considered to be good.

Although most nations felt the impact of the worldwide recession of 1980–1982, the current crisis facing nations with a large external debt and the financial institutions that are their creditors is fairly narrow in scope. The borrowing countries that are most heavily in debt still have great potential for economic development, albeit at a slower pace than the ambitious plans of the 1970s. The banks holding the debt have significant resources at their disposal that will permit them to restructure outstanding loans and to provide additional lending in future years. The recent Latin American debt holding of some major U.S. banks is shown in Table 14-3.

International Institutions

Although the IMF and the World Bank (International Bank for Reconstruction and Development) played relatively modest roles in the total lending to LDCs from 1973 to 1982, both institutions since then have been encouraged by governments and private financial institutions to play an expanded role in the current unsettled environment. At present, official international financial institutions provide 18 percent of the external debt of the LDCs.

IMF is not a lender of last resort since it has only limited funds obtained through contributions from member countries. By itself it cannot create either national or international currencies. Likewise, the World Bank has only its contributions plus what is raised through bond issues.

Table 14-3

Latin American Debt Held by Selected U.S. Banks,
December 31, 1983 (Millions of U.S. dollars)

	Loans	Nonper- forming Loans
Citibank	$10,132	$840
Bank of America	7,249	410
Chase Manhattan Bank	6,620	481
Manufacturers Hanover Trust Company	6,507	57
Morgan Guaranty Trust Company	4,324	160
Chemical Bank	3,569	103
Bankers Trust Company	2,555	54
Continental Illinois National Bank	2,130	149
First National Bank of Chicago	1,585	83
First Interstate Bank	1,344	90
Security Pacific National Bank	1,163	26

SOURCE: *The Wall Street Journal,* March 14, 1984 and April 2, 1984.

In mid-1983 it was proposed that IMF members increase their contributions to give the Fund more lending power to help out less developed nations. Congressional approval was required before the United States could increase its contribution. The additional contribution from the United States, which is the largest IMF contributor, was to be $8.5 billion. After weeks of Congressional debate, in which some members claimed the move was tantamount to a bailout of commercial banks, Congress finally, in November, 1983, passed a bill approving the additional contribution.

The Reagan Administration's official position in supporting the bill was stated by U.S. Deputy Secretary of the Treasury R. T. McNamara. He outlined the Administration's strategy for dealing with the next phase of the international debt problem. The strategy included five elements:[5]

1. Governments of industrialized nations should adopt policies to sustain noninflationary economic growth.
2. LDCs should follow sound economic policies and live within their means.
3. The IMF should be further strengthened.

---

5. Scott Pardee, "Prospects for LDC Debt and the Dollar," *Economic Review* (Federal Reserve Bank of Kansas, January 1984).

4. Continued commercial bank lending must be encouraged.

5. Bridge (temporary) financing should be available.

Official international financial institutions such as the Bank of International Settlements (BIS) have consistently made their lending contingent upon the borrowing country's willingness to accept advice concerning domestic macroeconomic policies, often including austerity measures. While the suggested programs have not been universally successful in helping borrowing countries avoid excessive inflation and repayment problems, the success rate has been impressive. On the other hand, in those countries receiving most of their externally borrowed funds from private institutions, there appears to have been an unsustainable sense of ability to repay.

THE CURRENT PROBLEM

The often-mentioned crisis facing the LDCs because they are having greater and greater difficulty servicing their external debts is, in fact, a potential crisis for all nations, public international financial institutions, and private financial institutions. Debtor nations with severe problems are few in number, and the private financial institutions involved compose a relatively small group. Nevertheless, an international credit collapse that initially involved only a few of the debtor nations and creditor institutions could have a catastrophic domino impact upon the world community. The implications of a financial collapse are so great that representatives of the developed countries, international financial institutions, private banks, debtor nations, and the oil-producing countries are coordinating their efforts to find viable solutions to the problem.

The current crisis has its roots in a variety of factors. The most important cause appears to be the depth and duration of the worldwide recession between 1980 and 1982. Most LDCs were slow to recover from the recession. The recession caused the demand for imports to decline in the developed countries. The impact on the exports of the LDCs, both oil-producing and non-oil-producing, was dramatic. Their balances of payments deteriorated rapidly because of reduced exports and increasingly expensive imports.

In October, 1979, the Federal Reserve System announced a monetary policy designed to control the rate of growth of the U.S. monetary supply. Interest rates rose rapidly in the United States. The high rates quickly spread to other money markets throughout the world, especially the Eurodollar market where the LDCs were heavy borrowers. Since many of the short-term loans were tied to the U.S. prime interest rate and the British LIBOR rate, the debt-servicing burden of loans rose rapidly. The oil-producing nations were the major source of funds in the Eurodollar market. These nations had less available cash because their reduced volume and lower price per barrel of crude oil resulted in smaller surpluses in their balances of payments. Therefore, the reduced flow of funds into the Eurodollar market put further pressure upon the already high rates of interest.

The inflationary expansion of the 1970s had lulled many bankers and government leaders into a false sense of security. The non-oil-producing LDCs suddenly found themselves facing a credit crunch. Those countries with especially high rates of inflation found external borrowing much more difficult to obtain. The private financial institutions changed from long-term lending with fixed interest rates to short-term lending at variable rates of interest.

Worsening balances of trade and increasing burdens of external debt forced the LDCs into a variety of restrictive actions. The actions included measures to reduce imports via quotas and outright bans on many items, exchange controls, and currency devaluations. Negotiations with private and public financial institutions were initiated for the purpose of rescheduling the external debt. Moreover, borrowing to meet interest payments became more widespread.

Rescheduling of Loans

Rescheduling their external debt may be an adequate solution for many of the debtor nations if their problem is primarily associated with the depth and duration of the worldwide recession. If the economic recovery in the United States is sustained and the value of the dollar remains high relative to other countries' currencies, the demand for imported goods from the LCDs will increase. Economic prosperity in the United States will eventually cause prices to rise for the raw materials and other products exported by the LDCs. The process of economic recovery and expanded exports would not be limited to economic activity in the United States. The return to prosperity would be expected to occur throughout the industrialized world, further stimulating the exports from the developing nations.

Higher prices for exports and increased volume of exports will improve the balances of payments of the LDCs. The improved balances of payments will provide them with the necessary foreign exchange to service their external debt, especially if the debt is spread over several years and the interest rates are fixed at levels that are more consistent with long-term historical rates than with the extremely high rates that prevailed in the early 1980s.

Although the IMF and the Federal Reserve are urging banks to reschedule loans and adjust interest rates, many banks are reluctant to do so for fear of establishing a precedent. However, since the banks are in a bind and do not want their loans to go into default, they are offering new loans with up-front loading. This permits the borrower to draw down large portions of the new loan in early stages of the new loan period. Thus, banks hope the borrower will use the new loan money to pay off interest payments on old loans that are in arrears and avoid forcing the bank to write off the previous loan. Even the noncollection of interest for 90 days forces a bank to declare a loan as nonaccrual, or nonperforming, and causes the bank to write down its profits.[6]

Since 1973, 30 different LDCs have rescheduled their external debt. The amounts were small until 1983 when $21.5 billion was rescheduled. Since then over $30 billion has been rescheduled annually. Table 14-4 gives a clear picture of the increasing loan-rescheduling activity of the past few years. Other sources indicate that during the period 1983–85, 31 countries approached banks for new financial arrangements affecting $140 billion of debt.[7]

Moreover, in mid-1983 a group of smaller Latin American countries, including Ecuador, Uruguay, and Bolivia, was discussing the feasibility of forming a "debtors' cartel" in the hope of obtaining more generous repayment terms from their foreign lenders. Actual or de facto repudiation of external debt may be a possible solution. In June, 1984, officials of

Table 14-4

Capital-Importing Developing Countries: Rescheduled Debt Service, 1980–1987
(In billions of U.S. dollars)

	1980	1981	1982	1983	1984	1985	1986	1987
Total	4.7	2.0	6.8	21.5	36.8	31.6	31.7	32.9
Africa	0.8	1.4	0.4	2.7	4.2	4.5	4.1	2.2
Europe	3.0	0.1	1.8	2.5	1.8	2.0	1.4	1.3
Western Hemisphere	0.9	0.5	4.4	16.3	30.6	23.0	23.0	27.2

SOURCE: *World Economic Outlook,* IMF (April 1986) 67.

---

6. When a bank puts a loan in the nonaccrual category it does not record as income the interest due, and it subtracts from the current profit any previous interest that was recorded but not collected.

7. *World Economic Outlook* (1986) 92.

seven nations—Brazil, Mexico, Argentina, Colombia, Venezuela, Peru, and Ecuador—that accounted for 80 percent of Latin America's foreign debt met to discuss solutions to the region's debt problems.

The hypothesis that economic recovery, lower interest rates, and rescheduled external debt that spreads payments over several years will eliminate the alleged crisis has a considerable amount of merit. It would reinforce the position of many that bankers did not act imprudently as they recycled the flow of petrodollars through the 1970s and early 1980s. Wharton Econometric Forecasting Association, however, indicated in 1983 that even with economic recovery LDC debt will be a problem for a number of years.[8]

Rescheduling Loans May Not Be the Solution

There are those who doubt that the external debt of at least some of the developing countries is manageable under any circumstances. Their position is that sovereign debt cannot be analyzed in the same way that private company debt is analyzed. Governments and government-owned companies are not motivated to restrict costs and stimulate revenues in order to generate profits from which external debt can be serviced.

Brazil and Argentina are often cited as debtor nations that may not be capable of ever paying their foreign creditors. It is pointed out that their governments have created extremely high rates of domestic inflation by continually increasing their money supplies in order to finance their many government programs and to support government-owned industries. Many years of inflation and expanded welfare programs designed to placate populations that might otherwise have been extremely resistant to unpopular military governments created a situation that by the early 1980s made it impossible to service their external debt. There were those who took the position that these countries did not have any intention of ever repaying their foreign debts. Perhaps the bank officers responsible for lending to these countries recognized that conditions were such that the loans could not be repaid. Yet banks continued to lend because the loans provided high interest earnings and an outlet for the large holdings of petrodollars.

To the extent that borrowed funds were used for current consumption, the borrowing countries did not create producing assets that would allow them to earn the foreign exchange needed to pay the interest and principal on the external debt. Under those circumstances the accepted formula for servicing the debt would be to greatly reduce inflation through restrictive

---

8. *U.S. News & World Report* (August 8, 1983) 68.

monetary and fiscal policies designed to reduce domestic consumption and thereby free resources for export. These restrictive measures would have to be accompanied by deep devaluations of their currencies in order to make their products competitive on the world markets.

Such draconian measures may be politically impossible in either Brazil or Argentina. In the early 1980s the IMF had not succeeded in getting either nation to accept its suggested reforms as a condition for new loans. Brazil's inflation rate continued to exceed 100 percent per year, and Argentina's inflation rate exceeded 400 percent in 1983. Some Latin American countries, including the Dominican Republic and Peru, were facing economic turmoil and social unrest arising from attempts to implement the economic sanctions required for IMF loans.

Should Banks Write Off Loans? A solution to the current debt might be to force the lending banks to write down or write off their loans since they appear to be uncollectible. In this way the bank officers and the stockholders would have to take direct responsibility for their decisions to lend to debtors who were unwilling to manage their external debt in ways that ensured their continued credit worthiness.

Even if the lending banks are forced to absorb the losses associated with writing off those huge debts, the borrowing countries would not be relieved of any responsibility for their domestic policies that contributed to the situation. The countries' credit ratings would be so poor that they could not borrow abroad until they demonstrated to the rest of the world that they could manage their affairs in a way that would lead to future solvency. Eliminating excessive inflation and creating a domestic economy that could sustain growth with a low level of unemployment might span an entire generation.

There are many who find such solutions to dealing with the external debt of the developing countries to be unacceptable, even though they may accept the validity of the position that some of the debtor nations will never be able to pay off the debt. Critics argue that the international political and economic environment makes such a solution unthinkable.

It is possible that the process of absorbing the losses associated with writing off the huge debts might force one or more of the world's large banks into bankruptcy. A single bankruptcy could have a domino effect that would spread quickly to many other banks. Any semblance of international monetary order would collapse. Worldwide depression might result, with its attendant economic miseries and the likelihood of military confrontations that could spread into a destructive world war.

Should Central Banks Absorb Loans? A suggested alternative to forcing the private banks to write off many of their loans to LDCs is for the central banks of the countries in which the commercial banks are located to purchase the potentially bad debts and absorb the losses. Obviously, the central banks could choose to purchase all or any portion of the questionable debt of the commercial banks. The amount purchased might be made dependent on the solvency problems of the individual banks.

To the extent that the central bank purchases the debts of the developing nations from the lending banks, the citizens of the nation in which the central bank is located must share the burden created by the imprudent lending policies of that nation's commercial bankers. The burden placed upon the citizens of the developed nations would probably manifest itself in the form of (1) higher rates of inflation because of an increase in the money supply; (2) a transfer of purchasing power to the nations whose debt is forgiven; (3) a subsidy, in the form of interest on the deposits that formed the basis of the loans to the OPEC countries; or (4) some combination of 1, 2, and 3.

Some Near Crises

After close scrapes by Mexico and Brazil to meet their interest payments in late 1983, it appeared that Argentina would be unable to meet its $500 million in interest payments due December 31, 1983. Fortunately, after weeks of discussion and manipulation involving its Latin American neighbors, numerous commercial banks, the IMF, and the U.S. Treasury, an eleventh-hour aid package was agreed to on Saturday, March 31, 1984, avoiding the pending crisis. It permitted Argentina to meet its late interest payments and averted write-offs by the commercial banks.

Mexico and Venezuela, although both heavily in debt, each loaned Argentina $100 million. Brazil and Colombia, also big debtors, each contributed $50 million to the aid package. A consortium of eleven U.S. and British banks loaned a total of $100 million. This $400 million along with $100 million from Argentina's foreign reserves gave it the needed $500 million. The United States, in turn, agreed to exchange $300 million for an equivalent amount of Argentine pesos so Argentina could pay off its loans from the four Latin American countries. The United States was repaid when Argentina obtained a pending loan from the IMF.

In order to obtain the IMF loan, however, Argentina agreed to curb its sky-high inflation, which was 400 percent in 1983. It agreed also to reduce its budget deficit, cut government spending, and take other austerity measures. These measures, limiting the growth of money supply and restraining wage increases, were unpopular with some government officials, labor leaders, and the general public.

Fortunately, the Argentine crisis on its external debt was averted by the rescue package. In addition to the IMF loan, loans with 11 international banks and numerous other banks were rescheduled, new loans were extended, and past due interest was paid. Since then Argentina has kept up with interest payments and met some principal payments.

After several months, however, it was apparent that the IMF plan of gradualism and austerity, accepted as part of the loan agreement, was ineffective in combatting inflation, which was soaring at an annual rate of 2,000–3,000 percent. Hoping to remove the inflation mentality of its citizens and prevent hyperinflation, Argentina in 1985, with IMF approval, adopted a bold new economic plan. It created a new currency called the "austral" by lopping three zeros off the peso. It froze prices, wages, and official exchange rates. It promised to stop printing money, to reduce government spending, and to balance the budget. The austral plan, as it is known, called for de-indexing wage and other contracts. In addition it sought to privatize some of the 350 state-owned and/or state-operated companies. Within a year the plan was successful in reducing inflation to a double-digit level and strides were being made with the other goals.

In 1986 Brazil adopted a similar plan to combat its rapid inflation. It created a new currency, the "cruzado," worth 1,000 cruzieros, instituted wage-price controls and abandoned automatic inflation adjustment measures. Also in that year, Brazil and its 700 creditor banks agreed on a package to restructure and reduce interest rates of $31 billion of its external debt. Early in 1987 the cruzado plan was acclaimed as a success by the Brazilians.

PROPOSALS TO EASE WORLD DEBT PROBLEMS

In July, 1985, the so-called Cartagena group, composed of 11 Latin American nations, met in Mexico to discuss ways and means of easing the problems of external debt. Suggestions included more new loans, capitalization of interest payments by adding them to existing loans, and reduction of interest rates.

The Baker Plan

In the fall of 1985 at a joint meeting of the IMF and World Bank in Seoul, Korea, U.S. Treasury Secretary James Baker proposed a three-point program to ease world debt problems. The Baker plan proposed that: (1) debtor nations would begin making serious policy changes to promote long-term growth in their economies; this would include steps to attract more foreign investment, reduce trade restrictions, and transfer costly state-run industries to the private sector; (2) at the same time, partly as an incentive, commercial banks would pledge in advance to provide debtor countries with up to $20–25 billion in new loans over the next three years; and (3) the World Bank and the Inter-American Development Bank would take wider roles reinforcing the IMF in overseeing the debtor nations' policies and guaranteeing new bank loans to help borrowers attract more money.

At the same time, Brazil, the Third World's largest borrower, was seeking relief from some of the IMF austerity measures previously imposed on it and threatening to trim its interest payments. (In early 1987, Brazil stopped paying on its loans.) It was about this time also that Peru, under its new president, Alan Garcia, announced among other measures that it would limit interest payments on its $14 billion external debt to 10 percent of its export income. At the time, Peru was $500 million in arrears on interest payments. Garcia stressed, as have other Latin American leaders, that domestic economic development has to come before debt repayments can be made. Later Peru unilaterally reduced the interest rate that it would pay on its existing loans. Subsequently, federal banking regulators required U.S. banks with loans to Peru to set aside special reserves on their Peruvian loans.

In early 1986 Bolivia's loans were considered to be nearly worthless. By 1985 Bolivia's price index had risen more than 8,000 percent and it had not made a payment on its $833 million debt to private banks in two years. In the spring of 1986, with its hyperinflation then running at a rate of more than 26,000 percent annually, one of the highest in world history, it took a two-inch stack of paper money to buy a chocolate bar. Unable to collect sufficient taxes to keep up with the rapid growth of government spending, it was running the presses daily and printing one, five, and ten million peso notes. Reminiscent of Germany, after World War I, when nearly worthless paper money was used by some for wall paper, Bolivia took drastic steps to stop inflation. It lifted controls on prices, interest rates, imports, and exports. It freed the official exchange rate and brought about a 93 percent devaluation of its currency. It stopped printing new money, cut government spending, froze public sector wages, and eliminated many government jobs. As a result of its success thus far, the IMF signed an agreement with Bolivia that helped reopen its credit lines with the World Bank and reschedule its debt. Moreover, Bolivia planned to reopen talks with its private bank lenders in late summer of 1986. It closed down or privatized many state-owned enterprises. Although the country continued in its depressionary condition, it stabilized its currency at 1.9 million to the U.S. dollar. By mid-1986 it had cut its inflation rate to 20 percent and had plans to replace the peso early in 1987 with the "condor" worth 1 million pesos.

In 1986 new trouble developed in Mexico, which had averted an earlier crisis in 1982. The steep decline in oil prices, on which Mexico relied heavily for export earnings and debt payments, the decline in the value of the peso, a flight of capital, and an expected 4 percent decline in real GNP for 1986 left the country in a financial squeeze. It was evident that Mexico would be unable to meet its mid-year debt obligations. Fearing that it might follow the Peruvian approach or even default, U.S. officials began working with Mexican authorities and others to seek a solution.

Mexico suggested loan restructuring, lower interest rates, modification of IMF austerity requirements, new commercial bank loans, and linkage of interest payments to oil prices. Mexico indicated it would need $15 billion in new loans from foreign creditors and interest rate concessions over the next three years in order to bring about sound economic recovery in 1987 and 1988. This would put them in a position to meet their debt obligations.

After several months of talks involving IMF and U.S. officials along with others, a $12.2 billion international package was agreed upon. It included an IMF loan of $1.6 billion with less austere terms to permit domestic growth, a World Bank loan of $1.9 billion to be used primarily for domestic development, an Inter-American Bank loan of $400 million, international export credits of $1.5 billion, U.S. farm credits of $800 million, and a recommended $6 billion in commercial bank loans to be subsequently negotiated.

In exchange for the loan package, Mexico agreed to reduce its budget deficit, tighten monetary policy, forestall the flight of capital, keep interest rates higher than inflation, and broaden its tax base. It also agreed to modernize, merge, or privatize 300 of its 500 state-owned corporations. As a further concession, it agreed to liberalize import restrictions and actively recruit more foreign capital and investments.

The Mexican agreement appeared to follow the Baker plan presented in Seoul months earlier. It involved changes by the debtor nation, pledges of loans by commercial banks, and involvement of the World Bank and the Inter-American Development Bank along with the IMF.

By early 1987, however, some disagreement still existed among commercial banks, both U.S. and foreign, that were expected to provide $6 billion in new loans. Several of them thought it was too much of the total package. Others were upset by the fact that some principal was to be repaid to certain banks, and practically all of them disagreed with the idea of interest rate concessions, including a rebate of some 1986 interest payments requested by Mexico.

Another jolt regarding LDC debt came in February, 1987 when Brazil suspended interest payments on $67 billion of its foreign bank debt. This forced some large American banks to place certain of their Brazilian loans on a non-accrual status and reduced first quarter profits of the banks.

Although the Baker plan seemed to have worldwide support, late in 1986 U.S. Senator Bradley criticized it as ineffective and incapable of solving Third World debt problems. He declared it to be equivalent to throwing good money after bad. Consequently, he introduced a federal bill calling for the United States and other creditors to forgive 9 percent of Third World debt and reduce interest rates on the remainder by 3 percentage points. He hoped the bill would be scheduled for Congressional hearings early in 1987. In the meantime, he suggested that President Reagan call an international summit to discuss the Bradley bill and other proposals for solving Third World debt problems.

Debt to Equity Swaps

One of the other proposals to ease world debt problems involves the switching of debt to equity. This method began being used more in 1986. When using this procedure, a commercial bank holding Third World debt sells some of the loans (through a brokerage firm) to a private investor at a discount. The buyer then presents the loan to the original borrower nation for repayment. The borrower pays off the loan in its own currency, rather than in U.S. dollars. The buyer of the loan then invests the money in its subsidiary or some other firm in that country. In this manner, the original borrower nation reduces some of its debt obligation, the lending bank gets some of its money back, the broker earns a fee, and the private investor has a politically acceptable way of making an investment in a Third World country.

In 1986 more than $2 billion in debt was swapped in this fashion, and another $5 billion is expected to be swapped in 1987. For example, in 1986 Chrysler Corporation pumped $100 million into its Mexican subsidiary using money obtained through the purchase of a Mexican loan at a 56 percent discount. One brokerage firm estimates that swaps could reach $50 billion annually in five to ten years, as the process becomes more acceptable.

CONCLUSION

From all indications, the problem of Third World or LDC debt, especially that of Latin American countries, will be with us for some years. In the short run, crises may be averted by rescheduling of debt and emergency borrowing from commercial banks, other nations, and international institutions. Although quick fixes, such as those provided for Argentina, Mexico, and others, have averted turmoil thus far, they do not provide a permanent solution to the problem.

The long-run solution involves a continuation of worldwide economic recovery and growth, including economic expansion in debtor nations. The adoption of stringent monetary and fiscal policies, a return to price stability, and an increase in exports are essential for the LDCs. On the part of lenders, longer-term rescheduling and lower interest rates are needed. Whether long-run measures can be implemented and take effect before a crisis of catastrophic proportions hits lenders and borrowers is a matter of conjecture. In mid-1987, several large U.S. banks set aside billions of dollars of reserves against their Third World loans.

QUESTIONS FOR DISCUSSION

1. Should LDCs be aided or left alone to work out their own loan problems?
2. Should the IMF increase its lending to LDCs to help them meet their loan commitments?
3. Should government agencies, commercial banks, and the IMF reschedule or restructure LDC loans?
4. Should interest rates on existing LDC loans be lowered?
5. What form of aid, if any, should be extended to LDCs?
6. Should the United States and other nations increase their contributions to the IMF still more to enhance IMF lending ability?
7. Should smaller LDC nations with heavy external debt form a "debtors' cartel" for the purpose of bargaining with lenders?
8. Should commercial banks consider many of their problem loans to LDCs as uncollectable and write them off as losses?
9. Do you favor the Bradley proposal as a solution to Third World debt problems?
10. Do you think swapping debt for equity is a viable solution for the Third World debt problem?

SELECTED READINGS

Anderson, Gerald H. "Solution to the International Debt Problem." *Economic Commentary*. Federal Reserve Bank of Cleveland (August 1, 1985).

Barth, James R., and Robert E. Keleher. "Financial Crises and the Role of the Lender of Last Resort." *Economic Review*. Federal Reserve Bank of Atlanta (January 1984).

Frieden, Jerry A., and David A. Lake. *International Political Economy*. New York: St. Martin's Press, 1987: Ch. 20 and 26.

Gotur, Padma. "Interest Rates and the Developing World." *Finance & Development* (December 1983).

International Monetary Fund Annual Report. Washington D.C.: International Monetary Fund, annually.

Kahn, Mohsin, and Malcolm Knight. "Sources of Payment Problems in LDCs." *Finance & Development* (December 1983).

Pardee, Scott E. "Prospects for LDC Debt and the Dollar." *Economic Review*. Federal Reserve Bank of Kansas City (January 1984).

Progress in LDC Debt. New York: Citicorp 1985.

"Third World Debt: A Way to Turn Debt from a Burden to a Boon." *Business Week* (December 22, 1986.)

"The Third World Debt Problem." *Economic Report of the President, 1984*. Washington, D.C.: U.S. Government Printing Office, 1984.

Watkins, Alfred J. *Till Debt Do Us Part*. Washington, D.C.: Roosevelt Center for American Policy Studies, 1986.

The World Bank Annual Report. Washington D.C.: The World Bank, annually.

World Economic Outlook. Washington D.C.: International Monetary Fund, annually.

15

THE POPULATION EXPLOSION CAN THE WORLD FEED ITSELF?

Most of the chapters in this book have dealt with economic problems that are somewhat specific in nature. In the majority of the chapters, the problem considered was placed in the context of the U.S. economy and had a fairly clear-cut economic dimension. As such, most of the problems were amenable to economic analysis; and while there may have been several alternative solutions presented to each of the problems, the solutions themselves were couched in economic terms.

Unfortunately, such is not the case with the question of the population explosion. The problem of population growth cuts across national boundaries, academic disciplines, and cultural and moral systems, and becomes more acute with the passage of time. It is virtually impossible to restrict the question of population solely to economic factors because of this fact.

Given the all-pervasive character of the population question, we must therefore widen our view somewhat in this chapter to include not just the U.S. economy but the world economy. In addition, we must admit the testimony of various academic disciplines, such as demography, the physical sciences (particularly the ecological aspects), geography, and history. We shall consciously omit only one body of learning that bears heavily on the question of population, that being theology.

THE POPULATION EXPLOSION

An explosion ordinarily describes a situation in which matter, under the impetus of some form of released energy, expands at an extremely rapid rate away from its center. When the word is used in connection with population, it usually connotes a rapid expansion of the world's population, an expansion that has proceeded at a faster rate than in earlier periods. Such is the case today. In 1986 the world's population expanded at an annual rate of 1.7 percent. Such a growth rate doubles the population in about 41 years. Some countries, notably those in South America and some sections of Asia, are growing at the rate of 3 percent per year, a rate that would double their populations in just 23 years.

Table 15-1 presents estimates of world population for certain check-point years 1 A.D. through 1986 and indicates the number of years necessary to double the world population for these years. Population doubling time has dropped from about 1,650 years to an estimated 41 years. Figure 15-1 contains a curve that portrays the historical growth in world population to date. Note that for much of humanity's recorded existence on earth, population growth has not been excessive. In fact, it has only been in the last 300 years or so that world population has increased at a rapid rate.

It is quite obvious from Table 15-1 and Figure 15-1 that the growth in world population has not conformed to anything resembling a constant growth rate. To the contrary, the world rate of population growth has been increasing. It should be equally obvious that this trend cannot continue indefinitely, because of finite resources and spatial constraints. In the long run, growth rates in world population must take on the shape of the S-curve

Table 15-1

Estimated Population of the World

Year (A.D.)	Population (Billions)	Number of Years to Double
1	0.25(?)	1,650(?)
1650	0.50	200
1850	1.1	80
1930	2.0	46
1986	4.9	41

SOURCE: Philip M. Hauser, ed., *The Population Dilemma* (Englewood Cliffs, NJ: Prentice-Hall, 1963), 10; and United Nations, *Demographic Yearbook* (New York: United Nations, annually).

Figure 15-1

Historical World Population Growth Curve

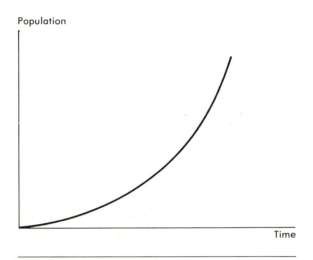

shown in Figure 15-2. This curve typifies the expected growth pattern for all living species. As the curve indicates, although population of a species first grows geometrically, it quickly levels off. This reflects the fact that a living species cannot continue to grow exponentially because overcrowding brings population growth to a halt.

In other words, in an environment with finite resource limitations, there is a maximum ceiling for human population growth. It is for this reason that it is inevitable that world population growth will conform to something approximating an S-shaped curve.

The Resource Factor

Mere evidence of rapid population growth is not evidence of a problem. Certain nations—Canada, Australia, and New Zealand, to name a few—are seriously inhibited in their economic growth rates by insufficient population. To draw any reasonable conclusions about rates of population growth, these rates must be related to other factors. Among the most important of these other factors are food, space, energy resources, and natural resources. Of these, food must certainly be rated of primary importance. Thus, we might couch the problem of population in terms of the available world food supply. Is the world supply of food currently sufficient to feed today's world population at the necessary minimal nutritional level? Is the

Figure 15-2

S-shaped Population Growth Curve

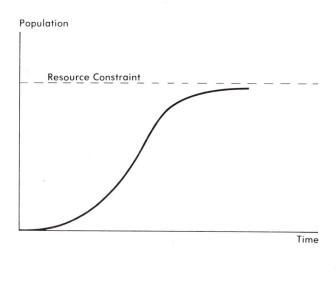

rate of growth in the world's food supply greater than, or at least equal to, the rate of growth in world population? If the answer to either of these questions is no, then the problem of a "food gap" is very real.

The Problem of Distribution. Even affirmative answers to these two questions do not remove the problem of excess population completely. Probing a little deeper into the question, we might ask, assuming that the world's food supply is adequate to feed the world population, whether this food supply is distributed among the nations of the world in an equitable manner so that all peoples have access to an adequate diet? If the answer to this question is negative, then necessarily the populations of some countries—or at least segments of these populations—are existing on substandard diets, while other countries enjoy surplus food supplies.

People and Their Environment. In the final analysis, the question of whether population is truly excessive or whether current growth trends in population are too high or too low can only be answered by examining the relationship between people and their environment. By *environment* we mean simply the sum total of all the available resources by which people seek to maintain themselves as a species. Since people exist in a finite (limited) environment, we must assume that there is some finite limit to the number of people the earth's environment can support at any given time.

The Population Explosion—Can the World Feed Itself?   **383**

All living organisms, plant and animal, tend toward a state of equilibrium with their environment. An ecological *equilibrium* state implies a situation in which the resources available to a given species are precisely equal to the amount of those resources necessary to maintain that species. Notice that we have used the phrase "tend toward" an equilibrium. A given species may or may not reach the equilibrium depending on the changing character of the variables that determine the equilibrium state.

The Disturbance of Equilibrium

People have conformed to this basic "law of nature" during most of their sojourn on the planet Earth. Only in very recent times (on the historical time scale) have people been able to offset this law. With the advent of the Industrial Revolution in the late eighteenth century, people began to change their environment drastically. They gained control of mechanical energy sources and broke the age-old dependence on animal or human energy. Thus, they were able to devote increasing amounts of time to devising ways to improve their own material welfare. Certainly the two most important developments affecting population growth that grew out of the Industrial Revolution were vastly improved agricultural methods and advances in medical knowledge. The first of these two developments gave rise to large increases in the available food supply. The second development gave rise to drastically lower death rates, first in the developed industrial nations and then in the less developed countries.

The unfortunate fact is, however, that these two developments have not proceeded apace in all nations. In the industrialized countries, where medical advances first began to significantly lower the death rate, the application of an advanced technology to agricultural methods generally has been successful in providing more than enough food for a burgeoning population. In those countries characterized as less developed because of their relative lack of industrial activity, however, the application of technologically advanced agricultural production methods has lagged seriously behind medical advances. As a consequence, food supplies have not kept pace with growing populations that have resulted from a declining death rate and a stable birth rate. This phenomenon seems to be at the heart of what is referred to as the "population explosion."

In recent years, as the pressure of increasing population has begun to press against available food supplies, particularly in the nonindustrialized countries of the world, a great deal of effort has been put forth to isolate the causes of and solutions to the population question. But before we turn to an examination of causes and tentative solutions, we should gain some insight into the historical development of our current population expansion.

THE HISTORICAL DEVELOPMENT OF POPULATION EXPANSION

We have discussed the concept of equilibrium between people and their environment, and we have noted that people managed to disrupt this equilibrium through the Industrial Revolution. Prior to the Industrial Revolution, the growth in the world population proceeded at the figurative snail's pace. The lives of primitive people were fully occupied with looking for their next meal. They could gather seeds and fruit. An individual might kill small animals; in combination, several people might kill large ones. Yet in very early years, people were as much hunted as they were hunters. Their food supply was uncertain. They had little or no protection against the elements. They had only the most primitive weapons. Their death rate in such a merciless environment must have been very high. The discovery of fire and the improvement of weapons made life somewhat less precarious, but it is unlikely that the world supported more than a few million humans until people began to control their environment. This control of the environment began with the cultivation of crops and the domestication of animals.

The Agricultural Revolution

The agricultural revolution was the first recorded disruption by people of the relationship between themselves and their environment. It permitted the human species to begin to increase in number for the first time. Ten thousand years later isolated societies still exist that have not participated in this revolution. Most of the human race, however, gradually abandoned the role of the hunter and became farmers or shepherds. The relatively assured food supply permitted a somewhat lower death rate and a consequent rise in population. This newly acquired control of plant and animal energy permitted at least some humans to direct their activities away from the never-ending chore of providing the necessities of life. Pyramids were built, philosophers began to question the origin and purpose of people and their environment, and new territories were discovered and exploited. Still, about three-fourths of the population were engaged in agriculture. Only about one-fourth of the people were surplus in that their efforts could, at least potentially, be invested in activities that tended to raise the standard of living.

The Industrial Revolution

The Industrial Revolution, with its associated improvements in food supply and medical knowledge, resulted in a sharp drop in the death rate. The consequent increase in population began in Western Europe around the end of the eighteenth century. From 1 A.D. until 1650, the average rate of increase in world population was about 150,000 per year. By the end of the nineteenth century, a new, sharply higher rate of increase was well under way. In England between 1800 and 1900, the population increased from 9 to 35 million, excluding the millions who emigrated. During the nineteenth century, the world population grew at an average annual rate of 4.5 million. Most of this increase occurred in Europe, the scene of the original Industrial Revolution, and in the new territories to which Europeans had migrated.[1]

The existence of newly discovered lands served as an escape valve for the swelling populations of Europe during the eighteenth and nineteenth centuries. Between 1800 and 1924, when the United States brought a halt to its open immigration policies, almost 60 million Europeans left Europe for new lands. This tremendous outflow of humanity had a twofold beneficial effect on the countries of Europe. First, it relieved population pressures that were swiftly building as a result of the Industrial Revolution; and second, the virgin soil of the new lands provided food for the emigrants as well as for those who stayed behind to work the shops and factories.

The Twentieth Century

The beginning of the present century witnessed a new development that was to have a tremendous effect on the rate of growth of the world's population. Those countries with the more advanced medical technology began to export that technology to countries that had not developed such medical care. Diseases that had previously ravaged entire populations were brought under control with a consequent dramatic drop in the death rates. The birth rates in such countries, however, continued to increase, and population began to grow rapidly. By 1946 the world was adding to its population at the rate of 22 million people a year. By the early 1950s the rate of increase had reached 30 million a year. It is now estimated that world population is growing at the rate of at least 81 million people a year.

THE DEMOGRAPHIC TRANSITION

In past agricultural societies, birth rates ranged from 35 to 50 per thousand, while death rates ranged from 30 to 40 per thousand. Such historical rates

---

1. Lord Boyd Orr, "Mankind's Supply of Food," in *Our Crowded Planet,* ed. Fairfield Osborn (London: George Allen & Unwin, 1963), 83.

would normally result in a natural rate of population growth of about 0.5 to 1 percent per year. However, agricultural societies did not grow at this rate because of wars, famine, and epidemics. After the Industrial Revolution, death rates in Europe declined sharply. Currently, many countries have death rates under 10 per thousand. Naturally, this drop in the death rate increased the rate of population growth during that time.

The excess of births over deaths is now approximately the same as it was at the beginning of the Industrial Revolution. But note the time lag. It took much more time for the birth rate to fall than the death rate. This lag is commonly called the *demographic transition*, and it is the main reason for the increased rate of world population growth in the last two centuries.

The European countries were able to increase their standards of living during this period of rapid population growth because they enjoyed the fruits of the Industrial Revolution. Technology improved agriculture to the point that 10–20 percent of the population could now produce more food than the total population could consume. This freed the energies of the balance of the population for the production of other goods and services.

The Lag in the Less Developed Nations

The primary concern with population growth today is that the less developed nations have reduced their death rates through the use of modern medical techniques, but they have neither fully participated in the Industrial Revolution nor reduced their birth rates. In short, the time lag between reduction in death rates and reduction in birth rates is still present—the demographic transition has not taken place. The resultant rising populations are pressing against available food supplies that are limited by obsolescent cultivation methods.

Certain sections of the world are still caught in the dilemma of falling death rates and stable birth rates at a relatively high level. Notice particularly in Table 15-2 the birth rates and death rates that have prevailed in Africa, Asia, and Latin America. Although the death rates are higher than other areas, these areas of the world are characterized by very high birth rates. The figures take on even greater importance when one considers the fact that these areas have experienced little or no decline in the birth rate, while the death rate has dropped sharply over time. A consequence of the time lag in the fall of birth rates is the relatively high rate of population growth in these areas, which are most often characterized by lack of industrialization and low agricultural crop yields.

The Aggregate Effect of the Demographic Transition Lag

To get some idea of the aggregate effect of the demographic transition, we should look at total population by area over time. Table 15-3 presents such

Table 15-2

Annual Rates of Increase in Population, 1986

	Annual Rate of Increase	Birth Rate per 1,000	Death Rate per 1,000
Africa	2.8%	45	16
Asia (except USSR)	1.8	28	10
North America	0.7	16	9
Latin America	2.3	31	8
Europe	0.3	13	10
Oceania	1.2	21	8
USSR	0.9	20	11
World Total	1.7	27	11

SOURCE: "Data Sheet," Population Reference Bureau, Inc. (1986).

a picture. If we combine the world annual rate of increase of about 1.7 percent with the world population of over 4.9 billion people, we can then project a world population in excess of 6.1 billion people around the turn of the twenty-first century. What are the chances that such a projection is valid? What factors can we expect to affect the validity of such a projection?

PROJECTIONS OF WORLD POPULATION

Any attempt to project world population into the future is necessarily fraught with difficulties. It is hard to imagine any event that would be

Table 15-3

Population of the World
(Millions of people)

	1920	1930	1940	1950	1960	1970	1980	1986
Africa	140	157	176	207	257	344	472	563
Americas	208	244	277	329	412	511	617	686
Asia	966	1072	1212	1384	1684	2056	2618	2876
Europe	329	356	381	395	426	462	484	493
Oceania	9	10	11	13	17	19	23	25
USSR	158	176	192	181	214	243	266	280
World total	1810	2015	2249	2509	3010	3635	4640	4942

SOURCE: United Nations, *Demographic Yearbook,* (1960 and annually); and "Data Sheet," Population Reference Bureau, Inc. (1986).

subject to more forces than the rate at which the human race reproduces itself. Nevertheless, if one wishes to forecast future world population, an attempt must be made to isolate those forces that seem to bear most significantly on the rate of human reproduction. An attempt to quantify these forces must then be made to get some idea of the dimension of their effect on population growth. In short, the population forecaster must try to construct a logical explanation of why population is growing at the rate it is in order to be able to predict what it will be in the future.

Even if the explanation does adequately explain past population growth, however, the forecaster must guarantee that the forces affecting the past population growth will continue to act similarly in the future. Many individuals over the past 200 years or so have offered explanations of population dynamics (change). None of them has ever offered a theory of population growth that has been universally acceptable to all. Perhaps the most notable commentator on the population question was Thomas Robert Malthus (1766–1834), an English clergyman and economist.

MALTHUSIAN THEORY

Essentially Malthus's theory was that in a given state of the art, population will outrun the means of subsistence if the food supply is based on the law of diminishing returns and if population grows by a biological urge. According to Malthusian doctrine, population shows a tendency to grow in a geometric progression (2, 4, 8, 16, 32, etc.), while food production increases only in arithmetic progression (2, 4, 6, 8, 10, etc). Eventually, population will outstrip food production, as shown in Figure 15-3.

At the point at which population increased to the limits of food supply, further population growth would be restrained by powerful "checks." These checks are summarized as moral restraint, vice, and misery; they are classified as either "positive checks," which increase the death rate, or "preventive checks," which reduce the birth rate.

Malthus's positive checks were simply the cruel aspects of life. He included overwork, insufficient food, poor habitation, overcrowded urban areas, unwholesome foods and medicines, sickness and disease, and war.[2] Malthus recommended self-restraint and delayed marriage as the only acceptable preventive checks. He considered contraception to be an unqualified vice and refused to consider it as an admissable alternative to the positive checks just enumerated.

---

2. Thomas Robert Malthus, *Population: The First Essay,* Foreward by Kenneth E. Boulding (Ann Arbor: University of Michigan Press, 1959), 38.

Figure 15-3

Geometric Growth in Population and
Arithmetic Growth in Food Supply

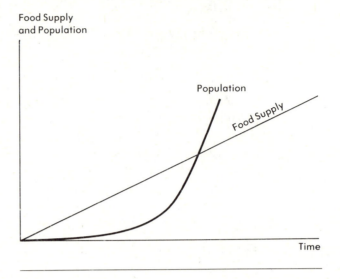

Malthus's writings on population have had a curious history. They have been both praised and criticized from the first, but they display an astounding ability to recover from criticism and remain at the forefront of writings on this question. It is literally impossible to pick up any book even vaguely concerned with population change without encountering some reference to his work.

Two basic criticisms have been made against Malthus's work. The first, which was particularly relevant in his time, was that his population theory was written as part of an argument against William Pitt's Poor Laws. Malthus believed that society could not effectively help the poor through charity since the poor would simply lose all incentive to work productively or to limit births. Thus, the number of poor would increase as a consequence, while the supply of food would not be affected. He wrote that the net result of charity to the poor would be that ". . .the same produce must be divided among a greater number, and consequently that a day's labor will purchase a smaller quantity of provisions, and the poor therefore in general will be more distressed."[3] One can well imagine the hue and cry that arose against this seemingly heartless position, which held that to assist the poor was actually to injure them in the long run.

---

3. Malthus, *Population,* 47.

The second major criticism against Malthus's theory is still heard today. He is charged with ignoring or not anticipating the tremendous increase in arable land in the newly discovered and exploited areas of the world and, more seriously, with not anticipating the rising productivity per acre that resulted from the Industrial Revolution. One can well ask, however, whether these events have completely contradicted Malthus's theory or whether they have merely postponed the attainment of a population equilibrium.

The Problem of Inherent Uncertainty

As can be seen from the Malthusian theory of population, long-range population forecasts are highly vulnerable to unpredictable forces. These forces can greatly alter the population growth rates for individual nations and for the world as a whole. For example, there is always the chance that the people of the world may accept a cheap and readily available method of artificial birth control. In addition, science may perfect a method of consciously controlling family size that is acceptable to the Roman Catholic Church and, thus, to a significant proportion of the world's population. On the more ominous side, previously unknown diseases may develop that could decimate large segments of the populations in countries unprepared to deal with them. Or, equally horrible to contemplate, a nuclear war could rid the world of perhaps one-half its population in the initial onslaught and the aftermath. These are all possibilities, good or evil, that would radically upset the current growth rate of world population and possibly bring it to a complete halt.

Thus far we have dealt almost exclusively with the broad forces that affect population growth. Birth rates, death rates, supplies of food, public health measures, all of these factors are constantly acting upon the rate at which the human race is changing in number. Admittedly, we have not dealt specifically with the forces that motivate people to have children at one rate rather than another. There are a number of theories that purport to explain why a given nation's birth rate is higher or lower than that of other nations at any given time or over time. Unfortunately, none of these theories completely explains variations in birth rates for all countries over all periods of time. Nor should we expect to find such a generalized theory. Given the large number of cultures and subcultures that the world has experienced, the vastly different educational levels that exist in various parts of the world, the different attitudes of the world's major religions toward the question of conscious birth control, and other very basic economic and sociological differences between peoples, it would be too much to expect one explanation of the rate of population growth to fit all regions of the world.

The Continued Increase in World Population

There now seems to be little doubt that population will exceed 6.1 billion at the turn of the next century, assuming that the necessary food supplies and shelter are forthcoming. But this is an important assumption. Can the people of the world continue to provide the necessities of life within this relatively short period? And further, will it be possible to raise the portion of the human race that is presently close to starvation to a nutritional level that is considered adequate by the standards presently existing in the wealthier developed nations? Finally, assuming that food supplies and shelter are produced that are sufficient to significantly raise the living standard of the people of the world, a remaining question must be raised: Will these supplies be distributed in an equitable manner to all of the world's people so that presently existing large pockets of hunger, sickness, and privation can be erased?

THE WORLD FOOD SUPPLY—FUTURE PROSPECTS

When dealing with the prospects for the world supply of food, and considering its adequacy for feeding the prospective future world population, we are confronted with something of a dilemma. World food supply and world population, when placed in juxtaposition, form a classic case of circular causality. Obviously, world population cannot grow faster than the food supply available to sustain it. On the other hand, it is extremely doubtful whether the food supply will increase fast enough to outpace the population and yield surpluses. Should world population growth outpace the rate of increase in food, then the only result can be eventual famine and death for a considerable portion of the world's people. What are the prospects of feeding a forecasted world population of 6.1 billion at the beginning of the next century?

The present outlook for feeding the world's population in future years is ambiguous. At a time when many nations of the world chronically face poverty, hunger, and starvation, other nations are in the envious position of being self-sufficient in food. In addition, some nations are self-sufficient in certain years, but food surpluses resulting from bumper crops are apparently insufficient to cover shortages in less productive years. In years of low crop yield, nations such as the Soviet Union and China turn to the world marketplace for food to avert disaster. In fact, the entry of the Soviet Union and China into the world food marketplace in 1972–1973 emphasized the delicate balance between food and population that exists throughout the world.

The world food crisis of 1972–1974 brought a greater awareness on the part of developed nations concerning the long-range challenge of feeding the world's population. Most experts concur that the crisis was symptomatic of the inadequate production and distribution of food throughout the world. But they disagree on the extent of the future food shortage problem. Those who are optimistic about our ability to provide adequate food supplies believe that the food crisis was the result of unfavorable factors that by chance occurred simultaneously during that period. In their view, the world will resume its long-term upward trend in agricultural production. Pessimists are likely to concede that the number, timing, and magnitude of the forces responsible for the crisis were atypical, but they hold that the return to more normal circumstances will not mean that the basic long-term problem of feeding the world's population has been solved.

Food-producing nations—particularly those such as the United States, which had artificially restricted the production of certain foods—re-examined their production policies. With an emphasis on increased production, harvest yields improved significantly and world food stocks were replenished to safe levels. However, it is recognized that shortfalls in some areas of the world are not only possible but likely, in which case the matter of an equitable distribution of export supplies will again become a major concern. Like world population, the world's food supply has become a matter of inherent uncertainty.

World Food Bank

The food crisis of the early 1970s emphasized the need for some sort of stop-gap solution to cope with such short-term food problems. One proposal that has gained wide support in many quarters is that of a world food bank. Proponents contend that the creation of an internationally managed world food bank would provide stability in the world food supplies and perhaps substantially reduce the pressures of international competition for available food supplies. It would also diminish the world's dependence on the United States for the production and distribution of excess food products.

Although there are many variations of the world food bank proposal, the plan of the United Nations' Food and Agriculture Organization (FAO) serves as a general model. This plan calls for all governments, including food importers as well as exporters, to maintain specific minimal levels of food stocks to meet international emergencies. Government representatives would meet regularly to review the adequacy of existing food stocks and take whatever action they deemed necessary. Developed nations would be assisted in establishing and maintaining the reserve crop necessary for self-protection against crop failures by international agencies such as the World Bank, the International Monetary Fund (IMF), and the FAO.

Despite the many complex political and economic problems associated with a world food bank plan, many believe that it offers the best promise of ameliorating short-term pressures on food products. However, it is recognized that the basic problem is of a long-term nature and that the race between food products and population must be ultimately dealt with by effectively increasing agricultural output or reducing population growth.

Potential versus Actual

Potentiality is not actuality. Because the arable land of the world is limited by geographical characteristics and climate, other productive inputs, such as farm equipment and particularly fertilizer, will have to be substituted to increase production to feed the world's growing population.

Investment. In recent years there has been a growing awareness on the part of less developed nations of the need for greater resource allocation to the farm sector. To a large extent this realization has been fostered by the fact that the necessity of importing food to feed increasing populations is costing valuable foreign exchange that could be better used in buying farm machinery and equipment. Greater effort is now being directed toward the construction of a chemical industry that would be capable of satisfying the tremendous need for fertilizers. The production potentiality of intensive application of fertilizers in the less developed countries is pointed up by the fact that over the past four decades output per acre in North America and Europe, where fertilizer and machinery are extensively used, has more than doubled, while in Asia, Africa, and Latin America it has risen by only approximately 10 percent. Obviously, if the latter areas begin to adopt the modern, highly productive techniques of agriculture, their farm output will increase dramatically.

The achievement of successful developmental breakthroughs in agricultural productivity is generally thought to be the responsibility of the developed nations. One such agricultural development may exist with hybrid wheat. Major seed producers in the United States are attempting to develop a commercially viable hybrid wheat that ultimately could bring abundance to granaries throughout the world. The new wheat could allow a farmer to increase wheat yields 25 percent or more immediately. As the hybrids are improved, yields could double within a few years. Since wheat is the world's largest crop, the benefits derived from hybrid wheat may significantly lessen hunger throughout the world.

In addition to wheat, agricultural researchers are trying to boost yields of soybeans but thus far have been frustrated in their attempts to do so. Agricultural scientists see this research as a major means of providing enough protein for the growing world population. World consumption of soybeans is increasing faster than the beans can be grown in the United States, which is a major world supplier. Although soybean production in the United States has increased over 100 percent since 1970, much of this increase has been achieved by increasing soybean acreage. Therefore, the discovery of a superproductive soybean is being given high research priority in the hope of providing sufficient soybeans to meet the world's protein needs.

Obviously, if the previously mentioned research proves successful, world farm output could increase dramatically. Those who are optimistic about the future see such efforts as a hopeful indication that in the not-too-distant future some of the developing nations of the world will become self-sustaining as far as food products are concerned.

Aquaculture. Another recent development has excited the imagination of many commentators on world food problems. Research scientists in the U.S. Department of Agriculture have developed a method for converting ocean fish that had previously been considered inedible into a palatable, high-protein food. This conversion process opens up almost inexhaustible supplies of a high-protein food that is badly needed in many of the world's food shortage areas. The potential food resources contained in the world's oceans have barely been touched up to the present time. This new method of exploiting them is a manifestation of the quickening interest in utilizing the virtually inexhaustible resources of the sea. So much interest in the potential of the sea as a source of food has developed that a new term, *aquaculture*, has emerged to describe scientific methods of harvesting the ocean's riches. Many experts feel that if the sea can be exploited scientifically as a source of food, the problem of shortages in the world food supply will be largely solved.

Redistribution of Food Supplies. It goes without saying that some areas of the world are better endowed with the prerequisites for agriculture than are others. Soil fertility, climate, length of the growing season, and the amount of arable land and available land will all influence the ability of any given nation to become self-sufficient in food production. Certain areas of Africa, for example, offer little hope of developing an efficient agricultural sector because of the lack of sufficient rainfall. This being the case, it becomes necessary for such nations to cultivate those resources that they do have in abundance (in the case of the African nations adjacent to the Sahara, this would be petroleum) and trade them to the more fortunately endowed agricultural nations in exchange for food. A famous case in point is Great Britain, a poorly endowed nation as far as agricultural resources are concerned that grew to economic greatness by utilizing its facility for the fabrication of goods.

It must be made quite clear, however, that most of the nations of the world do have the potential for developing a viable agricultural sector. In any discussion of the redistribution of food products from food surplus areas to food shortage areas, one always encounters the danger of emphasizing too heavily the responsibility the surplus areas have to the shortage areas. It is certainly true that it would be morally inexcusable for the surplus areas to refuse food to nations in need, but it is equally true that the persistent subsidizing of food needs by those nations with surplus food supplies may remove any incentive the shortage areas have to develop efficient agricultural sectors in their own economies.

Pessimistic Consensus

Not all experts agree that the long-run solution to feeding the world's growing population lies in increased investment or the redistribution of excess food. Pessimists point to several critical factors that they believe will undermine attempts to increase the supply of food at a rate required to sustain the population of the world.

They argue that the use of fertilizers and technology to increase per-acre yields may be reaching a point of diminishing returns, at which it costs more to increase production than the added production is worth. As for aquaculture, many oceans already have been overfished, as is evidenced by the fact that despite more time and money invested in ships and equipment, fish production has been declining since 1970.

The world is also running out of new lands to cultivate, and in many countries land is being taken out of production quite rapidly as urbanization expands. Fresh water for use in agricultural production is becoming scarce; most rivers that form convenient sources of fresh water already have been tapped.

A final factor is the growing affluence of the developed nations. In Western European and North American countries, the demand for food is increasing out of proportion to population. People consume more beef, poultry, eggs, and milk as they become more affluent, and grain that would be directly consumed in poor countries is indirectly consumed as food for livestock in developed countries.

Consequently, those who maintain this point of view contend that the only viable alternative to rapid increases in food supplies is to slow down world population growth rates. The question of conscious control of the birth rate and thus population size is a thorny one indeed, but it cannot be ignored in view of the increasing sentiment all over the world in favor of it.

POPULATION CONTROLS

Since rates of population growth are, in a broad sense, determined by the relationship between birth rates and death rates, it follows that measures to control population size must operate on one or the other or both of these variables. To a large extent, efforts over the last 50 years or so have concentrated on lowering death rates. A great deal of research effort and public health activity has been devoted to alleviating suffering and death in less developed nations whose populations regularly have been decimated by diseases that are historically endemic to the area. Probably the outstanding effort in this regard has been the worldwide effort to eradicate malaria through mosquito control and vaccines that impart immunity to this dread disease. Massive efforts to improve sanitation levels through the construction of sewage systems and water systems that would ensure the delivery of pure water have significantly increased the overall health levels of populations and thus reduced death rates. In Sri Lanka, for example, the expenditure of $2 per person on a public health campaign resulted in the reduction of the death rate by 75 percent in a ten-year period.[4]

These widespread efforts to decrease death rates have not been matched by measures that would induce similar downward trends in birth rates. The result, as we have seen, has been to accelerate population growth rates to the point that, in many countries, population is beginning to press heavily upon the available food resources. Only in the past 25–30 years has much attention been focused on the question of spiraling population growth, and only in the past 15–20 years has significant research been undertaken in an attempt to deal with the problem. This is not to say that the problem was completely ignored for a long period of time, but the commentators on population problems were largely voices in the wilderness during much of that earlier period.

Now there are indications of a growing awareness of the very real problems posed by rapid population growth the world over. This awareness is being accompanied by positive programs of action in many of the less developed nations. Efforts to disseminate knowledge and methods of birth control are proceeding apace in India and South America, as well as in other areas of rapid population growth. The effectiveness of such programs cannot be validly appraised, but the combination of money, time, and human effort that is being devoted to them seems to indicate that eventually they will begin to significantly lower birth rates.

---

4. Eugene R. Black, "Population Increase and Economic Development," in *Our Crowded Planet,* 67–68.

Of course, the obstacles to the success of such programs loom very large in many areas of the world. The relatively low educational level of the people, the social stigma attached to the small family in some areas, the active opposition to contraception that is present in predominantly Catholic countries, and finally, the almost overwhelming immensity of the task of spreading birth control information to literally billions of people give rise to serious questions as to whether birth control is really a solution to rapid population growth, at least in the all-important short run.

Because of these obstacles, the subject of birth control is a highly emotional one in many countries throughout the world. Consequently, the United Nations and other international bodies, such as the World Bank, have approached the problem of birth control with great caution. Recognizing the diversity of interests among nations, global organizations are seeking ways to present the case for birth control in terms of national social values. Interests are easy to antagonize, and emotions can become aroused at either the private or the public level. For example, the alarmist Malthusian-style position that focuses attention on rapid population growth in the less developed countries may be interpreted as being neoimperialistic in nature. If the less developed nations suspect an ulterior motive on the part of developed nations, distrust and possibly fear may result, which could lead to open opposition to any birth control program. Thus, international organizations must relate birth control to the national interests of the less developed nations and convincingly show that those nations will be the principal beneficiaries of any control program.

It is possible that such attempts to check the current rate of population increase will succeed, and in a much shorter period than is currently envisioned. Should this happen, and birth rates begin to decline toward a more normal relationship with death rates within the present generation, then much of the potential danger inherent in the aggregate population problem will be removed.

POPULATION TRENDS IN THE UNITED STATES

Up to this point we have confined our discussion of population questions primarily to a world context. We have examined the forces influencing population growth and food supply on a worldwide basis for the simple reason that such forces operate for the planet as a whole and only secondarily for particular nations. In this section, however, we shall turn our attention to the United States, that most fortunate of nations if judged by material standards, to give some insight into how its people will fare as their numbers increase in the coming years.

In late 1986, the Bureau of the Census released a preliminary report of the results of its 1986 survey concerning population data. The median age for first marriages in the survey was 23.1 for women and 25.7 for men. The median first-marriage age for women has remained higher in the 1980s than previously recorded. Average family size fell to a new low of 3.21 persons in 1986, down from 3.29 in 1980. The drop reflects the preference for fewer or no offspring. Average household size also reached a record low of 2.67 individuals.

Bureau of the Census Projections

The Bureau of the Census has published revised estimates of the population through the year 2025. These estimates were constructed under different sets of assumptions regarding the birth rate, death rate, and net immigration in the United States for the relevant time period. Table 15-4 shows these different projections.

Series 1 reflects the assumptions of 1.6 births per woman, an average life expectancy of 76.7 years, and a yearly net immigration of 250,000. These assumptions produce the lowest population figures of the four series. Series 2 is based on a birth rate of 1.9, an average life expectancy of 79.6, and an annual net immigration figure of 450,000. The third projection, Series 3, incorporates the highest set of assumptions with 2.3 births per woman, an average life expectancy of 83.3 years, and a net immigration of 750,000. Series 4 is identical to Series 2 except it assumes a net immigration figure of zero.

It is obviously impossible to predict which of the four estimates will ultimately be correct, if any. Since actual population in 1985 was about

Table 15-4

Projections of Total U.S. Population, 1990–2025
(In thousands)

Year	Series 1 (Lowest)	Series 2 (Middle)	Series 3 (Highest)	Series 4 (Zero immigration)
1990	245,753	249,657	254,122	245,764
1995	251,876	259,599	268,151	252,927
2000	258,096	267,955	281,542	258,412
2005	259,181	275,677	295,276	263,085
2010	261,482	283,238	310,006	267,468
2015	262,795	290,406	325,423	271,335
2020	262,695	296,597	340,762	274,118
2025	260,904	301,394	355,503	275,436

SOURCE: U.S. Bureau of the Census, *Current Population Reports,* Series P-25, No. 922.

239 million, even if we accept Series 1, the most conservative of the group, we can expect an increase of about 17 million people in this country by the year 2000 and an increase of about 21 million by the year 2025.

Note that of the four projections presented in Table 15-4, only Series 3 assumes a fertility rate higher than the zero population growth rate of 2.1. Series 4, however, is the only series to assume a net immigration figure of zero.

Zero Population Growth

In recent years many Americans concerned with population growth have advocated population control policies for this country. The basic argument for population control is that although the United States has a small population problem compared to other countries, a rapidly growing population increases environmental pollution of all kinds, thus resulting in a sharp reduction in the quality of life for more Americans. It is pointed out that each child born in the United States consumes 30 times more of the basic resources of the earth than the average child born in India. Therefore, the average American places inordinate stress on the world's irreplaceable natural resources, and it is difficult to envision any large-scale success in offsetting the increased use of such resources.

As a means of controlling population growth, a policy of zero population growth is advocated. According to this plan, population can be stabilized, not immediately, but over the next 50 years or so. To accomplish this feat, the total fertility rate of American women would have to be 2.1 for each generation to replace itself exactly. Thus, 1,000 women would have to bear 2,100 children so as to replace themselves with 1,000 women of childbearing age in the next generation. The extra 100 children allow for mortality and the fact that more than half of the children born each year are male.

If this childbearing rate were maintained over several generations, a stabilized population would result. (This, of course, ignores increases in population due to immigration.) Consequently, even if the zero population growth policy were adopted at the present time, the nation's population would increase significantly by the time stabilization was attained.

The overall impact of a zero population growth rate is uncertain, but major economic and social adjustments will undoubtedly be necessary. With much lower birthrates in recent years, many such changes are already occurring.

Population Characteristics. One of the most obvious results of a zero population growth rate is the fact that over time the average age of the nation's population will continue to rise. The Bureau of the Census reports that in 1984 the median age of the U.S. population was 31.2 years. The consequence of a policy of a zero population growth rate over a 50-year span or so would be a population with a median age of approximately 37.

The "graying" of America is already underway. Currently, there are 28 million Americans 65 years old and over, constituting 11.9 percent of the total U.S. population. Each day about 5,000 people reach the age of 65, and each day about 3,400 people 65 years and over die. The result is a net increase of 1,600 older people. This older segment of our population is the fastest growing segment of our nation's population and is expected to increase dramatically in the year 2010 when the post–World War II baby boomers enter the ranks of those 65 and over.

Labor Force. An older population brings about noticeable changes in the participation rate of the labor force. Female participation rates would swing upward as women have fewer children and thus are free to enter the labor force in larger numbers at an earlier age. On the average, the rate of participation for males would probably be unchanged, except perhaps in the 20–24 and 55-and-over age groups. Young males are expected to continue delaying their entry into the labor force due to the trend for more advanced education and training. The participation rate of males 55 years and older is more difficult to analyze because of the unknown effect of such factors as medical advances, work-leisure choices, technological advances, and early retirement.

Supporters of a zero population growth rate contend that the labor force of the future will be of greater quality and thus more productive. Stabilized population growth should lead to better medical facilities and treatment. The result would be a healthier and more efficient work force. Workers would also be better educated and skilled. Critics respond to such purported advantages of zero population growth by indicating that in all likelihood the aforementioned increases in productivity could readily occur with a growing population as well as with a stabilized one.

Environment. If population growth were stabilized, society, according to zero population growth supporters, would become much more livable because environmental pollution would be drastically reduced. This view is based on the fact that population growth and density are key factors in accounting for our present rate of environmental decay. Critics claim this view is oversimplified. Zero population growth, they argue, would make only a minor dent in solving the environmental crisis, for it is affluence, not population, that is the major factor contributing to pollution. Even with a stabilized population, our demands for the kind of goods and services that are responsible for much of the environmental pollution would increase rapidly in the future because of increased affluence. Zero population growth, in itself, would not serve to stem the tide of pollution because even with the attainment of zero population growth, a much greater amount of resources would have to be allocated to the public sector to launch a meaningful attack on environmental decay.

Public Sector. Skeptics question the effect that an older population would have on the public sector. Older persons have one-half of the income of their younger counterparts. Many of our nation's poor became poor only after reaching the age of 65. For one in five elderly persons, Social Security provides at least 90 percent of their income. But the Social Security System is already under fiscal duress in attempting to provide an adequate level of income for the elderly. The System relies on large numbers of working taxpayers to finance the old age and retirement needs of those who are no longer in the work force. Real concerns exist as to whether the System can survive with a population increasingly topheavy with older people.

Also, critics doubt the ease with which resources can be easily adapted to other uses. Their concern rests, to a large extent, with the number of institutions that have been built or are in the process of being completed on the basis of an expanding population. Schools, hospitals, and other institutions would have to be put to effective use in some other capacity and in a manner in which there is a net benefit to society. The problem would probably be difficult for many of the individuals who have been specially trained, such as teachers and nurses. What kind of a transition would these people be likely to make? The transition, undoubtedly, could be made over a long period of time, but not without some hardships to individuals and to society as a whole.

Private Sector. The attainment of a stable population in the next century would necessitate adjustments in the private business sector. But in general, national output and income per capita should be considerably higher than would be the case if higher population growth rates were to prevail. Higher incomes with a stabilized population would lead to increased sales for business people, for it is the number of spending customers, not total population, that is critical in determining the level of business activity. It is also recognized that an older population would affect various industries differently. Detrimentally affected industries, however, would have more than ample time to adjust to predictable changes that are slow in materializing. Many of these firms could diversify their product lines without undue difficulty.

Opponents argue that the degree of transition would be of much greater magnitude than is supposed. Some industries would be hit almost immediately and would have little time to adjust. Those industries that produce goods and services for preschool children, such as dolls, toys, tricycles, and diapers, would feel the impact first. Soon thereafter, the impact would be felt by those producing goods for the elementary school group. One would expect the demand for toys, games, bicycles, school equipment, clothing, and children's furniture to fall off appreciably. Eventually, as the number of teenagers decreased, producers of records, motorcycles, and perhaps automobiles would be affected. As the years passed, the diminished rate of new family formations would affect the housing, appliance, and durable goods industries. Through all these phases producers of services, as well as goods, would also be affected. For example, in the field of medicine, obstetricians would be first affected, followed by pediatricians and then orthodontists. As a result of lower birth rates throughout the 1970s, many of these industries have experienced declining sales activity.

Should the momentum of the zero population growth movement continue, population growth rates would continue to decline. The biggest question for demographers is whether the reduced birth rates of the present reflect a position of long-run stabilization or whether they are merely short-term experiences.

Immigration

The previous discussion centered on the possible effects of a stabilized population based on maintaining birth rates of 2.1 per women. However, the possibility exists that the realization of a stabilized population in the foreseeable future could be frustrated by continued high rates of immigration.

Immigration policy tends to reflect existing economic, social, and political conditions, not only in the United States but also in other countries of the world. Statistically, immigration figures are not well documented and in the case of illegal immigrants are only "best estimates." Records indicate that the United States has received more immigrants than any other nation in history. More than 52 million have arrived since 1820. In more recent years, the United States has accepted more than twice as many legal immigrants for permanent resettlement as the rest of the world's countries combined. Only a few nations, including Canada, Israel, Australia, and Argentina currently accept immigrants.

In 1985, an estimated 570,000 legal immigrants entered the United States. In addition to those legally entering the country, 60,000 arrived as refugees and another 750,000 entered illegally. Thus, the number of immigrants entering the United States in 1985 was in the vicinity of 1.3 million people. In 1985, total immigration accounted for approximately 26 percent of the nation's population growth. By the early part of the next century, assuming a fertility rate of 2.1, immigration's share of annual population growth may exceed 50 percent. To grasp the magnitude of this figure, a historical frame of reference might be useful. Not since the first 20 years of this century, when the United States opened its doors to large numbers of Southern and Eastern Europeans, have immigration totals been in excess of 1 million per year.

Immigration figures could possibly exceed the 1 million mark in the foreseeable future. For one reason, illegal immigration is not sensitive to changes in legal immigration quotas. The estimate of at least 750,000 illegal immigrants per year is likely to continue, although a decline in the number of refugees admitted to the United States is probable. The best estimate is that in 1985 approximately 6 million people resided in the country as illegal aliens.

After nearly a decade of trying to pass legislation dealing with illegal aliens, Congress passed an immigration bill in 1986. In an effort to curtail the number of illegal aliens entering the country, the 1986 law imposes stiff fines and even jail terms on individuals who knowingly hire illegal aliens. Employers must ensure that all new employees have either a U.S. passport, a U.S. birth certificate, or a Social Security card plus a state-issued identification card. Illegal aliens who can prove they entered the United States prior to January 1, 1982 or who have worked on U.S. farms for at least 90 days will be granted resident status for 18 months. After that they can earn permanent resident status, and after five years they can become U.S. citizens. Although the law will be difficult and expensive to enforce, it is thought to be a reasonable first step in confronting the problem.

A second reason for projecting continued high immigration is the fact that a large number of refugees from Cuba, Haiti, and elsewhere will have been in this country long enough to apply for citizenship. Upon being granted citizenship, former refugees will be able to bring in spouses, minor children, and parents. Their number will be in addition to the annual quotas for legal immigrants.

Regardless of where immigrants originate or whether they cross our borders as legal immigrants, illegal immigrants, or refugees, population projections for the next century that are based on low birth rates and low immigration rates will fall short of the mark. Immigration rates rule out a stationary U.S. population in the foreseeable future.

Surpluses, Social Obligations, and Survival

Our discussion of population growth and the United States cannot end here. Precisely because we are so well endowed with the goods of the world and because we have taken advantage of our good fortune, we are in a very vulnerable position. There is no deeper envy than that felt by the hungry person outside the house of the well fed. Thus, Americans are the envy of a considerable portion of the world's people simply because we are among the world's best-fed people. Envy is a destructive emotion, and unless it can be satisfied, the consequences are likely to be disastrous. Thus, the United States cannot be satisfied simply to feed, clothe, and house its own increasing population; it must also be prepared to aid materially those areas of the world that are not so well endowed.

Probably the most important export of the United States in the coming years will not be its surplus food supplies, but rather technology and capital. While the United States could perhaps increase its output of food enough to feed the hungry of the world in the short term, it is doubtful whether it could long sustain them, given the rate of population increase. The only viable long-run solution lies in the development of efficient agricultural systems in those countries that are not self-sufficient in food production. It is in the development of such systems that the United States can be of greatest service, both to the world's people and to itself.

THE ALTERNATIVES BEFORE US

Since the question of world population is multifaceted and complex and since we have been able merely to scratch the surface of the problem, perhaps it would be best to consider a series of alternative approaches for dealing with population pressures. Admittedly, some of the proposed methods for dealing with the problem of population pressure do not touch upon all facets of the problem, but by presenting several alternative proposals we should be able to highlight virtually all of its aspects.[5]

Ultimately, the rate of growth of world population rests upon the rate at which children are born and the rate at which the population as a whole is dying. As a corollary to this statement, control of the rate of growth in world population must work upon either of these rates or both.

History, ecology, and reason tell us that eventually population growth will return to an equilibrium with the growth in the economic environment. There seems to be little doubt that the world's population is currently growing faster than the environment, especially in the developing nations. It is impossible to predict at this time at what point stability will be reached, but the fundamental choice seems to be between lowering the birth rate and allowing the death rate to rise. If one objects to limiting births, one must be prepared to favor increasing the death rates unless another development, such as the Industrial Revolution, is believed to be imminent in the developing nations. Until such a revolution comes to pass, we shall be forced to witness and participate in a remorseless return to equilibrium.

---

5. We do not pretend to have presented an exhaustive treatment of this issue. Some aspects such as the psychological, the political, and the theological have not been covered at all; our only defense is that this book is concerned primarily with the economic issues.

The question is not whether the equilibrium is to be reestablished, but by what means this is to be achieved.

A Significant Increase in the Death Rate

There are relatively few serious advocates of a return to higher death rates. Our entire medical research effort is oriented in precisely the opposite direction. Nations throughout the world are involved in public health programs designed to prevent premature deaths caused by diseases arising from unsanitary living conditions, disease-carrying insects, or outright ignorance of the lifesaving capabilities of modern medicine. A few economists have suggested that developing nations postpone public health expenditures and, in this way, prevent the drastic fall in death rates that always accompanies the undertaking of such expenditures. These economists suggest that such a measure would hold the population fairly steady and allow resources that would have to go to feed a burgeoning population to be allocated to investment goods that would quicken the pace of economic development. Aside from the moral aspects of this suggestion, one can question the effectiveness of a stable population if it continues to be wracked by disease because of the absence of public health expenditures. One can also question the possibility of economic development based upon a disease-ridden work force.

Another proposal that might fit in this category is forced abortion. A few countries, such as the People's Republic of China, have adopted this method, but there seems to be a great deal of opposition to such a remedy throughout most of the world. The idea of taking the life of an unborn infant runs directly against the concept of the worth and dignity of the individual that pervades most cultures. It is somewhat doubtful whether such a remedy would ever find widespread acceptance on a worldwide basis, although that does not mean it is impossible.

There does not seem to be much hope for a solution to population pressures by increasing the death rate. People have been exposed to the opposite kind of effort for too long to consciously permit efforts to shorten life expectancy. Thus, we must look elsewhere for a solution.

Emigration

The movement of masses of people from overcrowded areas to places where living room is available has long been one solution to population pressure. The emigration from Europe in the eighteenth and nineteenth centuries was one of the relief valves following the rapid population growth generated by the Industrial Revolution. It is questionable, however, whether emigration is still a viable solution in the last quarter of the twentieth century. The monetary costs involved in moving literally millions of people from, say,

Latin America to Canada would be prohibitive, to say nothing of the almost impossible human difficulties that would be involved.

As in the case of an increasing death rate, emigration does not seem to be a reasonable short-run solution to overpopulation. The monetary costs alone would be so large that a fraction of their total spent to increase agricultural productivity in low-crop-yield areas might well result in a much more efficient expenditure of funds. Such a possibility leads directly to the third alternative solution to the problem of excess population.

Rise in Production to Meet Growing Needs

This alternative seems to present the greatest possibilities for a short-run solution to the population problem because most of the increase in food production has taken place in the agriculturally advanced economies. Many of these nations have been burdened with heavy surpluses of food products, while at the same time experiencing lower birth rates. Less developed nations, on the other hand, have barely been able to keep pace with the needs of their rapidly growing populations.

We have already discussed the difficulties of raising agricultural production in the food shortage areas of the world. We would emphasize once again, however, that the exporting of surplus food from the haves to the have-nots does not seem to be a viable long-run solution to the problem. The emphasis must shift to the exporting of agricultural expertise and capital to allow the nations with food shortages to develop their own efficient agriculture. Certainly, this does not mean that all exports of food should be immediately cut off. Such a move would likely prove disastrous for some of the less developed countries. But unless there is a massive effort on the part of all the nations of the world to develop self-sufficient agricultures in food-deficient countries, by the year 2000 we could well witness that most terrible of all population checks, famine.

Thus, rapid increases in agricultural productivity hold the most immediate hope for the problem of excessive population precisely because, most often, populations are larger than the food supply available to feed them. This fact is spreading rapidly among nations faced with the problem, and the growing emphasis on the development of efficient agricultures is cause for hope that the world need not face widespread famine.

Birth Control

With this alternative probably rests one important long-run solution to the problem of excessive population. There is some finite limit to the number of people that can inhabit the earth, if only because of spatial and resource restrictions. World population cannot grow without end; an equilibrium must eventually be reached between the people inhabiting the earth and the resources available to sustain them. It is reasonable to suggest, then, that world population should stabilize at a point somewhat short of the absolute limit imposed by the above restrictions. If we have any concern at all for what has been termed the quality of life, then this suggestion must be admitted. By the quality of life, we mean the manner in which people live—the composite of material and nonmaterial things that are available to use and to enjoy during an individual's sojourn on earth. Food, clothing, shelter, elbow room, the right to acquire knowledge or enjoy beauty, the right to be left alone, all of these things depend in large part on the number of people that inhabit the world at any given time.

Certainly, the most-discussed method of controlling population is birth control. Whether such control is exercised by the use of artificial methods including sterilization, or whether it results from later marriage, abstinence, or some natural method, the results are the same—a decrease in the number of children.

But birth control is not the panacea that it might seem. There are serious difficulties involved in the implementation of birth control programs, especially on a national scale, and many experts on the subject remain to be convinced that such programs can significantly affect the birth rate. The reason for this uncertainty is readily seen. Birth control concerns what is probably the most intimate and personal facet of human relationships, and the problems in rigidly ordering this relationship make practically all other difficulties pale by comparison. But this is not to say that birth control does not hold out real hope for stabilizing population growth in the long run. As the educational and economic level of the world's people rises, and as methods of birth control acceptable to all groups are developed, it is likely that more people will accept this solution.

CONCLUSION

With this study of the alternative solutions to the population explosion, we hope that the reader has acquired some grasp of the immensity and complexity of the problems facing us as a result of rapid population growth. To be aware of and to understand a problem are the first steps along the road to its solution.

Ultimately, the hope for a solution to this problem lies in the intelligence and ingenuity of people. We became the dominant species on earth precisely because we alone among the vastly diversified life forms on earth have such powers. To deny that we will use them in the solution of this problem is to deny the one characteristic that differentiates us from all other life forms. We are not prepared to make such a denial, and thus we can predict that an acceptable solution to the problem of population pressures will be found and implemented in the not-too-distant future.

QUESTIONS FOR DISCUSSION

1. What political and economic obstacles would have to be overcome to organize a world food bank such as the one the United Nations proposes?
2. What political implications do you foresee if the people in areas with food shortages are forced by rapid population growth to remain at the margin of subsistence?
3. Do you think the less developed nations will be able to develop efficient agricultural systems with help from the more developed nations?
4. To what extent should the United States concern itself with the world's supply of nonrenewable resources?
5. What factors have accounted for the actual shape of the world population growth curve in the past several centuries?
6. Under what conditions should the United States limit exports of food to foreign nations?
7. Would you agree with the assertion that the United States is capable of feeding the world's population for the foreseeable future?
8. Can you think of any short-term solutions for the famine and starvation currently taking place in less developed nations?
9. Is it likely that the less developed nations will adopt a policy of zero population growth during the next decade?

SELECTED READINGS

Brown, Lester, *et al. State of the World.* New York: W.W. Norton & Co., 1986.

"The Disappearing Border." *U.S. News and Weekly Report* (August 19, 1985): 30–35.

The Global 2000 Report to the President: Entering the Twenty-first Century. Washington, D.C.: U.S. Government Printing Office, 1981.

Gupte, Pranay. *The Crowded Earth: People and the Politics of Population.* New York: W.W. Norton & Co., 1985.

Murdock, William W. *The Poverty of Nations: The Political Economy of Hunger and Population.* Baltimore: Johns Hopkins University Press, 1980.

U.S. Population: Where Are We; Where We're Going. Washington, D.C.: Population Reference Bureau, Inc., June 1982.

World Development Report ——. Washington, D.C.: The World Bank, annually.

INDEX

paying twice for, 193-194
Social Security Act, 178-181
   amendments of 1972, 181-183
   amendments of 1977, 188-189
   amendments of 1983, 189-193
   hospital and health care provisions of,
     180-181
   Title 18 of, 231
   Title 19 of, 232
Soil bank program, 104, 113
Soil Conservation and Domestic Allotment
   Act, 103
Solid waste disposal, and air pollution, 252,
   253, 254
Special Drawing Rights (SDRs), 340-341
Stagflation, 43
Structural inflation, 32-33
Structural unemployment, 10-11, 21-26
Subsidiary Coinage Act of 1853, 326
Sulfur dioxide, as pollutant, 256
Sulfuric acid, as pollutant, 256
Superregional banks, 139
Supplemental security income (SSI), 181
Supplementary Security Income (SSI), 232
Supplementary unemployment benefits
   (SUB), 11
Supply and demand analysis
   and crime, 292-293
   and farm products, 95-96
   and health care industry, 238-240
   and inflation, 33-34
Supply shocks, 33
Survey method, of measuring unemployment,
   14-16
   criticism of, 17-18
Synthetic fuels, 82

Target prices, agriculture, 105
Tax credit program, targeted jobs, 26
Tax Equity and Fiscal Responsibility Act, 26
Technological unemployment, 11, 21-26
Terms of trade, 361
Thermal inversion, 252
Third Liberty Bond Act, 208
Third World debt, 358-378

Thrift institutions, 120-121, 128-129, 130
Timber resources, 302
Time deposit interest ceilings, 125
Time utility, 60
Trade deficits, and value of dollar, compari-
   son of, 352
Transportation, and air pollution, 252, 253
Treasury bills, as part of national debt,
   209-210
Treasury notes, as part of national debt,
   210-211

Underemployment, 16
Unemployed
   age and race of, 12-13
   hard-core, 14
   sex and marital status of, 13-14
Unemployed persons, definition of, 15
Unemployment, 1-28
   alleviating, 20-26
   comparison with other nations, 19
   and crime, 289-290
   cyclical, 11, 21
   duration of, 11-12
   and employment, 3-4
   and inflation, 34-45
   measurement of, 14-19
   problems of, 9-10
   structural, 10-11, 21-26
   technological, 11, 21-26
Urbanization, and air pollution, 251
Usury laws, 127-128
Utility, 59-60

Venezuela, external debt of, 364, 373
Victimless crime, policy toward, 274, 294-
   296
Victims of crime, policy toward, 293-294
Violent crime, 274, 275, 278, 280-281

Wage and price controls, 39-43
Wage-price spiral, 32
Water consumption, 308, 309
Water resources, 307, 308-315